ASTRONOMY
DEMYSTIFIED

Other Titles in the McGraw-Hill Demystified Series

STAN GIBILISCO • *Physics Demystified*
RHONDA HUETTENMUELLER • *Algebra Demystified*
STEVEN KRANTZ • *Calculus Demystified*

ASTRONOMY DEMYSTIFIED

STAN GIBILISCO

McGraw-Hill

New York   Chicago   San Francisco   Lisbon   London
Madrid   Mexico City   Milan   New Delhi   San Juan
Seoul   Singapore   Sydney   Toronto

The McGraw·Hill Companies

Cataloging-in-Publication Data is on file with the Library of Congress

Copyright © 2003 by The McGraw-Hill Companies, Inc. All rights reserved. Printed in the United States of America. Except as permitted under the United States Copyright Act of 1976, no part of this publication may be reproduced or distributed in any form or by any means, or stored in a data base or retrieval system, without the prior written permission of the publisher.

    4 5 6 7 8 9 0   AGM/AGM   0 9 8 7 6 5 4

ISBN 0-07-138427-8

The sponsoring editor for this book was Scott Grillo, the editing supervisor was Steven Melvin, and the production supervisor was Pamela Pelton.

Quebecor/Martinsburg was printer and binder.

McGraw-Hill books are available at special quantity discounts to use as premiums and sales promotions, or for use in corporate training programs. For more information, please write to the Director of Special Sales, McGraw-Hill, 2 Penn Plaza, New York, NY 10121-2298. Or contact your local bookstore.

 This book is printed on recycled, acid-free paper containing aminimum of 50% recycled, de-inked fiber.

DEDICATION

To Tim, Samuel, and Tony from Uncle Stan

CONTENTS

PREFACE

This book is for people who want to learn basic astronomy without taking a formal course. It also can serve as a supplemental text in a classroom, tutored, or home-schooling environment. I recommend that you start at the beginning of this book and go straight through.

In this book, we'll go on a few "mind journeys." For example, we'll take a tour of the entire Solar System, riding hybrid space/aircraft into the atmospheres and, in some cases, to the surfaces of celestial bodies other than Earth. Some of the details of this trip constitute fiction, but the space vehicles and navigational mechanics are based on realistic technology and astronomical facts.

This book is about astronomy, not cosmology. A full discussion of theories concerning the origin, structure, and evolution of the Universe would constitute a full course in itself. While the so-called Big Bang theory is mentioned, arguments supporting it (or refuting it) are beyond the scope of this volume. The fundamentals of relativity theory are covered; these ideas are nowhere near as difficult to understand as many people seem to believe. Space travel and the search for extraterrestrial intelligence are discussed as well.

This book contains an abundance of practice quiz, test, and exam questions. They are all multiple-choice and are similar to the sorts of questions used in standardized tests. There is a short quiz at the end of every chapter. The quizzes are "open book." You may (and should) refer to the chapter texts when taking them. When you think you're ready, take the quiz, write down your answers, and then give your list of answers to a friend. Have your friend tell you your score but not which questions you got wrong. The answers are listed in the back of the book. Stick with a chapter until you get most of the answers correct.

This book is divided into several major sections. At the end of each section is a multiple-choice test. Take these tests when you're done with the respective sections and have taken all the chapter quizzes. The section tests are "closed book." Don't look back at the text when taking them. The questions are not as hard as those in the quizzes, and they don't require that you memorize trivial things. A satisfactory score is three-quarters of the answers correct. Again, answers are in the back of the book.

There is a final exam at the end of this course. The questions are practical and are easier than those in the quizzes. Take this exam when you have finished all the sections, all the section tests, and all the chapter quizzes. A satisfactory score is at least 75 percent correct answers.

With the section tests and the final exam, as with the quizzes, have a friend tell you your score without letting you know which questions you missed. In that way, you will not subconsciously memorize the answers. You might want to take each test and the final exam two or three times. When you have gotten a score that makes you happy, you can check to see where your knowledge is strong and where it is not so keen.

I recommend that you complete one chapter a week. An hour or two daily ought to be enough time for this. Don't rush yourself; give your mind time to absorb the material. But don't go too slowly either. Take it at a steady pace, and keep it up. In that way, you'll complete the course in a few months. (As much as we all wish otherwise, there is no substitute for *good study habits.*) When you're done with the course, you can use this book, with its comprehensive index, as a permanent reference.

Suggestions for future editions are welcome.

Stan Gibilisco

ACKNOWLEDGMENTS

Illustrations in this book were generated with CorelDRAW. Some clip art is courtesy of Corel Corporation, 1600 Carling Avenue, Ottawa, Ontario, Canada K1Z 8R7.

I extend thanks to Linda Williams, who helped with the technical editing of the manuscript for this book.

PART ONE

The Sky

CHAPTER 1

Coordinating the Heavens

What do you suppose prehistoric people thought about the sky? Why does the Sun move differently from the Moon? Why do the stars move in yet another way? Why do star patterns change with the passing of many nights? Why do certain stars wander among the others? Why does the Sun sometimes take a high course across the sky and sometimes a low course? Are the Sun, the Moon, and the stars attached to a dome over Earth, or do they float free? Are some objects farther away than others?

A thousand generations ago, people had no quantitative concept of the sky. In the past few millennia, we have refined astronomical measurement as a science and an art. Mathematics, and geometry in particular, has made this possible.

Points on a Sphere

It is natural to imagine the sky as a dome or sphere at the center of which we, the observers, are situated. This notion has always been, and still is, used by astronomers to define the positions of objects in the heavens. It's not easy to specify the locations of points on a sphere by mathematical means. We can't wrap a piece of quadrille paper around a globe and make a rectangular coordinate scheme work neatly with a sphere. However, there are ways to uniquely define points on a sphere and, by extension, points in the sky.

MERIDIANS AND PARALLELS

You've seen globes that show lines of *longitude* and *latitude* on Earth. Every point has a unique latitude and a unique longitude. These lines are actually half circles or full circles that run around Earth.

The lines of longitude, also called *meridians*, are half circles with centers that coincide with the physical center of Earth (Fig. 1-1A). The ends of these arcs all come together at two points, one at the *north geographic pole* and the other at the *south geographic pole*. Every point on Earth's surface, except for the north pole and the south pole, can be assigned a unique longitude.

The lines of latitude, also called *parallels*, are all full circles, with two exceptions: the north and south poles. All the parallels have centers that lie somewhere along Earth's *axis of rotation* (Fig. 1-1B), the line connecting the north and south poles. The *equator* is the largest parallel; above and below it, the parallels get smaller and smaller. Near the north and south poles, the circles of latitude are tiny. At the poles, the circles vanish to points.

All the meridians and parallels are defined in units called *degrees* and are assigned values with strict upper and lower limits.

DEGREES, MINUTES, SECONDS

There are 360 degrees in a complete circle. Why 360 and not 100 or 1000, which are "rounder" numbers, or 256 or 512, which can be divided repeatedly in half all the way down to 1?

No doubt ancient people noticed that there are about 360 days in a year and that the stellar patterns in the sky are repeated every year. A year is like a circle. Various familiar patterns repeat from year to year: the general nature of the weather, the Sun's way of moving across the sky, the lengths of the days, the positions of the stars at sunset. Maybe some guru decided that 360, being close to the number of days in a year, was a natural number to use when dividing up a circle into units for angular measurement. Then people could say that the stars shift in the sky by 1 degree, more or less, every night. Whether this story is true or not doesn't matter; different cultures came up with different ideas anyway. The fact is that we're stuck with degrees that represent 1/360 of a circle (Fig. 1-2), whether we like it or not.

For astronomical measurements, the degree is not always exact enough. The same is true in geography. On Earth's surface, 1 degree of latitude rep-

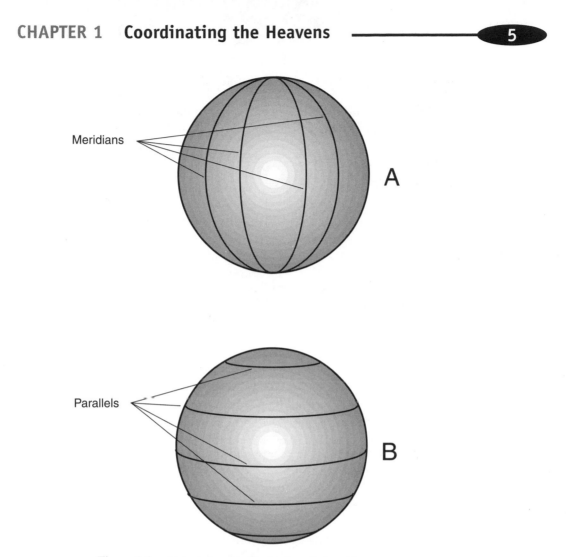

Figure 1-1.   At *A*, circles of longitude, also called meridians.
At *B*, circles of latitude, also called parallels.

resents about 112 kilometers or 70 miles. This is okay for locating general regions but not for pinpointing small towns or city blocks or individual houses. In astronomy, the degree may be good enough for locating the Sun or the Moon or a particular bright star, but for dim stars, distant galaxies, nebulae, and quasars, smaller units are needed. Degrees are broken into *minutes of arc* or *arc minutes*, where 1 minute is equal to ¹⁄₆₀ of a degree. Minutes, in turn, are broken into *seconds of arc* or *arc seconds*, where 1 second is equal to ¹⁄₆₀ of a minute. When units smaller than 1 second of arc are needed, decimal fractions are used.

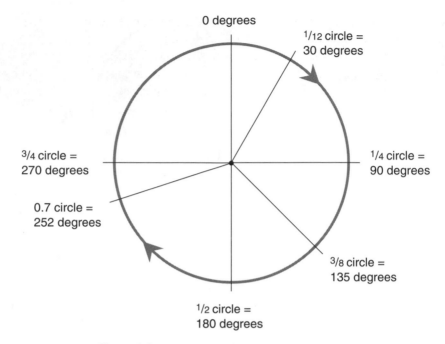

Figure 1-2.   There are 360 degrees in a complete circle.

Let's take a close look at how latitude and longitude coordinates are defined on the surface of Earth. It will help if you use a globe as a visual aid.

LATITUDE

In geography classes you were taught that latitude can range from 90 degrees south to 90 degrees north. The north geographic pole is at 90 degrees north, and the south geographic pole is at 90 degrees south. Both the poles lie on the Earth's axis. The equator is halfway between the poles and is assigned 0 degrees latitude. The *northern hemisphere* contains all the north-latitude circles, and the *southern hemisphere* contains all the south-latitude circles.

As the latitude increases toward the north or south, the circumferences of the latitude circles get smaller and smaller. Earth is about 40,000 kilometers (25,000 miles) in circumference, so the equator measures about 40,000 kilometers around. The 45-degree-latitude circle measures about 28,000 kilometers (17,700 miles) in circumference. The

60-degree-latitude circle is half the size of the equator, or 20,000 kilometers (12,500 miles) around. The 90-degree-latitude "circles" are points with zero circumference. Every latitude circle lies in a geometric plane that slices through Earth. All these planes are parallel; this is why latitude circles are called *parallels*. Every parallel, except for the poles, consists of infinitely many points, all of which lie on a circle and all of which have the same latitude.

There is no such thing as a latitude coordinate greater than 90 degrees, either north or south. If there were such points, the result would be a redundant set of coordinates. The circle representing "100 degrees north latitude" would correspond to the 80-degree north-latitude circle, and the circle representing "120 degrees south latitude" would correspond to the 60-degree south-latitude circle. This would be confusing at best because every point on Earth's surface could be assigned more than one latitude coordinate. At worst, navigators could end up plotting courses the wrong way around the world; people might mistakenly call 3:00 P.M. the "wee hours of the morning"!

An ideal coordinate system is such that there is a one-to-one correspondence between the defined points and the coordinate numbers. Every point on Earth should have one, and only one, ordered pair of latitude-longitude numbers. And every ordered pair of latitude/longitude numbers, within the accepted range of values, should correspond to one and only one point on the surface of Earth. Mathematicians are fond of this sort of neatness and, with the exception of paradox lovers, dislike redundancy and confusion.

Latitude coordinates often are designated by abbreviations. Forty-five degrees north latitude, for example, is written "45 deg N lat" or "45°N." Sixty-three degrees south latitude is written as "63 deg S lat" or "63°S." Minutes of arc are abbreviated "min" or symbolized by a prime sign ('). Seconds of arc are abbreviated "sec" or symbolized by a double prime sign ("). So you might see 33 degrees, 12 minutes, 48 seconds north latitude denoted as "33 deg 12 min 48 sec N lat" or as "33°12′48″N."

As an exercise, try locating the above-described latitude circles on a globe. Then find the town where you live and figure out your approximate latitude. Compare this with other towns around the world. You might be surprised at what you find when you do this. The French Riviera, for example, lies at about the same latitude as Portland, Maine.

LONGITUDE

Longitude coordinates can range from 180 degrees west, down through zero, and then back up to 180 degrees east. The zero-degree longitude line, also called the *prime meridian*, passes through Greenwich, England, which is near London. (Centuries ago, when geographers, lexicographers, astronomers, priests, and the other "powers that were" decided on the town through which the prime meridian should pass, they almost chose Paris, France.) The prime meridian is also known as the *Greenwich meridian*. All the other longitude coordinates are measured with respect to the prime meridian. Every half-circle representing a line of longitude is the same length, namely, half the circumference of Earth, or about 20,000 kilometers (12,500 miles), running from pole to pole. The *eastern hemisphere* contains all the east-longitude half circles, and the *western hemisphere* contains all the west-longitude half circles.

There is no such thing as a longitude coordinate greater than 180 degrees, either east or west. The reason for this is the same as the reason there are no latitude coordinates larger than 90 degrees. If there were such points, the result would be a redundant set of coordinates. For example, "200 degrees west longitude" would be the same as 160 degrees east longitude, and "270 degrees east longitude" would be the same as 90 degrees west longitude. One longitude coordinate for any point is enough; more than one is too many. The 180-degree west longitude arc, which might also be called the 180-degree east-longitude arc, is simply called "180 degrees longitude." A crooked line, corresponding approximately to 180 degrees longitude, is designated as the divider between dates on the calendar. This so-called *International Date Line* meanders through the western Pacific Ocean, avoiding major population centers.

Longitude coordinates, like their latitude counterparts, can be abbreviated. One hundred degrees west longitude, for example, is written "100 deg W long" or "100°W." Fifteen degrees east longitude is written "15 deg E long" or "15°E." Minutes and seconds of arc are used for greater precision; you might see a place at 103 degrees, 33 minutes, 7 seconds west longitude described as being at "103 deg 33 min 7 sec W long" or "103°33′07″W."

Find the aforementioned longitude half circles on a globe. Then find the town where you live, and figure out your longitude. Compare this with other towns around the world. As with latitude, you might be in for a shock. For example, if you live in Chicago, Illinois, you are further west in longitude than every spot in the whole continent of South America.

Celestial Latitude and Longitude

The latitude and longitude of a celestial object is defined as the latitude and longitude of the point on Earth's surface such that when the object is observed from there, the object is at the *zenith* (exactly overhead).

THE STARS

Suppose that a star is at x degrees north celestial latitude and y degrees west celestial longitude. If you stand at the point on the surface corresponding to $x°N$ and $y°W$, then a straight, infinitely long geometric ray originating at the center of Earth and passing right between your eyes will shoot up into space in the direction of the star (Fig. 1-3).

As you might guess, any star that happens to be at the zenith will stay there for only a little while unless you happen to be standing at either of the

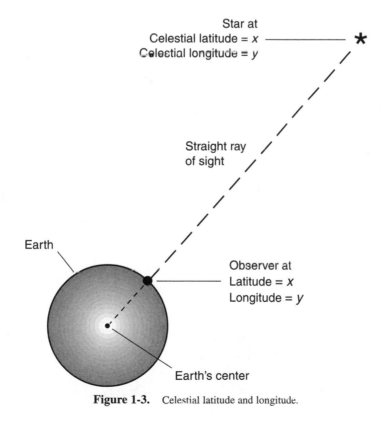

Figure 1-3.   Celestial latitude and longitude.

geographic poles (not likely). Earth rotates with respect to the stars, completing a full circle approximately every 23 hours and 56 minutes. In a few minutes, a star that is straight overhead will move noticeably down toward the western horizon. This effect is exaggerated when you look through a telescope. The greater the magnification, the more vividly apparent is the rotation of Earth.

The next time you get a chance, set up a telescope and point it at some star that is overhead. Use the shortest focal-length eyepiece that the telescope has so that the magnification is high. Center the star in the field of view. If that star is exactly overhead, then its celestial latitude and longitude correspond to yours. For example, if you're on the shore of Lake Tahoe, your approximate latitude is 39°N and your approximate longitude is 120°W. If you have a telescope pointing straight up and a star is centered in the field of view, then that star's celestial coordinates are close to 39°N, 120°W. However, this won't be the case for long. You will be able to watch the star drift out of the field of view. Theoretically, a star stays exactly at a given celestial longitude coordinate (x, y) for an infinitely short length of time—in essence, for no time at all. However, the celestial latitude of each and every star remains constant, moment after moment, hour after hour, day after day. (With the passage of centuries, the celestial latitudes of the stars change gradually because Earth's axis wobbles slowly. However, this effect doesn't change things noticeably to the average observer over the span of a lifetime.)

WHAT'S THE USE?

The celestial longitude of any natural object in the sky (except those at the north and south geographic poles) revolves around Earth as the planet rotates on its axis. No wonder people thought for so many centuries that Earth must be the center of the universe! This makes the celestial latitude/longitude scheme seem useless for the purpose of locating stars independently of time. What good can such a coordinate scheme be if its values have meaning only for zero-length micromoments that recur every 23 hours and 56 minutes? This might be okay for the theoretician, but what about people concerned with reality?

It turns out that the celestial latitude/longitude coordinate system is anything but useless. Understanding it will help you understand the more substantial coordinate schemes described in the next sections. And in fact, there is one important set of objects in the sky, a truly nuts-and-bolts group of hardware items, all of which stay at the same celestial latitude and longitude as viewed from any fixed location. These are the geostationary satellites, which lie in a human-made ring around our planet. These satellites orbit several thousand kilometers above the equator, and they revolve right along with Earth's rotation (Fig. 1-4).

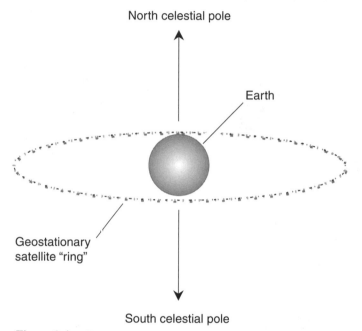

North celestial pole

Earth

Geostationary satellite "ring"

South celestial pole

Figure 1-4.   Geostationary satellites are all at 0 degrees celestial latitude, and each has a constant celestial longitude.

When it is necessary to point a dish antenna, such as the sort you might use to receive digital television or broadband Internet signals, at a geostationary "bird," the satellite's celestial coordinates must be known, in addition to your own geographic latitude and longitude, with great accuracy. The celestial latitude and longitude of a geostationary satellite are constant for any given place on Earth. If a satellite is in a geostationary orbit precisely above Quito, Ecuador, then that is where the "bird" will stay, moment after moment, hour after hour, day after day.

An Internet user fond of broadband and living in the remote South American equatorial jungle might use a dish antenna to transmit and receive data to and from a "bird" straight overhead. The dish could be set to point at the zenith and then left there. (It would need a hole near the bottom to keep it from collecting rain water!) A second user on the shore of Lake Tahoe, in the western United States, would point her dish at some spot in the southern sky. A third user in Tierra del Fuego, at the tip of South America, would point his dish at some spot in the northern sky (Fig. 1-5). None of the three dishes, once positioned, would ever have to be moved and, in fact, should never be moved.

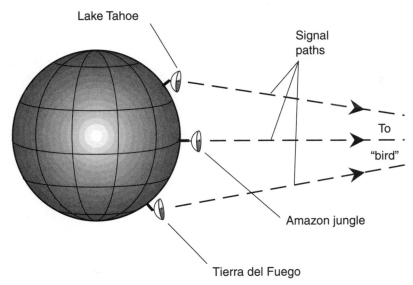

Figure 1-5.   A geostationary satellite has constant celestial latitude and longitude, so dish antennas can be aimed at it and then left alone.

If you're astute, you'll notice that although the geostationary satellite is directly above the equator, its celestial latitude is zero only with respect to observers located at the equator. If viewed from north of the equator, the satellite shifts a little bit into the southern celestial hemisphere; when observed from south of the equator, the satellite shifts slightly into the northern celestial hemisphere. The reason for this is *parallax*. The satellite is only a few thousand kilometers away, whereas the stars, whose celestial latitudes remain fixed, are trillions and quadrillions of kilometers distant. This is why the signal paths in Fig. 1-5 aren't exactly parallel. On a small scale, the phenomenon of parallax allows us to perceive depth with binocular vision. On a large scale, parallax is used to measure the distance to the Sun, the Moon, the other planets in the solar system, and even a few of the nearer stars.

The Az/El System

For centuries, navigators and casual observers have used a celestial coordinate system that is in some ways simpler than latitude/longitude and in other ways more complicated. This is the so-called *azimuth/elevation* scheme. It's often called *az/el* for short.

COMPASS BEARING

The *azimuth* of a celestial object is the compass bearing, in degrees, of the point on the horizon directly below that object in the sky. Imagine drawing a line in the sky downward from some object until it intersects the horizon at a right angle. The point at which this intersection occurs is the azimuth of the object. If an object is straight overhead, its azimuth is undefined.

Azimuth bearings are measured clockwise with respect to geographic north. The range of possible values is from 0 degrees (north) through 90 degrees (east), 180 degrees (south), 270 degrees (west), and up to, but not including, 360 degrees (north again). This is shown in Fig. 1-6A. The azimuth bearing of 360 degrees is left out to avoid ambiguity, so the range of possible values is what mathematicians call a *half-open interval*. Azimuth bearings of less than 0 degrees or of 360 degrees or more are reduced to some value in the half-open interval (0°, 360°) by adding or subtracting the appropriate multiple of 360 degrees.

ANGLE RELATIVE TO THE HORIZON

The *elevation* of an object in the sky is the angle, in degrees, subtended by an imaginary arc running downward from the object until it intersects the horizon at a right angle. This angle can be as small as 0 degrees when the object is on the horizon, or as large as 90 degrees when the object is directly overhead. If the terrain is not flat, then the horizon is defined as that apparent circle halfway between the zenith and the *nadir* (the point directly below you, which would be the zenith if you were on the exact opposite side of the planet).

Elevation bearings for objects in the sky are measured upward from the horizon (Fig. 1-6B). Such coordinates are, by convention, not allowed to exceed 90 degrees because that would produce an ambiguous system. Although you might not immediately think of them, elevation bearings of less than 0 degrees are possible, all the way down to −90 degrees. These

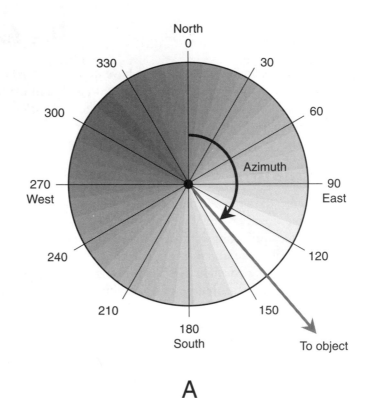

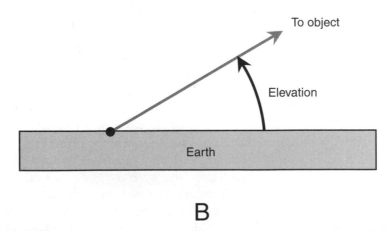

Figure 1-6A.   Azimuth is the compass bearing. The observer is shown as a black dot.

Figure 1-6B.   Elevation is the angle above the horizon. The observer is shown as a black dot.

bearings represent objects below the horizon. While we can't see such objects, they are there nevertheless. At night, for example, the Sun has a negative elevation. Technically, elevation bearings always have values within the closed interval [−90°, 90°].

SKY MAPS ON THE WEB

Various Internet sites provide up-to-the-minute maps of the sky for stargazers. One excellent site can be found by pointing your browser to *Weather Underground* at the following URL

http://www.wunderground.com

and then clicking on the link that says "Astronomy." From there, it's a simple matter of following the online instructions.

Some star maps are drawn so that the sky appears as it would if you lie on your back with your head facing north and your feet facing south. Thus west appears on your right, and east appears on your left (Fig. 1-7a). Others are drawn so that the sky appears as it would if your head were facing south and your feet were facing north, so west appears on your left and east appears on your right (Fig. 1-7b). Points having equal elevation form concentric circles, with the zenith (90 degrees) being a point at the center of the map and the horizon (0 degrees) being a large circle representing the periphery of the map. Simplified sets of *grid lines* for such az/el maps are shown in both illustrations of Fig. 1-7.

These maps show the Sun and the pole star Polaris as they might appear at midafternoon from a location near Lake Tahoe (or anyplace else on Earth at the same latitude as Lake Tahoe). The gray line represents the path of the Sun across the sky that day. From this you might get some idea of the time of year this map represents. Go ahead and take an educated guess! Here are two hints:

• The Sun rises exactly in the east and sets exactly in the west.

• The situation shown can represent either of two approximate dates.

Right Ascension and Declination

There are two points in time every year when the Sun's elevation, measured with respect to the center of its disk, is positive for exactly 12 hours and

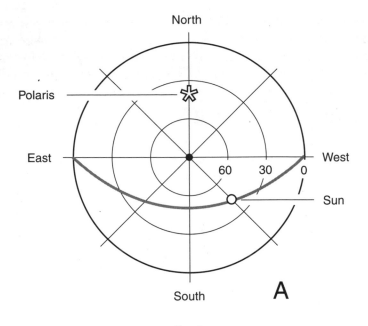

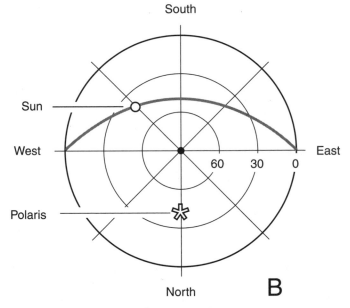

Figure 1-7.   Az/el sky maps for viewer lying flat, face-up.
At *A*, top of head facing north; at *B*, top of head facing south.

negative for exactly 12 hours. One of these time points, the *vernal equinox*, occurs on March 21, give or take about a day; the other, the *autumnal equinox*, occurs on September 22, give or take about a day. At the equinoxes, the Sun is exactly at the celestial equator; it rises exactly in the east and sets exactly in the west, assuming that the observer is not at either of the geographic poles.

The crude celestial maps of Fig. 1-7 show the situation at either of the equinoxes. That is, the date is on or around March 21 or September 22. You can deduce this because the Sun rises exactly in the east and sets exactly in the west, so it must be exactly at the celestial equator. At the latitude of Lake Tahoe, the Sun is 39 degrees away from the zenith (51 degrees above the southern horizon) at high noon on these days. Polaris is 39 degrees above the northern horizon all the time. The entire heavens seem to rotate counterclockwise around Polaris.

THE VERNAL EQUINOX

What's this about the Sun being above the horizon for exactly 12 hours and below the horizon for exactly 12 hours at the equinoxes? The stars in the heavens seem to revolve around Earth once every 23 hours and 56 minutes, approximately. Where do the 4 extra minutes come from?

The answer is that the Sun crosses the sky a little more slowly than the stars. Every day, the Sun moves slightly toward the east with respect to the background of stars. On March 21, the Sun is at the celestial equator and is located in a certain position with respect to the stars. This point among the stars is called, naturally enough, the *vernal equinox* (just as the date is called). It represents an important reference point in the system of celestial coordinates most often used by astronomers: *right ascension* (RA) and *declination* (dec). As time passes, the Sun rises about 4 minutes later each day relative to the background of stars. The *sidereal* (star-based) day is about 23 hours and 56 minutes long; the *synodic* (sun-based) day is precisely 24 hours long. We measure time with respect to the Sun, not the stars.

Declination is the same as celestial latitude, except that "north" is replaced by "positive" and "south" is replaced by "negative." The south celestial pole is at dec = -90 degrees; the equator is at dec = 0 degrees; the north celestial pole is at dec = $+90$ degrees. In the drawings of Fig. 1-7, the Sun is at dec = 0 degrees. Suppose that these drawings represent the situation on March 21. This point among the stars is the zero point for right ascension (RA = 0 h). As springtime passes and the Sun follows a higher and higher course across the sky, the declination and right ascension both increase for a while. Right ascension is measured eastward along the celestial equator from the March equinox in units called *hours*. There are 24 hours of right ascension in a complete circle; therefore, 1 hour (written 1 h or 1^h) of RA is equal to 15 angular degrees.

THE SUN'S ANNUAL "LAP"

As the days pass during the springtime, the Sun stays above the horizon for more and more of each day, and it follows a progressively higher course across the sky. The change is rapid during the early springtime and becomes more gradual with approach of the *summer solstice*, which takes place on June 22, give or take about a day.

At the summer solstice, the Sun has reached its northernmost declination point, approximately dec = +23.5 degrees. The Sun has made one-quarter of a complete circuit around its annual "lap" among the stars and sits at RA = 6 h. This situation is shown in Fig. 1-8 using the same two az/el coordinate schemes as those in Fig. 1-7. The gray line represents the Sun's course across the sky. As in Fig. 1-7, the time of day is midafternoon. The observer's geographic latitude is the same too: 39°N.

After the summer solstice, the Sun's declination begins to decrease, slowly at first and then faster and faster. By late September, the autumnal equinox is reached, and the Sun is once again at the celestial equator, just as it was at the vernal equinox. Now, however, instead of moving from south to north, the Sun is moving from north to south in celestial latitude. At the autumnal equinox, the Sun's RA is 12 h. This corresponds to 180 degrees.

Now it is the fall season in the northern hemisphere, and the days are growing short. The Sun stays above the horizon for less and less of each day, and it follows a progressively lower course across the sky. The change is rapid during the early fall and becomes slower and slower with approach of the *winter solstice*, which takes place on December 21, give or take about a day.

At the winter solstice, the Sun's declination is at its southernmost point, approximately dec = −23.5 degrees. The Sun has made three-quarters of a complete circuit around its annual "lap" among the stars and sits at RA = 18 h. This is shown in Fig. 1-9 using the same two az/el coordinate schemes as those in Figs. 1-7 and 1-8. The gray line represents the Sun's course across the sky. As in Figs. 1-7 and 1-8, the time of day is

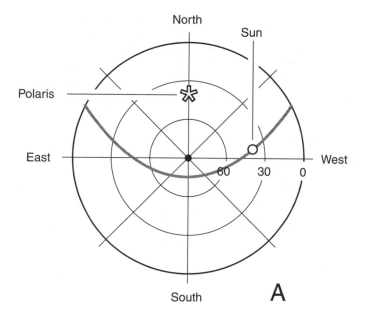

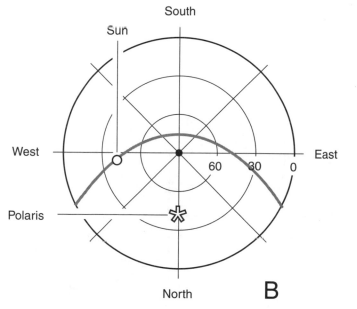

Figure 1-8.   Az/el sky maps for midafternoon at 39 degrees north latitude on or around June 21.

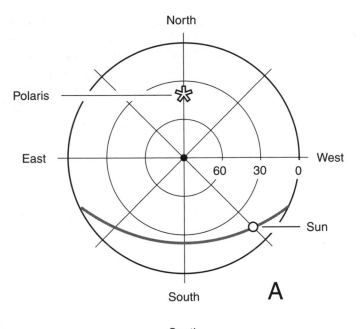

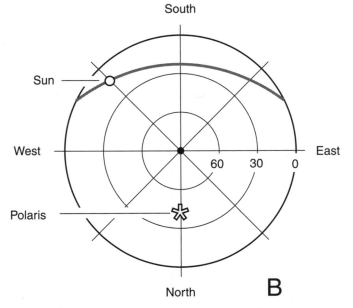

Figure 1-9.   Az/el sky maps for midafternoon at 39 degrees north latitude on or around December 21.

midafternoon. The observer hasn't moved either, at least in terms of geographic latitude; this point is still at 39°N. (Maybe the observer is in Baltimore now or in the Azores. Winter at Lake Tahoe can be rough unless you like to ski.)

After the winter solstice, the Sun's declination begins to increase gradually and then, as the weeks pass, faster and faster. By late March, the Sun reaches the vernal equinox again and crosses the celestial equator on its way to warming up the northern hemisphere for another spring and summer. The "lap" is complete. The Sun's complete circuit around the heavens takes about 365 solar days plus 6 hours and is the commonly accepted length of the year in the modern calendar. In terms of the stars, there is one extra "day" because the Sun has passed from west to east against the far reaches of space by a full circle.

THE ECLIPTIC

The path that the Sun follows against the background of stars during the year is a slanted celestial circle called the *ecliptic*. Imagine Earth's orbit around the Sun; it is an ellipse (not quite a perfect circle, as we will later learn), and it lies in a flat geometric plane. This plane, called the *plane of the ecliptic*, is tilted by 23.5 degrees relative to the plane defined by Earth's equator. If the plane of the ecliptic were made visible somehow, it would look like a thin gray line through the heavens that passes through the celestial equator at the equinoxes, reaching a northerly peak at the June solstice and a southerly peak at the December solstice. If you've ever been in a planetarium, you've seen the ecliptic projected in that artificial sky, complete with RA numbers proceeding from right to left from the vernal equinox.

Suppose that you convert the celestial latitude and longitude coordinate system to a Mercator projection, similar to those distorted maps of the world in which all the parallels and meridians show up as straight lines. The ecliptic would look like a sine wave on such a map, with a peak at +23.5 degrees (the summer solstice), a trough at −23.5 degrees (the winter solstice), and two nodes (one at each equinox). This is shown in Fig. 1-10. From this graph, you can see that the number of hours of daylight, and the course of the Sun across the sky, changes rapidly in March, April, September, and October and slowly in June, July, December, and January. Have you noticed this before and thought it was only your imagination?

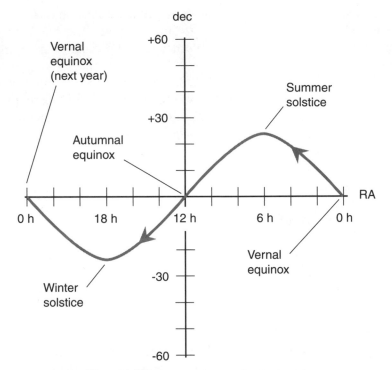

Figure 1-10.   The ecliptic (gray line) is the path
that the Sun follows in its annual "lap around the heavens."

Quiz

Refer to the text if necessary. A good score is 8 correct. Answers are in the back of
the book.

1. Which of the following points or objects seems to stay fixed in the sky all the
   time?
   (a) The vernal equinox
   (b) The background of stars
   (c) The Sun
   (d) Polaris

2. What is the declination represented by the south celestial pole?
   (a) 18 h
   (b) 0°
   (c) −90
   (d) It cannot be defined because it changes with time.

3. What is the celestial longitude of the winter solstice?
   (a) 18 h
   (b) 0°
   (c) −90
   (d) It cannot be defined because it changes with time.

4. A point is specified as having a celestial latitude of 45°30′00″N. This is equivalent to how many degrees as a decimal fraction?
   (a) 45.5°
   (b) 45.3°
   (c) 30.00°
   (d) There is not enough information to tell.

5. How many hours of right ascension correspond to one-third of a circle?
   (a) 3 h
   (b) 6 h
   (c) 12 h
   (d) None of the above

6. How many sidereal days are there in one full calendar year?
   (a) Approximately 366.25
   (b) Approximately 365.25
   (c) Approximately 364.25
   (d) It depends on the celestial latitude of the observer.

7. One second of arc represents what fraction of a complete circle?
   (a) 1/60
   (b) 1/3600
   (c) 1/86,400
   (d) 1/1,296,000

8. The celestial latitude/longitude frame of reference
   (a) is fixed relative to the background of stars.
   (b) is fixed relative to the Sun.
   (c) is fixed relative to geostationary satellites.
   (d) is not fixed relative to anything.

9. Which of the following represents an impossible or improperly expressed coordinate value?
   (a) RA = 12 h
   (b) dec = +100
   (c) 103 deg 00 min 20 sec W long
   (d) 23°S

10. Azimuth is another name for
    (a) right ascension.
    (b) compass bearing.
    (c) celestial latitude.
    (d) celestial longitude.

CHAPTER 2

Stars and Constellations

We may never know exactly what the common people of ancient times believed about the stars. We can read the translations of the works of the scribes, but what about the shepherds, the nomads, and the people in the ages before writing existed? They must have noticed that stars come in a variety of brightnesses and colors. Even though the stars seem to be scattered randomly (unless the observer knows that the Milky Way is a vast congregation of stars), identifiable star groups exist. These star groups do not change within their small regions of the sky, although the vault of the heavens gropes slowly westward night by night, completing a full circle every year. These star groups and the small regions of the sky they occupy are called *constellations*.

Illusions and Myths

We know that the constellations are not true groups of stars but only appear that way from our Solar System. The stars within a constellation are at greatly varying distances. Two stars that look like they are next to each other really may be light-years apart (a *light-year* is the distance light travels in a year) but nearly along the same line of sight. As seen from some other star in this part of the galaxy, those two stars may appear far from each other in the sky, maybe even at celestial antipodes (points 180 degrees apart on the celestial sphere). Familiar constellations such as Orion and Hercules are their true selves only with respect to observers near our Sun. Interstellar travelers cannot use the constellations for navigation.

COSMIC CIRCUS

Many of the constellations are named for ancient Greek gods or for people, animals, or objects that had special associations with the gods. Cassiopeia is the mother of Andromeda. Orion is a hunter; Hercules is a warrior; Draco is a dragon. There are a couple of bears and a couple of dogs. There is a sea monster, most likely a crazed whale, who almost had Andromeda for supper one day. There is a winged horse, a pair of fish whose tails are attached, a bull, a set of scales used to mete out cosmic justice, a goat, twin brothers who look after ocean-going vessels, and someone pouring water from a jug that never goes empty. The Greeks saw a lot of supernatural activity going on in the sky. They must have thought themselves fortunate that they were far enough away from this lively circus so as not to be kept awake at night by all the growling, barking, shouting, and whinnying. (If the ancient Greeks knew what really happens in the universe, they would find the shenanigans of their gods and animals boring in comparison.)

The people of Athens during the age of Pericles saw the exact same constellations that we see today. You can look up in the winter sky and recognize Orion immediately, just as Pericles himself must have. The constellations, as you know them, will retain their characteristic shapes for the rest of your life and for the lives of your children, grandchildren, and great-grandchildren. In fact, the constellations have the same shapes as they did during the Middle Ages, during the height of the Roman Empire, and at the dawn of civilization, when humans first began to write down descriptions of them. The positions of the constellations shift in the sky from night to night because Earth revolves around the Sun. They also change position slightly from century to century because Earth's axis slowly wobbles, as if our planet were a gigantic, slightly unstable spinning top. But the essential shapes of the constellations take millions of years to change.

By the time Orion no longer resembles a hunter but instead perhaps a hunched old man or a creature not resembling a human at all, this planet likely will be populated by beings who look back in time at us as we look back at the dinosaurs. As life on Earth evolves and changes shape, and as the beings on our planet find new ways to pass through life, so shall the mythical deities of the sky undergo transformations. Perhaps Orion will become his own prey, and Draco will be feeding Hercules his supper every evening.

SKY MAPS

In this chapter, the general shapes of the better-known constellations are shown. To see where these constellations are in the sky from your location this evening, go to the *Weather Underground* Web site at the following URL:

http://www.wunderground.com

Type in your ZIP code or the name of your town and state (if in the United States) or your town and country and then, when the weather data page for your town comes up, click on the "Astronomy" link. There you will find a detailed map of the entire sky as it appears from your location at the time of viewing, assuming that your computer clock is set correctly and data are input for the correct time zone.

Celestron International publishes a CD-ROM called *The Sky*, which shows stars, planets, constellations, coordinates, and other data for any location on Earth's surface at any time of the day or night. This CD-ROM can be obtained at hobby stores that sell Celestron telescopes.

Circumpolar Constellations

Imagine that you're stargazing on a clear night from some location in the mid-northern latitudes, such as southern Europe, Japan, or the central United States. Suppose that you sit down and examine the constellations on every clear evening, a couple of hours after sunset, for an entire year. Sometimes the Moon is up, and sometimes it isn't. Its phase and brightness affect the number of stars you see even on the most cloud-free, haze-free nights. But some constellations stand out enough to be seen on any evening when the weather permits. The constellations near the north celestial pole are visible all year long. The following subsections describe these primary constellations.

STAR BRIGHTNESS

In this chapter, stars are illustrated at three relative levels of brightness. Dim stars are small black dots. Stars of medium brilliance are larger black dots. Bright stars are circles with black dots at their centers. But the terms *dim, medium,* and *bright* are not intended to be exact or absolute. In New York City, some of the *dim* stars shown in these drawings are invisible, even under good viewing conditions, because of scattered artificial light. After your eyes have had an hour to adjust to the darkness on a moonless, clear night in the mountains of Wyoming, some of the *dim* stars in these illustrations will look fairly bright. The gray lines connecting the stars (reminiscent of dot-to-dot children's drawings) are intended to emphasize the general shapes of the constellations. The lines do not, of course, appear in the real sky, although they are often shown in planetarium presentations and are commonly included in sky maps.

POLARIS

One special, moderately bright star stays fixed in the sky all the time, day and night, season after season, and year after year. This star, called *Polaris*, or *the pole star*, is a white star of medium brightness. It can be found in the northern sky at an elevation equal to your latitude. If you live in Minneapolis, for example, Polaris is 45 degrees above the northern horizon. If you live on the Big Island of Hawaii, it is about 20 degrees above the horizon. If you live in Alaska, it's about 60 to 65 degrees above the horizon. At the equator, it's on the northern horizon. People in the southern hemisphere never see it.

Polaris makes an excellent reference for the northern circumpolar constellations and in fact for all the objects in the sky as seen from any location in Earth's northern hemisphere. No matter where you might be, if you are north of the equator, Polaris always defines the points of the compass. Navigators and explorers have known this for millennia. You can use the pole star as a natural guide on any clear night.

URSA MINOR

Polaris rests at the end of the "handle" of the so-called little dipper. The formal name for this constellation is *Ursa Minor*, which means "little bear." One might spend quite a while staring at this constellation before getting the idea that it looks like a bear, but that is the animal for which it is named, and whoever gave it that name must have had a reason. Most constellations are named for animals or mythological figures that don't look anything like them, so you might as well get used to this. The general shape of Ursa Minor is shown in Fig. 2-1. Its orientation varies depending on the time of night and the time of year.

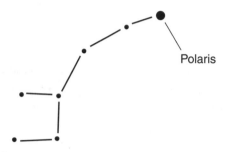

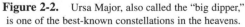

Figure 2-1.   Ursa Minor is commonly known as the "little dipper."

URSA MAJOR

The so-called big dipper is formally known as *Ursa Major*, which means "big bear." It is one of the most familiar constellations to observers in the northern hemisphere. In the evening, it is overhead in the spring, near the northern horizon in autumn, high in the northeastern sky in winter, and high in the northwestern sky in summer. It, like its daughter, Ursa Minor, is shaped something like a scoop (Fig. 2-2). The two stars at the front of

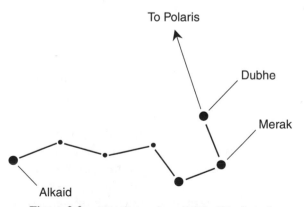

Figure 2-2.   Ursa Major, also called the "big dipper," is one of the best-known constellations in the heavens.

the scoop are *Dubhe* and *Merak*, and are called the *pointer stars* because they are roughly aligned with Polaris. If you can find the big dipper, look upward from the scoop five or six times the distance between Dubhe and Merak, and you will find Polaris. To double-check, be sure that the star you have found is at the end of the handle of the little dipper.

CASSIOPEIA

One of the north circumpolar constellations is known for its characteristic M or W shape (depending on the time of night and the time of year it is viewed). This is *Cassiopeia*, which means "queen." Ancient people saw this constellation's shape as resembling a throne (Fig. 2-3), and this is where the idea of royalty came in. In the evening, Cassiopeia is low in the north-northwestern sky in the spring, moderately low in the north-northeastern sky in the summer, near the zenith in the fall, and high in the northwestern sky in the winter.

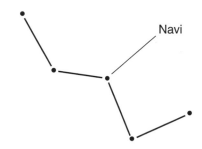

Figure 2-3.   Cassiopeia, also called the "queen," looks like the letter M or the letter W. To the ancients, it had the shape of a throne.

CEPHEUS

As the queen sits on her throne, she faces her spouse, the king, the constellation whose formal name is *Cepheus*. This constellation is large in size but is comprised of relatively dim stars. For this reason, Cepheus is usually obscured by bright city lights or the sky glow of a full moon, especially when it is near the horizon. It has the general shape of a house with a steeply pitched roof (Fig. 2-4). In the evening, Cepheus is near the northern horizon in the spring, high in the north-northeast sky in the summer, nearly overhead in the fall, and high in the northwestern sky in the fall.

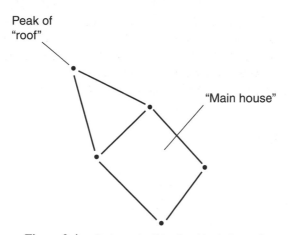

Figure 2-4. Cepheus, the "king," resides in front of Cassiopeia's throne. It has the shape of a house with a steep roof.

DRACO

One of the largest circumpolar constellations, obscure to casual observers on account of its long, winding shape, is *Draco*, the dragon. With the exception of *Eltanin*, a star at the front of the dragon's "head," this constellation is made up of comparatively dim stars (Fig. 2-5). Draco's "tail" wraps around the little dipper. The big dipper is in a position to scoop up the dragon tail first. In the evening, Draco is high in the northeastern sky in springtime, nearly overhead in the summer, high in the north-northwest sky in the fall, and near the northern horizon in the winter.

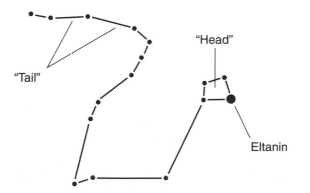

Figure 2-5. Draco, the "dragon," has a long, sinuous shape with an obvious "head" and "tail."

PERSEUS

Perseus is another circumpolar constellation with an elongated, rather complicated shape (Fig. 2-6). A mythological hero, Perseus holds the decapitated head of Medusa, a mythological female monster with hair made of snakes and a countenance so ugly that anyone who looked on it was turned into stone. Perseus is low in the northwestern sky in springtime, half above and half below the northern horizon in the summer, high in the northeastern sky in the fall, and nearly overhead in the winter.

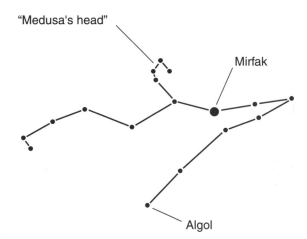

Figure 2-6.   Perseus is a mythological hero who holds the severed head of Medusa. Its star Algol varies in brilliance because it is actually two stars that eclipse each other as they orbit around a common center of gravity.

OTHER CIRCUMPOLAR CONSTELLATIONS

There are other, lesser constellations that remain above the horizon at all times. These can be found on star maps. The further north you go, the more circumpolar constellations there are. If you were to go all the way to the north pole, all the constellations would be circumpolar. The stars would all seem to revolve around the zenith, completing one full circle every 23 hours and 56 minutes. Conversely, the further south you go, the fewer circumpolar constellations you will find. At the equator, there are none at all; every star in the sky spends half the sidereal day (about 11 hours and 58 minutes) above the horizon and half the sidereal day below the horizon.

We haven't discussed what happens if you venture south of the equator. But we will get to that in the next chapter.

Constellations of Spring

Besides the circumpolar constellations, there are certain star groups that are characteristic of the evening sky in spring in the northern hemisphere. The season of spring is 3 months long, and even if you live in the so-called temperate zone, your latitude might vary. Thus, to get ourselves at a happy medium, let's envision the sky in the middle of April, a couple of hours after sunset at the latitude of Lake Tahoe, Indianapolis, or Washington, D.C. (approximately 39° N).

LIBRA

Libra is near the east-southeastern horizon. It has the general shape of a trapezoid (Fig. 2-7) if you look up at it facing toward the east-southeast. Libra is supposed to represent the scales of justice. This constellation is faint and once was considered to be part of Scorpio, the scorpion.

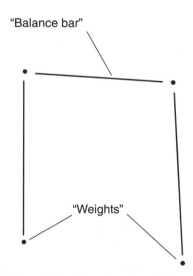

Figure 2-7.   Libra, the scales of justice.

VIRGO

Virgo, the virgin, is fairly high in the southeastern sky. It has an irregular shape, something like a letter Y with a hooked tail (Fig. 2-8) if you look up at it facing toward the southeast. Virgo contains the bright star *Spica*.

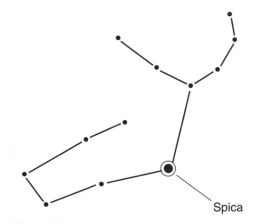

Spica

Figure 2-8.   Virgo, the virgin, holds a staff of wheat.

LEO

Leo, the lion, is just south of the zenith. This constellation is dominated by the bright star *Regulus*. If you stand facing south and crane your neck until you're looking almost straight up, you might recognize this constellation by its Sphinx-like shape (Fig. 2-9).

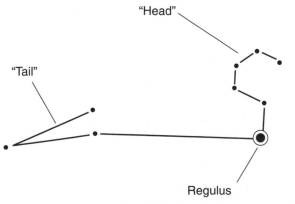

"Head"

"Tail"

Regulus

Figure 2-9.   Leo, the lion, resembles the Sphinx.

CANCER AND CANIS MINOR

Cancer, the crab, stands high in the southwestern sky. If you stand facing southwest and look up at an elevation about 70 degrees, you'll see a group of stars that resembles an upside-down Y (Fig. 2-10). In ancient mythology, Cancer was the cosmic gate through which souls descended to Earth to occupy human bodies. Next to Cancer is *Canis Minor*, the little dog, which contains the prominent star *Procyon*.

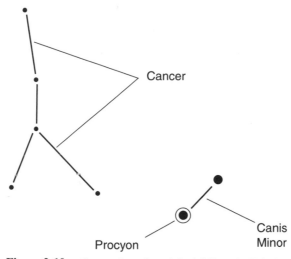

Figure 2-10.   Cancer, the crab; and Canis Minor, the little dog.

GEMINI

Gemini is moderately high in the western sky. This constellation has the general shape of a tall, thin, squared-off letter U if you stand facing west and look up at it (Fig. 2-11). At the top of the U are the prominent stars *Castor* and *Pollux*, named after the twin sons of Zeus, the most powerful of the ancient Greek gods. If you use your imagination, you might see Castor facing toward the left, with Pollux right behind him. The bright stars must be their left eyes.

AURIGA

Just to the right of Gemini is the constellation *Auriga*. It has the shape of an irregular pentagon (Fig. 2-12) as you face west-northwest and look

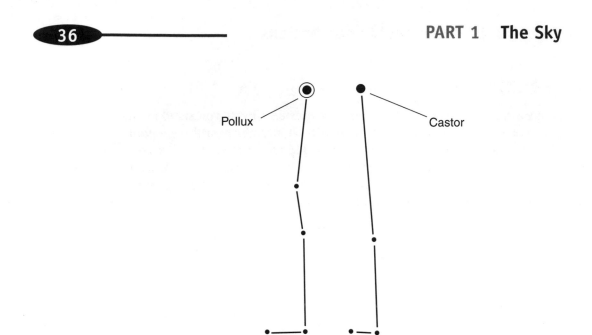

Figure 2-11.   Gemini. The stars Pollux and Castor represent the twins.

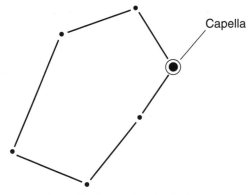

Figure 2-12.   Auriga, the charioteer.

upward about 30 degrees from the horizon. Auriga contains the bright star *Capella*. In ancient mythology, this constellation represented the king of Athens driving a four-horse chariot. Presumably, Capella is the king, and the four lesser stars are the horses.

HERCULES

Turn around and look toward the east-northeast, just above the horizon. You will see a complex of stars, none of them bright, forming a trapezoid with limbs (Fig. 2-13). This is the constellation *Hercules*, representing a man of legendary strength and endurance. In the sky, he faces Draco, the dragon. These two cosmic beings are engaged in a battle that has been going on for millennia and will continue to rage for ages to come. Who will win? No one knows, but eventually, as the stars in our galaxy wander off in various directions, both these old warriors will fade away. Hercules contains one of the most well-known star clusters in the heavens, known as *M13* (this is a catalog number).

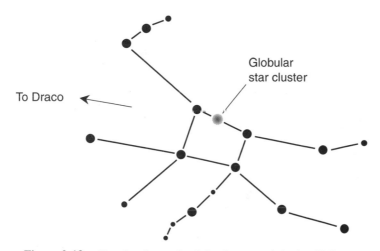

Figure 2-13. Hercules, the warrior, is in a long, cosmic battle with Draco.

CORONA BOREALIS

Looking slightly higher in the sky, just behind Hercules, you will see a group of several stars that form a backward C or horseshoe shape. These stars form the constellation *Corona Borealis*, the northern crown (Fig. 2-14). This is the head ornament that was worn by Ariadne, a princess of Crete. According to the legends, the Greek god Dionysis threw the crown up into the sky to immortalize the memory of Ariadne.

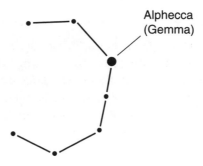

Figure 2-14.   Corona Borealis, the crown worn by Ariadne, princess of Crete.

BOOTES

Just above the northern crown you will see a bright star at an elevation of about 45 degrees, directly east or a little south of east. This is *Arcturus*. If you use your imagination, you might see that this star forms the point where a fish joins its tail; the fish appears to be swimming horizontally (Fig. 2-15). This constellation, *Bootes*, does not represent a fish but a herdsman. His job, in legend, is to drive Ursa Major, the great bear, forever around the north pole.

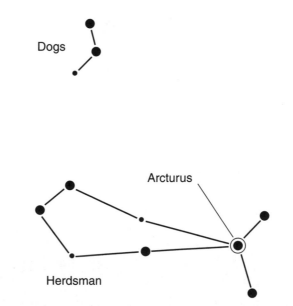

Figure 2-15.   Bootes, the herdsman, and his hunting dogs, Canes Venatici.

CANES VENATICI

Between Bootes and Ursa Major there is a group of three rather dim stars (shown in Fig. 2-15 along with their master). These are Bootes' hunting dogs, *Canes Venatici*, who snap at the heels of Ursa Major and keep the big bear moving. One might argue that according to myth, our planet owes its rotation, at least in part, to a couple of cosmic hounds.

CORVUS, CRATER, AND HYDRA

A large portion of the spring evening sky is occupied by three constellations consisting of relatively dim stars. These are *Corvus*, also known as the crow, *Crater*, also called the cup, and *Hydra*, the sea serpent or water snake (Fig. 2-16). It is not too hard to imagine how Hydra got its name, and one might with some effort strain to imagine Crater as a cup. But Corvus is a fine example of a constellation that looks nothing like the mythological creature or object it represents.

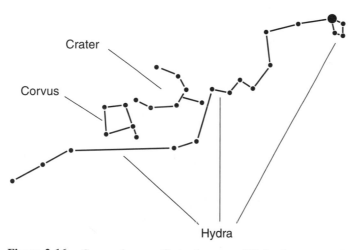

Figure 2-16.   Corvus, the crow; Crater, the cup; and Hydra, the sea serpent.

Constellations of Summer

Now let's look at the sky in the middle of July, a couple of hours after sunset at the latitude of Lake Tahoe, Indianapolis, or Washington, D.C.

(approximately 39° N). Some of the spring constellations are still visible. All the circumpolar constellations are still there, but they appear to have rotated around Polaris one-quarter of a circle counterclockwise from their positions in the spring. Other spring constellations that are still visible, though they have moved toward the west the equivalent of about 6 hours, include Virgo, Libra, Bootes, Canes Venatici, Corona Borealis, and Hercules. New star groups have risen in the east, and old ones have set in the west. Here are the prominent new constellations of summer.

CAPRICORNUS

Near the horizon in the east-southeast sky is a group of stars whose outline looks like the main sail on a sailboat. This constellation is *Capricornus* (often called *Capricorn*), the goat (Fig. 2-17). This goat has the tail of a fish, according to the myths, and dwells at sea. On its way to heaven after the death of the body, the human soul was believed to pass through this constellation; it is 180 degrees opposite in the celestial sphere from Cancer, through which souls were believed to enter this world.

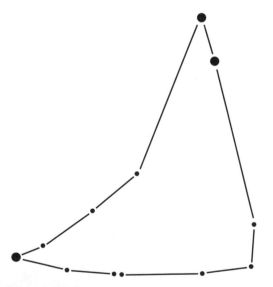

Figure 2-17. Capricornus, the goat, has the tail of a fish.

SAGITTARIUS

In the east-southeast sky, to the right and slightly above Capricornus, you will see a constellation whose outline resembles a teapot (Fig. 2-18). This is *Sagittarius*, the centaur. What, you might ask, is a centaur? You ought to know if you have read mythology or seen a lot of movies or television; it is a creature with the lower body of a horse and the chest, head, and arms of a human being. The centaur carries a bow and arrow with which to stun evil or obnoxious creatures. Sagittarius lies in the direction of the densest part of the Milky Way, the spiral galaxy in which our Solar System resides.

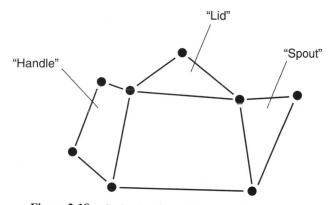

Figure 2-18.   Sagittarius, the centaur, is shaped like a teapot as viewed from northern-hemispheric temperate latitudes.

SCORPIUS

A huge and hapless scorpion, forever on the verge of feeling the bite of Sagittarius's arrow, sits in the southern sky, extending from near the horizon to an elevation of about 30 degrees (Fig. 2-19). This is *Scorpius* (also called *Scorpio*). This constellation is one of the few that bears some resemblance to the animal or object it represents. The eye of the scorpion is the red giant star *Antares*, which varies in brightness.

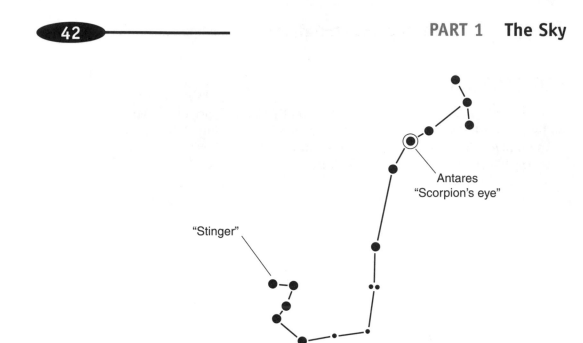

Figure 2-19.   Scorpius, the scorpion, contains the red star Antares.

THE SUMMER TRIANGLE

If you stand facing east and look up near the zenith, you will see the bright star *Vega*, flanked by a small parallelogram of dimmer stars. The quadrilateral forms the constellation *Lyra*, representing the lyre played by the mythical musician Orpheus. Below and to the left of Vega is another bright star, *Deneb*, that is at the tip of the tail of *Cygnus*, the swan. If you are at a dark location away from city lights on a moonless summer evening, you might imagine this bird soaring along the Milky Way that stretches from the north-northeastern horizon all the way to the southern horizon. Off to the right of these is a third bright star, *Altair*. This is part of the constellation *Aquila*, the eagle that pecks eternally at the liver of Prometheus as part of his punishment for stealing fire from the gods. Vega, Deneb, and Altair stand high in the east on summer evenings and comprise the well-known *summer triangle* (Fig. 2-20).

OPHIUCHUS AND SERPENS

In the southern sky, centered at the celestial equator, is the constellation *Ophiuchus*, the snake bearer. This poor soul holds a snake, the constellation *Serpens*, that stretches well to either side. You might imagine that Ophiuchus has a meaningless job, but nothing could be further from the truth. Ophiuchus must keep a tight hold on Serpens (Fig. 2-21), for if that

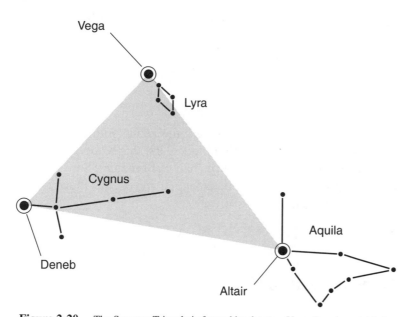

Figure 2-20.   The Summer Triangle is formed by the stars Vega, Deneb, and Altair, in the constellations Lyra, Cygnus, and Aquila (the lyre, the swan, and the eagle).

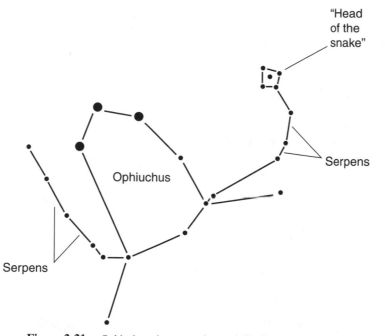

Figure 2-21.   Ophiuchus, the serpent bearer, holds Serpens, the snake. The snake's head is but a small distance from the back of unsuspecting Bootes.

snake gets away, it will easily be able to reach and bite Bootes, the herdsman. If that were to happen, Bootes and his dogs, Canes Venatici, would stop driving Ursa Major around Polaris, and Earth would stop spinning!

COMA BERENICES

About halfway between the horizon and the zenith in the west-northwest sky, you will see a fuzzy blob. With binoculars, this resolves into a cluster of stars known as *Coma Berenices* (the hair of Berenice). Some people mistake this group of stars for the Pleiades. However, the Pleiades are best observed in the winter.

Constellations of Autumn

Now imagine that it is an evening in the middle of October, a couple of hours after sunset, and you are at the latitude of Lake Tahoe, Indianapolis, or Washington, D.C. (approximately 39° N). New constellations have risen in the east, and old ones set in the west. The sky is looked after by new custodians as the nights grow longer. Here are constellations we have not described before that now occupy prominent positions in the sky.

PISCES AND ARIES

High in the southeast you will see *Pisces*, the two fish, and *Aries*, the winged ram (Fig. 2-22). Legend has it that Pisces were joined or tied together at their tails long ago, and to this day they are flailing about in that unfortunate condition. Aries has fleece of gold, and for this reason, the ram is sought after by a cosmic spirit called *Jason* and his cohorts called the *Argonauts*.

CETUS

Somewhat below and to the right of Pisces is *Cetus*, the whale, also considered a sea monster in some myths (Fig. 2-23). The variable star *Mira* is sometimes visible in the belly of the whale. Cetus is supposed to have been sent to swallow Andromeda, but this mission did not succeed. Cetus con-

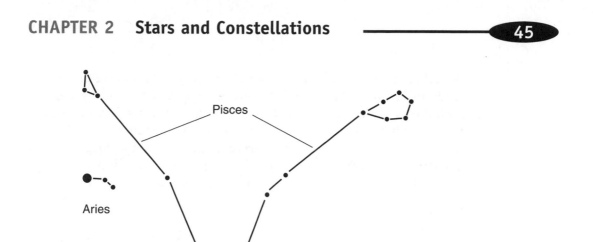

Figure 2-22.   Pisces, the fishes, and Aries, the ram.

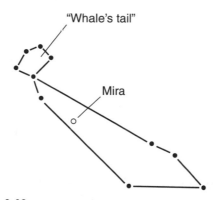

Figure 2-23.   Cetus, the whale, contains the variable star Mira.

tains one star, called *Tau Ceti*, believed to be a good candidate for having a solar system similar to ours.

PEGASUS AND ANDROMEDA

Nearly at the zenith there is a square consisting of four medium-bright stars. This is the body of *Pegasus*, the winged horse. Toward the northeast, *Andromeda*, representing a princess, rides the horse alongside the Milky

Way (Fig. 2-24). Andromeda had been chained to a rock and left out for Cetus to devour as the tide came in, but she was rescued by Perseus. Andromeda contains a spiral galaxy similar to our Milky Way but is more than 2 million light-years away. This galaxy can be seen as a dim blob by people with keen eyesight; with a massive telescope at low magnification, it resolves into a spectacular object. When photographed over a period of hours, it takes on the classic appearance of a spiraling disk of stars.

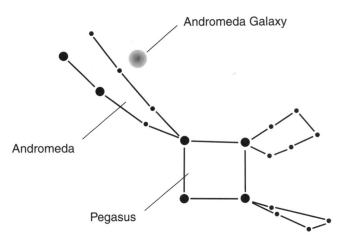

Figure 2-24.   Pegasus, the winged horse, and Andromeda, the Ethiopian princess who married Perseus. The Andromeda Galaxy is shown as a fuzzy dot.

AQUARIUS

Beneath Pegasus, in the southern sky, you will see *Aquarius*, the water bearer. This is not an easy constellation to envision as any sort of human figure; it more nearly resembles an exotic, long-necked bottle or a tree branch (Fig. 2-25). Aquarius supposedly brings love and peace as well as water.

PISCIS AUSTRINUS AND GRUS

Low in the southern sky is *Piscis Austrinus*, also called *Piscis Australis*. This is the southern fish and contains the bright star *Formalhaut* (Fig. 2-26). At the middle temperate latitudes in North America, Piscis Austrinus manages

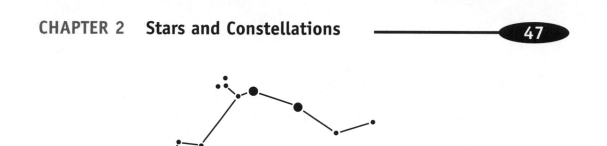

Figure 2-25.   Aquarius, the water-bearer, traverses the southern sky on autumn evenings.

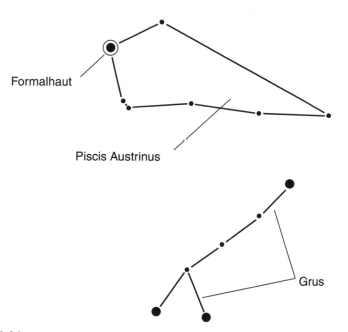

Figure 2-26.   Piscis Austrinus, the southern fish, and Grus, the crane. At latitudes higher than about 45 degrees north, Grus never rises above the southern horizon.

to rise only a few degrees above the horizon. Further north, in Europe and in England, it barely emerges at all. Immediately to the south of it is *Grus*, the crane. This constellation is not visible in the northern temperate extremes, although it can be seen on dark nights in most of the United States.

Constellations of Winter

Finally, let's get our jackets on and look at the evening sky in the middle of January. Some of the autumn constellations can still be seen. The circumpolar constellations have rotated around Polaris by yet another quarter circle and are now 90 degrees clockwise relative to their positions in the spring. Here are the prominent new constellations of winter as they appear from the latitude of Lake Tahoe, Indianapolis, or Washington, D.C. a few hours after suppertime.

CANIS MAJOR AND LEPUS

The southern portion of the winter evening sky is dominated by Canis Major, the big dog, and Lepus, the rabbit (Fig. 2-27.) Canis Major is easy to spot because of the brilliant white star, *Sirius*, that appears in the south-southeast. This is the brightest star in the whole sky, and its name in fact means "scorching." Because it is contained in Canis Major, Sirius is often called the Dog Star.

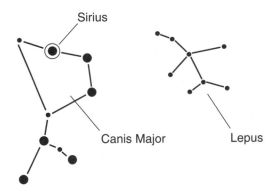

Figure 2-27.   Canis Major, the big dog, and Lepus, the rabbit.

ORION

Somewhat above and to the right of Sirius you will see another winter landmark, *Orion*, the hunter. It's not hard to imagine how ancient people saw a human form in this constellation (Fig. 2-28). Three stars in the middle of Orion represent the hunter's belt, from which hangs a knife or sword consisting of several dimmer stars. If you look at Orion's sword with good binoc-

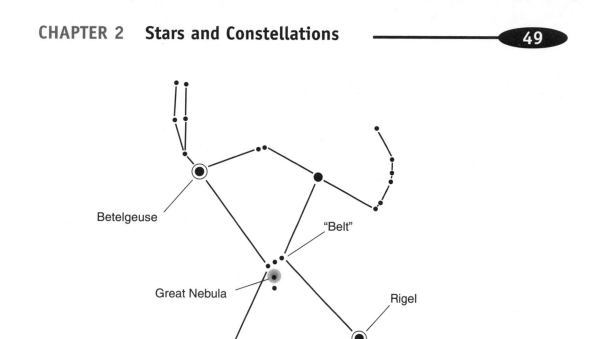

Figure 2-28.   Orion, the hunter, is one of the best-known constellations. It contains a nebula that is visible with good binoculars.

ulars or a wide-aperture telescope at low magnification, you will see the *Great Nebula in Orion*, a vast, glowing cloud of gas and dust in which new stars are being born. Orion contains two bright stars of its own, *Betelgeuse* (also spelled *Betelgeux*), a red giant, and *Rigel*, a blue-white star.

TAURUS AND THE PLEIADES

Above Orion, only a few degrees from the zenith in the southern sky on winter evenings, is *Taurus*, the bull (Fig. 2-29). This constellation contains the bright star *Aldebaran*, which represents the eye of the bull. Near Taurus is a group of several stars known as the *Pleiades*, or seven sisters (although there are really far more than seven of them). When seen through binoculars, these stars appear shrouded in gas and dust, indicating that they are young and that new members are being formed as gravity causes the material to coalesce.

ERIDANUS

Beginning at the feet of Orion and winding its way to the southern horizon and thence into unknown realms is a string of relatively dim stars. This

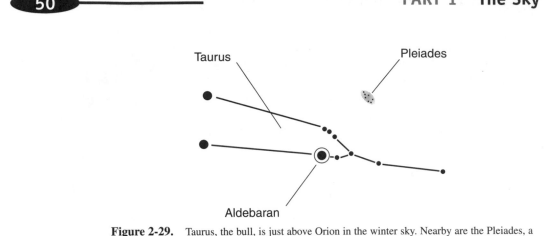

Figure 2-29. Taurus, the bull, is just above Orion in the winter sky. Nearby are the Pleiades, a loose cluster of stars.

constellation is *Eridanus*, the river. It, like Cetus, the whale, contains a star that is thought by many scientists to have a solar system like ours. That star, known as *Epsilon Eridani*, has been the subject of science fiction stories for this reason.

Quiz

Refer to the text if necessary. A good score is 8 correct. Answers are in the back of the book.

1. As the seasons progress, the constellations appear to gradually turn counterclockwise around the north celestial pole from night to night because
   (a) Bootes and Canes Venatici chase Ursa Major around Polaris.
   (b) the sidereal day is slightly shorter than the solar day.
   (c) Earth rotates on its axis.
   (d) the galaxy spirals around its center.

2. If you live in the northern hemisphere, the elevation of Polaris above the horizon, in degrees, is about the same as
   (a) your latitude.
   (b) 90 degrees minus your latitude.
   (c) the elevation of the Sun in the sky at noon.
   (d) nothing in particular; its elevation changes as the seasons pass.

3. Tau Ceti is considered a special star because
    (a) it revolves around Polaris in a perfect circle.
    (b) it is inside our solar system.
    (c) it is in a constellation all by itself.
    (d) some astronomers think that it might have a solar system like ours.

4. Earth slowly wobbles on its axis, causing the constellations to
    (a) change shape slightly from year to year.
    (b) gradually converge on Polaris.
    (c) shift position in the sky slightly from century to century.
    (d) follow the plane of the ecliptic.

5. People in the time of Julius Caesar saw constellations whose individual shapes
    were
    (a) the same as they are now.
    (b) somewhat different than they are now.
    (c) almost nothing like they are now.
    (d) nothing at all like they are now.

6. The constellation Andromeda is well known because it contains
    (a) the brightest star in the whole sky.
    (b) all the planets at one time or another.
    (c) the north celestial pole.
    (d) a spectacular spiral galaxy.

7. Orion is a landmark constellation in the northern hemisphere
    (a) all year round.
    (b) during the winter.
    (c) only north of about 45 degrees latitude.
    (d) because it contains the brightest two stars in the sky.

8. Coma Berenices is sometimes mistaken for
    (a) the sword of Orion.
    (b) Ursa Major.
    (c) the Andromeda galaxy.
    (d) the Pleiades.

9. The pole star, Polaris, is part of
    (a) Canis Major.
    (b) Pegasus.
    (c) Ursa Minor.
    (d) no constellation; it stands by itself.

10. The stars Vega, Altair, and Deneb dominate the sky
    (a) in the circumpolar region.
    (b) during the northern hemisphere summer.
    (c) during spring, summer, and fall, respectively.
    (d) No! These are not stars but constellations.

CHAPTER 3

The Sky "Down Under"

Suppose that you live in Charleston, South Carolina. Imagine that you go to bed, and in the morning, you awaken in Sydney, Australia, and do not know that you have been transported to a location south of the equator. It is a clear morning. The Sun appears to rise normally enough, but after awhile you notice something strange. Rather than progressing generally south and west, as the Sun always does during the early morning hours in the northern hemisphere, the Sun moves north and west. At high noon it sits squarely in the northern sky. The weather is improbable, everyone has a strange accent, and the phone numbers are weird. You look at the cover of the phone book or tune into a local radio station, and the mystery is solved—except, of course, for how you got transported halfway around the world without remembering any of the trip.

If you normally live in Sydney and some morning you awaken in Charleston, a similar surprise awaits you. In fact, if you come from "down under" and are transported to America by surprise, you'll be every bit as jarred as an American who is transported to Australia.

Southern Coordinates

Southern celestial coordinates are similar to northern celestial coordinates. They operate according to the same mathematics. The main difference is that the two coordinate hemispheres are mirror images of one another. While the northern heavens seem to rotate counterclockwise around the north celestial pole, the southern Sun, Moon, planets, and stars seem to rotate clockwise around the south celestial pole.

Recall your middle-school algebra class. Imagine the cartesian coordinate plane. The northern hemisphere is akin to the first and second quadrants, where the y values are positive; the southern hemisphere is cousin to the third and fourth quadrants, where the y values are negative. No particular quadrant is preferable to or more special than any of the other three. So it is with Earth and sky. Fully half the points on the surface of Earth are south of the equator. It is no more unusual in this world for the Sun to shine from the north at high noon than it is for the Sun to shine from the south. Only the extreme polar regions experience conditions that most people would call truly strange, where the Moon or Sun can stay above the horizon for days or weeks at a time, circling the points of the compass.

SOUTHERN AZ/EL

In the southern hemisphere, azimuth bearings are measured clockwise with respect to geographic north, just as they are in the northern hemisphere. However, an alternative system can be used; *azimuth* can be defined as the angle clockwise relative to geographic south. In this latter system, the range of possible values is from 0 degrees (south) through 90 degrees (west), 180 degrees (north), 270 degrees (east), and up to, but not including, 360 degrees (south again). This is shown in Fig. 3-1. The bearing of 360 degrees is omitted to avoid ambiguity, just as is the case with the north-based system. You will never hear of an azimuth angle less than 0 degrees nor more than 360 degrees, at least not in proper usage.

The *elevation* of an object in the sky is the angle, in degrees, subtended by an imaginary arc running away from the object until it intersects the horizon at a right angle. For visible objects over flat terrain, this angle can be as small as 0 degrees when the object is on the horizon or as large as 90 degrees when the object is directly overhead. This is exactly the same scheme as is used in the northern hemisphere. Elevation bearings are measured upward from the horizon to 90 degrees and downward below the horizon to −90 degrees. If the horizon cannot be seen, then it is defined as that apparent circle halfway between the zenith and the nadir.

SOUTHERN RA/DEC

There are two points in time every year when the Sun's elevation, measured with respect to the center of its disk, is positive for exactly 12 hours and negative for exactly 12 hours. One of these time points occurs on March 21,

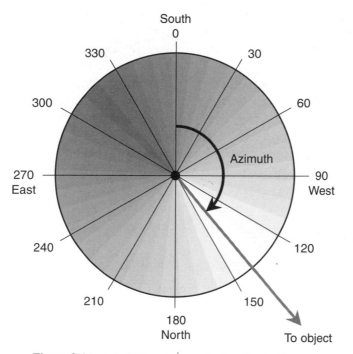

Figure 3-1. Azimuth based on a southerly point of reference.

give or take about a day; the other occurs on September 22, give or take about a day. At the equinoxes, the Sun is exactly at the celestial equator; it rises exactly in the east and sets exactly in the west, assuming that the observer is not at either of the geographic poles. The names *vernal* and *autumnal*, as used in the northern hemisphere, are not really correct in the southern hemisphere because the seasons are reversed compared with those in the north. Thus it is best to speak of the *March equinox* and the *September equinox*.

The crude celestial maps of Fig. 3-2 show the situation at either equinox. That is, the date is on or around March 21 or September 22. You can deduce this because the Sun rises exactly in the east and sets exactly in the west, so it must be at the celestial equator. At the latitude of Sydney, the Sun is 35 degrees away from the zenith (55 degrees above the northern horizon) at high noon on either of these days. The south celestial pole, which unfortunately has no well-defined sentinel star, as is the case for the northern hemisphere, is 35 degrees above the southern horizon all the time. The heavens seem to rotate clockwise around the south celestial pole. In the drawing at *A*, imagine yourself lying flat on your back, with your head fac-

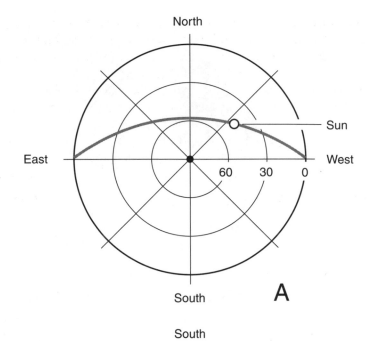

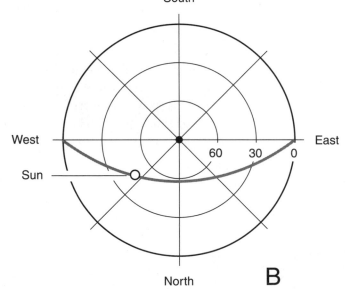

Figure 3-2.   Az/el sky maps for midafternoon at 35 degrees south latitude on or around the March or September equinoxes. At *A*, top of head facing north; at *B*, top of head facing south.

ing north and your feet facing south. In the drawing at *B*, imagine yourself rotated 180 degrees, that is, with your head facing south and your feet facing north. Either orientation is valid astronomically, and you will find star maps that use either scheme.

Every day the Sun moves slightly toward the east with respect to the background of stars. At the March equinox, the Sun is at the celestial equator and is located in a certain position with respect to the stars. This represents the reference point for *right ascension* (RA) and *declination* (dec). As time passes, the Sun rises about 4 minutes later each day relative to the background of stars. The *sidereal* (star-based) day is about 23 hours and 56 minutes long; the *synodic* (sun-based) day is precisely 24 hours long. In the southern hemisphere, the Sun's motion relative to the stars is from left to right.

In the drawings of Fig. 3-2, the Sun is at dec = 0 degrees. Suppose that these drawings represent the situation at the March equinox. This point among the stars is the zero point for right ascension (RA = 0 h). As autumn passes and the Sun follows a lower and lower course across the sky, the declination and right ascension both increase for a while. Remember that right ascension is measured in hours, not in degrees. There are 24 hours of right ascension in a circle, so 1 hour (written 1 h or 1^h) of RA is equal to 15 degrees.

THE SUN'S ANNUAL "LAP" IN THE SOUTH

Let us begin following the Sun during the course of the year starting at the March equinox. As the days pass during the months of April, May, and June, the Sun stays above the horizon for less and less of each day, and it follows a progressively lower course across the sky. The change is rapid in the first days after the equinox, and becomes more gradual with the approach of the *June solstice*, which takes place on around June 22 give or take a day. This might be called the "winter solstice," but again, to avoid confusion with northern-hemisphere-based observers who call it the "summer solstice," it is better to name the month in which it occurs.

At the June solstice, the Sun has reached its northernmost declination point, approximately dec = +23.5 degrees. The Sun has made one-quarter of a complete circuit around its annual "lap" among the stars and sits at RA = 6 h. This situation is shown in Fig. 3-3 using the same two az/el coordinate schemes as those in Fig. 3-2. The gray line represents the Sun's course across the sky. As in Fig. 3-2, the time of day is midafternoon. The observer's geographic latitude is the same too: 35°S.

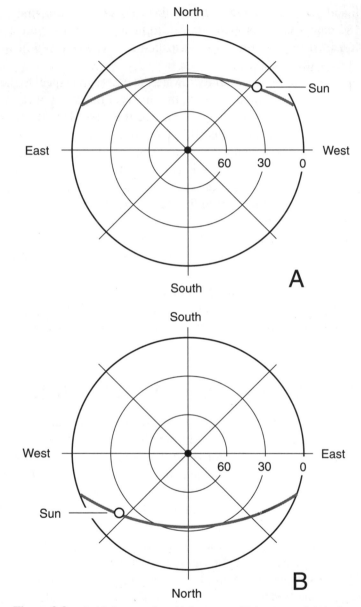

Figure 3-3. Az/el sky maps for midafternoon at 35 degrees south latitude on
or around June 21.

After the June solstice, the Sun's declination begins to decrease, slowly at first and then faster and faster. By late September, the other equinox is reached, and the Sun is once again at the celestial equator, just as it was at the March equinox. But now, instead of moving from south to north, the Sun is moving from north to south in celestial latitude. At the September equinox, the Sun's RA is 12 h. This corresponds to 180 degrees.

Now it is the spring season in the southern hemisphere, and the days are growing long. The Sun stays above the horizon for more and more of each day, and it follows a progressively higher course across the sky. The change is rapid during September and October and becomes slower and slower with the approach of the *December solstice*, which takes place on December 21, give or take a day.

At the December solstice, the Sun's declination is at its southernmost point, approximately dec = −23.5 degrees. The Sun has gone through three-quarters of its annual "lap" among the stars, and sits at RA=18 h. This is shown in Fig. 3-4 using the same two az/el coordinate schemes as those in Figs. 3-2 and 3-3. The gray line represents the Sun's course across the sky. As in Figs. 3-2 and 3-3, the time of day is midafternoon. The observer hasn't moved either, at least in terms of geographic latitude; this point is still at 35°S.

After the December solstice, the Sun's declination begins to increase gradually and then, as the weeks pass, faster and faster. By late March, the Sun reaches an equinox again and crosses the celestial equator on its way to forsaking the southern hemisphere for another autumn and winter. The "lap" is complete.

Mirrored Myths

The Greeks didn't name the southern circumpolar constellations, but many of the star groups near the equator, as seen from "down under," are the same ones that the Greeks made famous. The only difference is that they are all upside down.

SKY MAPS

In this chapter, the general shapes of the better-known southern constellations are shown. To see where these constellations are in the sky from your

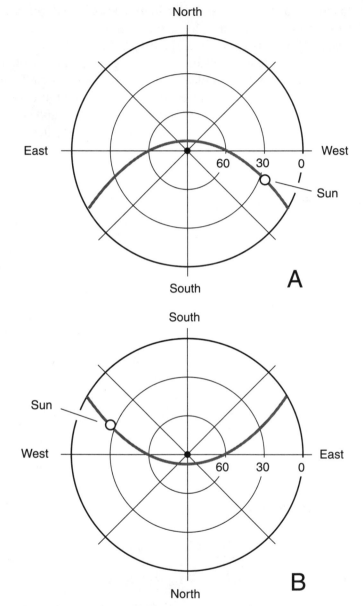

Figure 3-4.   Az/el sky maps for midafternoon at 35 degrees south latitude on or around December 21.

location this evening, go to the *Weather Underground* Web site at the following URL:

http://www.wunderground.com

Type in the name of your town and country, and then, when the weather data page for your town comes up, click on the "Astronomy" link. There you will find a detailed map of the entire sky as it appears from your location at the time of viewing, assuming that your computer clock is set correctly and data are input for the correct time zone.

Southern Circumpolar Constellations

From the latitude of 35°S, the circumpolar constellations encompass much of the sky. At some time or other during the year, it is possible to see more of the sky at lower latitudes (closer to the equator) than at higher latitudes (closer to the poles). If you live in Sydney, Buenos Aires, or Cape Town, you have a slight advantage in this respect over your counterparts who live in Minnesota and a bigger advantage over people in Scotland. However, the portion of the sky that stays above the horizon, no matter what time of the year you stargaze in the evening sky, becomes smaller as you go closer to the equator. Observers in chillier climes get to see more circumpolar constellations but less of the complete celestial sphere; people in warmer places get to see more of the celestial sphere, but they have to choose the proper times to see specific constellations near the pole.

STAR BRIGHTNESS

In this chapter, as in Chapter 2, stars are illustrated at three relative levels of brightness. Dim stars are small black dots. Stars of medium brilliance are larger black dots. Bright stars are circles with black dots at their centers. But the terms *dim*, *medium*, and *bright* are not intended to be exact or absolute. In downtown Sydney, some of the dim stars shown in these drawings are invisible, even under good viewing conditions, because of scattered artificial light. After your eyes have had an hour to adjust to the darkness on a moonless, clear night in the outback, some of the dim stars in these illustrations will be easy to see. The gray lines connecting the stars are included in the diagrams only to emphasize the general shapes of the constellations.

THERE IS NO SOUTHERN POLARIS

We need a time of reference for our circumpolar observations, and mid-April is as good a time as any. Imagine that you are in the countryside near Sydney or Cape Town or Buenos Aires and that you go outdoors to stargaze at around 10:00 P.M. Assume that the sky is clear, there is no haze, and the Moon is below the horizon so that its light does not interfere with stargazing. You know that the south celestial pole is 35 degrees above the southern horizon. You search for a significant star, or at least a constellation, to mark the spot using the "fist rule." (Hold your right arm out straight and make a tight fist. Point the knuckles toward your right. The top of your fist is about 10 angular degrees from the bottom.) You find the southern horizon using a compass or your knowledge of the area and proceed three and a half fists up into the sky. There is nothing significant. The south polar region is devoid of bright or even moderately bright stars. This caused some trouble for mariners who ventured south of the equator. They needed a convenient way to locate the south celestial pole.

CRUX AND MUSCA

As you stand facing toward the south, you will see, high in the sky, a group of four stars forming a kitelike shape. This is *Crux*, more commonly called the *southern cross*. Just below it, somewhat dimmer, is a star group shaped somewhat like a ladle. This is *Musca* or *Musca Australis*, the *southern fly*. Look at these two constellations carefully, and make educated guesses as to their centers (Fig. 3-5). The center of Crux is easy to decide on, but the center of Musca is a little tougher. Pick a point on the handle of the ladle, just above the scoop. These two constellation-center points are separated by about 10 degrees of arc, a fact that you can verify by the fist rule. Now go two fists down toward the southern horizon from the center point of Musca. This will give you a point close to the south celestial pole.

TRIANGULUM AND APUS

Just below and to the left of Crux, there is a group of three stars that form a triangle. It's almost a perfect equilateral triangle, with the apex at the bottom

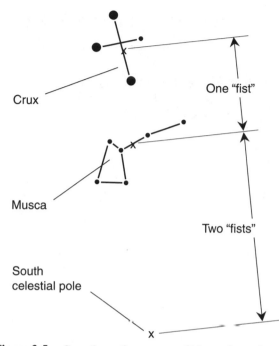

Figure 3-5.   Crux, the southern cross, and Musca, the southern fly,
as observed in mid-April from the latitude of Sydney, Australia.

and the base on top (Fig. 3-6). This is *Triangulum Australis*, the southern triangle, often called simply *Triangulum*. Below this constellation and to its right
is *Apus*, the bird of paradise. It is a small, dim constellation and at this time of
the year looks something like a lawn mower (if you have a good imagination).

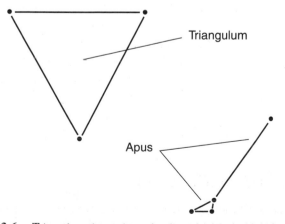

Figure 3-6.   Triangulum, the southern triangle; and Apus, the bird of paradise.

PAVO

Below Triangulum, in the south-southeast sky, is a large group of dim stars. This is the constellation *Pavo*, the peacock (Fig. 3-7). This is an inappropriate name for such an inconspicuous group of stars. However, you might, with some effort, imagine a large, fan-shaped set of tail feathers, with the bird, at this time of year, appearing upright.

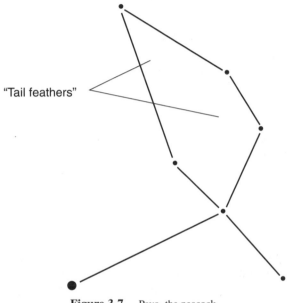

Figure 3-7.   Pavo, the peacock.

TUCANA

Low in the southern sky, almost grazing the horizon, is *Tucana*, the toucan (Fig. 3-8). At this time of year, you can imagine this dim group of stars as having the shape of a bird lying on its side with its beak pointed down and to the right. This constellation contains the *Small Magellanic Cloud*, a satellite of our Milky Way galaxy.

HYDRUS, RETICULUM, AND MENSA

In the south-southwest, below and to the right of the south celestial pole, there are three constellations of note: *Hydrus*, the little snake, *Reticulum*, the

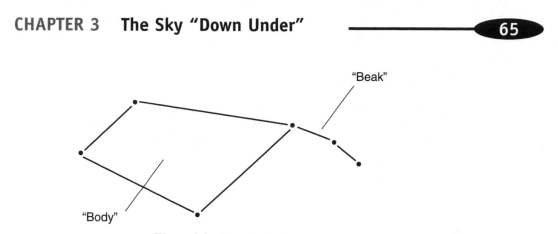

Figure 3-8.   Toucana, the toucan.

net, and *Mensa*, the table. These are shown in Fig. 3-9. It is hard to imagine how any of these star groups got the names they got. Mensa contains part of the *Large Magellanic Cloud*, another satellite of our galaxy similar to the Small Magellanic Cloud. Both of the Magellanic Clouds are named after the explorer Magellan, whose ship made the famous round-the-world trip hundreds of years ago and who sailed through the far southern oceans.

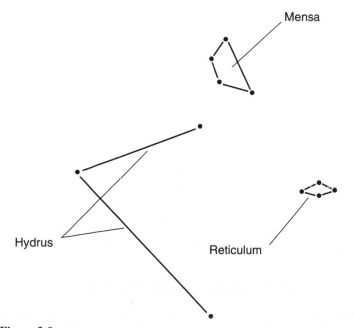

Figure 3-9.   Hydrus, the little snake; Reticulum, the net; and Mensa, the table.

VOLANS AND CARINA

Above and to the right of the celestial pole are *Volans*, the flying fish, and *Carina*, the keel or ship (Fig. 3-10). Carina is noteworthy because it contains the yellowish white star *Canopus*, which is the second brightest nighttime star after Sirius.

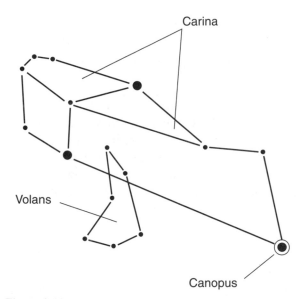

Carina

Volans

Canopus

Figure 3-10.   Volans, the flying fish, and Carina, the keel. Carina contains the well-known bright star Canopus.

CHAMELEON AND OCTANS

The two constellations closest to the south celestial pole are *Chameleon*, the lizard, and *Octans*, the octant. The stars in these groups are so dim that unless you are out in the country away from city lights, you will not see them. Also, if the Moon is near full phase and is above the horizon, its scattered light might wash these constellations out. At this time of the year, Octans in the evening sky appears as a tall, slender triangle immediately to the east of the celestial pole, and Chameleon is near and above it (Fig. 3-11). Apus, the bird of paradise, appears in this group too, centered about 12 degrees (a little more than one fist) above and to the left of the pole.

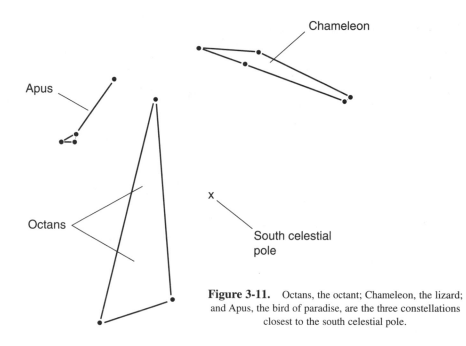

Figure 3-11.   Octans, the octant; Chameleon, the lizard; and Apus, the bird of paradise, are the three constellations closest to the south celestial pole.

Constellations of the Southern Autumn

Imagine that it is still the same night and still the same time (10:00 P.M.), and you turn your attention toward the east, north, and west, that part of the sky not confined to the vicinity of the celestial pole. Constellations in this part of the sky rise and set; they are not always above the horizon. As is the case in the northern hemisphere, the farther from the pole a constellation is located, the more time it spends each day below the horizon. A star at the equator spends exactly half the sidereal day, or about 11 hours and 58 minutes, above the horizon and half the time out of sight "beneath Earth." Ultimately, for observers at 35 degrees south latitude, constellations with declinations of more than +55 degrees (within 35 degrees of the north celestial pole) never make it above the northern horizon.

LIBRA

Libra is a group of faint stars representing the scales of justice. It is high in the northeastern sky. It has the general shape of a trapezoid or diamond at this time of the year (Fig. 3-12).

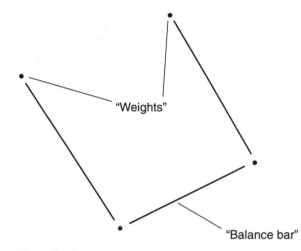

Figure 3-12.   Libra, the scales of justice, looks like a diamond.

VIRGO

Virgo, the virgin, is fairly high in the north-northeast sky. As viewed from this angle, it is shaped rather like a scorpion (Fig. 3-13). Virgo contains the bright star *Spica*.

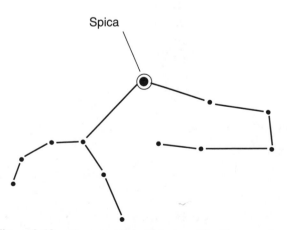

Figure 3-13.   Virgo, the virgin, looks something like a scorpion.

LEO

Leo, the lion, is high in the northwest sky. It bears no resemblance to a resting lion or Sphinx, as it does when looked at from north of the equator. Instead, its shape more nearly resembles that of a mangled coat hanger (Fig. 3-14) or a laundry iron held upside down. The bright star *Regulus* dominates.

CANCER AND CANIS MINOR

Cancer, the crab, is low in the northwest sky (Fig. 3-15). Next to Cancer is *Canis Minor*, the little dog, which contains the prominent star *Procyon*. In ancient Greek mythology, souls were said to enter the world by passing down from the heavens through Cancer.

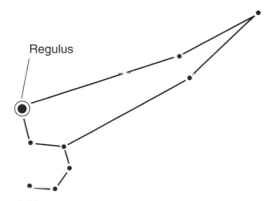

Figure 3-14. Leo, the lion, lacks his regal nature south of the equator.

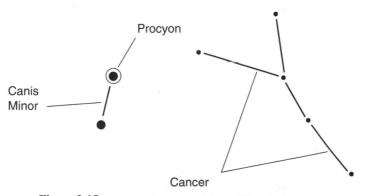

Figure 3-15. Cancer, the crab, and Canis Minor, the little dog.

CORONA BOREALIS

Low in the northeastern sky, near the horizon, is a group of several stars that form an inverted-U or Greek letter omega shape. These stars form the constellation *Corona Borealis*, the northern crown (Fig. 3-16). This constellation is dominated by the moderately bright star *Alphecca*, also known as *Gemma*.

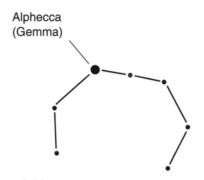

Figure 3-16.   Corona Borealis, the northern crown.

BOOTES AND CANES VENATICI

Just to the left of the northern crown you will see a brilliant, twinkling star at an elevation of about 20 degrees in the northeast or north-northeast sky. This is *Arcturus*. If you use your imagination, you might see that this star forms the point where a fish joins its tail (Fig. 3-17). The fish seems to be swimming straight downward. This is *Bootes*, the herdsman. Just to the left of Bootes, near the northern horizon, is a group of three rather dim stars. These are Bootes' canine companions, *Canes Venatici*.

CORVUS, CRATER, AND HYDRA

A large portion of the autumn evening sky is occupied by three constellations consisting of relatively dim stars. These are *Corvus*, the crow, *Crater*, the cup, and *Hydra*, the sea serpent or water snake (Fig. 3-18). Hydra stretches from low in the northwest, nearly through the zenith, to high in the eastern sky. Corvus and Crater are both high in the north, just below Hydra.

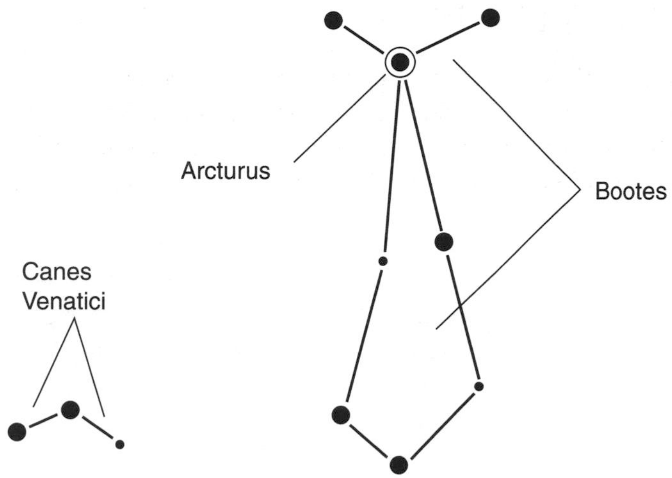

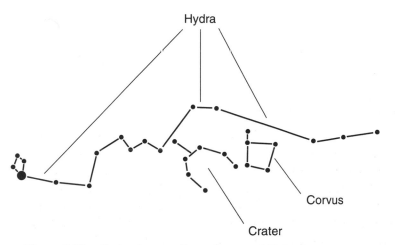

Figure 3-17.   Bootes, the herdsman, and Canes Venatici, his dogs.

Figure 3-18.   Corvus, the crow; Crater, the cup; and Hydra, the sea serpent.

Constellations of the Southern Winter

Now imagine that it is the middle of July—the dead of the southern-hemispheric winter—and that you are outdoors at around 10:00 P.M. The circumpolar constellations are all still above the horizon, but they have rotated 90 degrees clockwise around the pole from their positions in April. The noncircumpolar constellations have moved from east to west. As you look generally away from the circumpolar sky, you should be able to make out the following groups of stars, which are also visible from the northern temperate latitudes at this time of year.

HERCULES

Near the northern horizon, or just a little west of due north, is a moderately dim group of stars forming a trapezoid with limbs (Fig. 3-19). This is *Hercules*, the warrior. His nemesis, *Draco*, is mostly out of sight below the horizon. The well-known globular cluster *M13* is in this constellation, although from the southern temperate latitudes the viewing is somewhat less favorable than it is from northern locations.

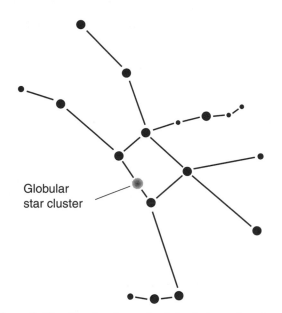

Globular
star cluster

Figure 3-19.   Hercules, the warrior, is low in the northern sky on southern-hemispheric winter evenings.

CAPRICORNUS

High in the eastern sky is *Capricornus* (also called *Capricorn*), the goat (Fig. 3-20). This goat has the tail of a fish, according to the myths, and dwells at sea. Ancient Greek mythology held that on its way to heaven after death of the body, the human soul would pass through this constellation.

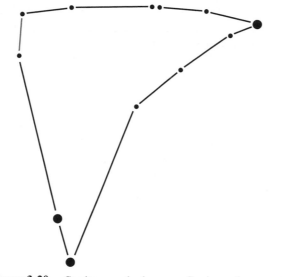

Figure 3-20.   Capricornus, also known as Capricorn, the sea goat.

SAGITTARIUS

Near the zenith, you will see Sagittarius, the centaur (Fig. 3-21). Sagittarius lies in the direction of the densest part of our galaxy. If it were not for interstellar dust, which is concentrated along the plane of the Milky Way, this constellation and those near it would be obscured by the brilliance of the galactic core.

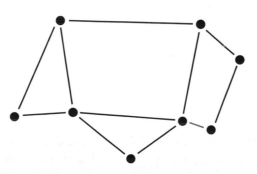

Figure 3-21.   Sagittarius, the centaur, is near the zenith on southern-hemispheric winter evenings.

SCORPIUS

Just to the west of Sagittarius, also near the zenith, is *Scorpius* (also called *Scorpio*), the scorpion (Fig. 3-22). This constellation is one of the few that bears some resemblance to the animal or object it represents. The eye of the scorpion is the red giant star *Antares*, which varies in brightness.

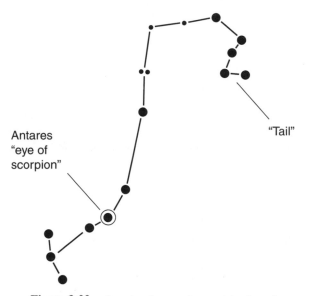

Antares
"eye of
scorpion"

"Tail"

Figure 3-22.   Scorpius, the scorpion, contains the red star Antares, and is just to the west of Sagittarius.

OPHIUCHUS AND SERPENS

High in the northwestern sky are the constellations *Ophiuchus*, the snake bearer, and *Serpens*, the snake (Fig. 3-23). As with most of the other constellations near the celestial equator, these two are inverted with respect to their appearance as seen from the northern hemisphere.

LYRA, CYGNUS, AND AQUILA

Low in the northern sky you will see the bright star *Vega*, flanked by a small parallelogram of dimmer stars. The quadrilateral forms the constellation

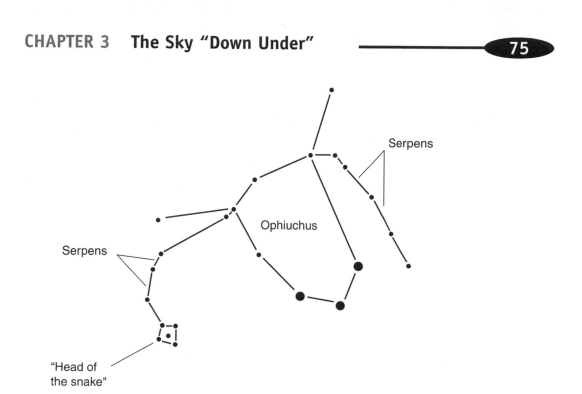

Figure 3-23.   Ophiuchus, the serpent bearer, holds Serpens, the snake.

Lyra, the lyre. To the right of Vega, grazing the horizon, is another bright star, *Deneb*, that is at the tip of the tail of *Cygnus*, the swan. Above and slightly to the right of Deneb is a third bright star, *Altair*. This is part of the constellation *Aquila* (Fig. 3-24). In the northern hemisphere, these three stars stand high in the sky and are sometimes called the *summer triangle*. However, in the southern hemisphere they have no special distinction apart from their relative brilliance.

Constellations of the Southern Spring

Now imagine that it is around 10:00 P.M. in the middle of October and that you are at the latitude of Sydney, Buenos Aires, or Cape Town (approximately 35°S). New constellations have risen in the east, and old ones have set in the west. Here are constellations we have not described before that now occupy prominent positions in the sky.

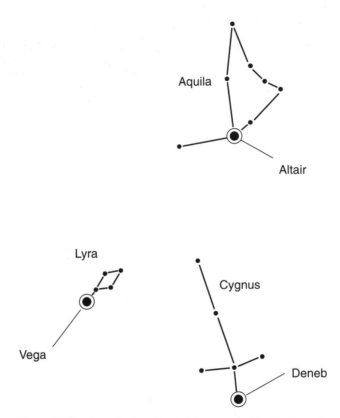

Figure 3-24.   Lyra, the lyre; Cygnus, the swan; and Aquila, the eagle.
These constellations are marked by the bright stars Vega, Deneb, and Altair.

PISCES AND ARIES

In the north you will see *Pisces*, the two fish. In the north-northeastern sky is *Aries*, the winged ram (Fig. 3-25). The fish of Pisces are, according to mythology, joined at their tails, and Aries has fleece of gold.

CETUS

High in the northeastern sky is *Cetus*, the whale or sea monster (Fig. 3-26). The variable star *Mira* is sometimes visible in this constellation. The star *Tau Ceti* is thought to be a candidate for having a solar system similar to ours and possibly an earthlike planet.

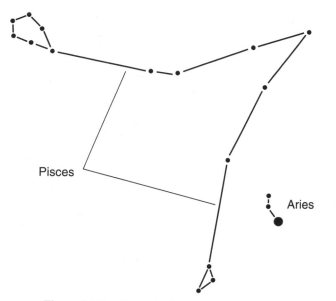

Figure 3-25.   Pisces, the fishes, and Aries, the ram.

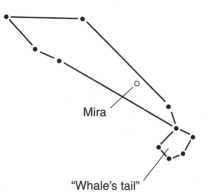

Figure 3-26.   Cetus, the whale, contains the variable star Mira.

PEGASUS AND ANDROMEDA

Low in the north-northwestern sky is *Pegasus*, the winged horse. Near the northern horizon, *Andromeda*, representing a princess, rides the horse alongside the Milky Way (Fig. 3-27). Andromeda contains a spiral galaxy similar to our own galaxy, but it is 2,200,000 light-years away, about 20 times the diameter of the Milky Way's spiral disk. The *Great Nebula in*

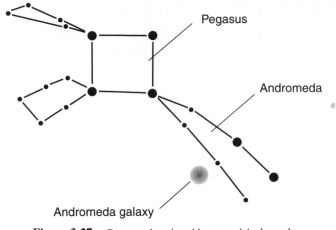

Figure 3-27.   Pegasus, the winged horse, and Andromeda,
the Ethiopian princess who married Perseus.

Andromeda, as it was originally called, is too near the horizon, as viewed from the southern temperate latitudes, to present itself well to casual observers.

AQUARIUS

High in the northwest sky, you will see *Aquarius*, the water bearer (Fig. 3-28). Aquarius supposedly brings love and peace. In ancient mythology, this constellation was seen as a person pouring water from a jug.

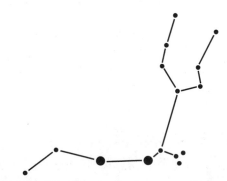

Figure 3-28.   Aquarius, the water-bearer, traverses the
northwestern sky on spring evenings in the southern hemisphere.

PISCIS AUSTRINUS AND GRUS

High in the western sky, nearly at the zenith, is *Piscis Austrinus*, also called *Piscis Australis*. This is the southern fish and contains the bright star *Formalhaut* (Fig. 3-29). Immediately to the south of it is *Grus*, the crane.

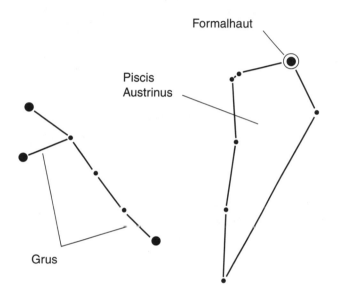

Figure 3-29.   Piscis Austrinus, the southern fish, and Grus, the crane, are near the zenith on southern-hemisphere spring evenings.

Constellations of the Southern Summer

Finally, let's look at the late-evening sky in the middle of January. Here are the prominent new constellations of the southern-hemispheric summer as they appear from the latitude of Sydney, Buenos Aires, or Cape Town around 10:00 P.M. local time.

CANIS MAJOR AND LEPUS

Just north of the zenith are *Canis Major*, the big dog, and *Lepus*, the rabbit (Fig. 3-30). Canis Major is easy to spot because of the brilliant white star, *Sirius*, nearly overhead at this latitude in the evening at this time of the

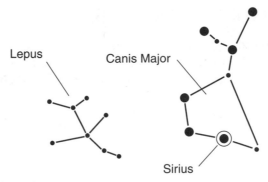

Figure 3-30.   Canis Major, the big dog, and Lepus, the rabbit.

year. Sirius is also called the *Dog Star* and is the brightest star in the sky except for the Sun.

ORION

Somewhat below and to the left of Sirius, but still high in the sky, is *Orion*, the hunter (Fig. 3-31). If you look at Orion's central region with good binoculars or a wide-angle telescope, you will see the *Great Nebula in Orion*, a vast, glowing cloud of gas and dust in which new stars are being born. Orion contains two bright stars, *Betelgeuse* (also spelled *Betelgeux*), a red giant, and *Rigel*, a blue-white star.

TAURUS AND THE PLEIADES

Low in the north-northwest sky is *Taurus*, the bull (Fig. 3-32). This constellation contains the bright star *Aldebaran*. Below Taurus is a group of several stars known as the *Pleiades*. At this latitude, the Pleiades are less spectacular than they are as seen from the northern hemisphere, but on an especially dark night, with a good wide-angle telescope, their splendor shines through. From extreme southern latitudes, the Pleiades never rise above the northern horizon.

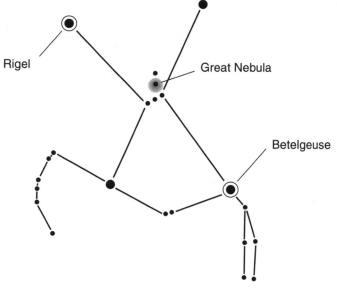

Figure 3-31.   Orion, the hunter, contains a nebula and
two well-known bright stars.

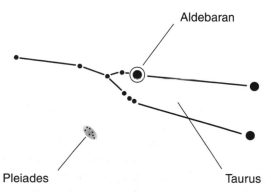

Figure 3-32.   Taurus, the bull, and the Pleiades, also known as the
Seven Sisters (although there are really far more than seven of them).

ERIDANUS

Beginning at the feet of Orion and winding its way high across the western sky into the circumpolar region is a string of relatively dim stars. This constellation is *Eridanus*, the river. It, like Cetus, the whale, contains a star, *Epsilon Eridani*, that is thought by many scientists to have a solar system like ours.

GEMINI

Gemini is low in the north-northeast sky. This constellation has the general shape of a long, thin, backward letter C if you stand facing north and look up at it (Fig. 3-33). At the right-hand extreme are two relatively bright stars, *Castor* and *Pollux,* named after the twin sons of the mythological Greek god Zeus.

AURIGA

Grazing the northern horizon is the constellation *Auriga*. It contains the bright star *Capella* (Fig. 3-34). From extreme southern latitudes, Capella

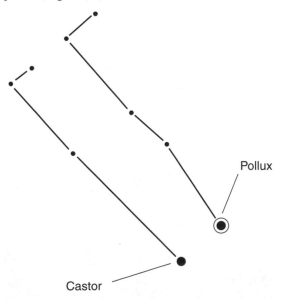

Figure 3-33.   Gemini contains the stars Castor and Pollux, and appears low in the north-northeast sky in the southern-hemispheric summer.

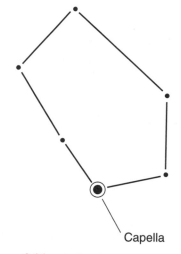

Figure 3-34.   Auriga, the charioteer, contains
the bright star Capella.

never makes it above the northern horizon, although from the more northerly temperate regions and from the southern tropics it is a brilliant, twinkling
landmark in the sky.

Quiz

Refer to the text if necessary. A good score is 8 correct. Answers are in the back of
the book.

1. In the southern hemisphere, elevation is measured in the same way as it is in
   the northern hemisphere, with the following exception:
   (a) It is a negative angle rather than a positive angle.
   (b) It is expressed in radians rather than in degrees.
   (c) It is measured with respect to the south pole rather than the north pole.
   (d) There are no exceptions.

2. Which of the following stars is also known as "south Polaris"?
   (a) Formalhaut
   (b) Capella

   (c) Sirius
   (d) None of the above

3. The spring equinox in the southern hemisphere occurs in which month?
    (a) March
    (b) June
    (c) September
    (d) December

4. Suppose that it is a few days after the summer solstice according to people who
    live in Sydney, Australia. The Sun's declination is
    (a) slowly decreasing.
    (b) rapidly decreasing.
    (c) slowly increasing.
    (d) rapidly increasing.

5. At a latitude of 35°S, the angular radius of the south circumpolar region is
    (a) 70 degrees of arc.
    (b) 55 degrees of arc.
    (c) 35 degrees of arc.
    (d) impossible to determine, not enough data are given.

6. Orion is a landmark constellation primarily in which season south of the equa-
    tor?
    (a) Winter
    (b) Summer
    (c) Spring
    (d) Fall

7. At a latitude of 55°S, the northern pole star Polaris would be approximately
    (a) 35 degrees above the horizon.
    (b) 35 degrees below the horizon.
    (c) 55 degrees above the horizon.
    (d) 55 degrees below the horizon.

8. A circumpolar constellation as viewed from Cape Town, South Africa, is
    (a) Ursa Minor.
    (b) Octans.
    (c) Ursa Major.
    (d) Draco.

9. A star that lies on the celestial equator as seen from 45°N would lie approxi-
    mately where as seen from 45°S?
    (a) On the celestial equator
    (b) Near the north celestial pole
    (c) Near the south celestial pole
    (d) The answer cannot be determined from the information given.

10. As the night progresses for an observer in Buenos Aires, Argentina, the south
    circumpolar stars seem to
    (a) revolve counterclockwise around the south celestial pole.
    (b) revolve clockwise around the south celestial pole.
    (c) rise in the east and set in the west.
    (d) never rise above the horizon.

CHAPTER 4

The Moon and the Sun

It has not been easy for humanity to develop consistent theories about what happens in the sky. If there were no Moon, it would have been more difficult. If the telescope had never been invented, the puzzle would have been tougher still. The Moon goes through obvious changes even to the most casual observer, but the reasons for these changes were not obvious to most people 50 generations ago. Neither the Sun nor the Moon is a smooth globe. Both have complicated surfaces. The Moon has craters, mountains, plains, and cliffs. The Sun has a mottled surface that is often strewn with spots. The Sun and Moon together perform a cosmic dance that, once in a while, puts on a show to rival anything else in nature.

The Moon

Earth has countless natural satellites—meteors captured by gravity and orbiting in all manner of elliptical paths. The only natural satellite of significance and the only one that can be detected without powerful observing aids, however, is the Moon. It's interesting that we have never come up with a better name for Earth's Moon; we speak about the moons of Jupiter and the moons of Saturn, and then we call our own Queen of the Night "the Moon." It is as if someone had a daughter and named her "Daughter." Sometimes the Moon is called "Luna," but that name conjures up visions of madness and worship and is not used by astronomers.

DOUBLE PLANET?

The Moon orbits Earth at a distance of about 381,000 kilometers (237,000 miles). Sometimes it's a little closer, and sometimes it's a little farther away. The Moon's diameter is 27.2 percent that of Earth, roughly 3480 kilometers (2160 miles). That's large for a moon relative to its parent planet. The Earth-Moon system is sometimes considered a double planet, and some astronomers think the pair formed that way. But Earth is 81 times more massive than the Moon, and the Moon has essentially no atmosphere. Thus, in planetary terms, the Moon is a dull place.

Perhaps you have seen drawings of the Earth-Moon system and have come to envision the Moon as much closer to Earth than is actually the case. (The drawings in this chapter, except for Fig. 4-1, are examples of such misleading data.) There is a reason for this distortion. If the Earth-Moon system were always drawn true to scale, the illustration would be of little use for most instructive purposes. Earth is a bit less than 12,800 kilometers (7,930 miles) in diameter, and the Moon is about 381,000 kilometers (237,000 miles) away on average. That's 30 Earth diameters. If drawn to scale, the Earth-Moon system would look like Fig. 4-1. Think of the Earth and the Moon as pieces of fruit. Suppose that Earth is a 10-centimeter-diameter grapefruit and the Moon is a 27-millimeter-diameter plum (4 inches and 1 inch across, respectively). To make a scale model, you must set the two fruits 3 meters (10 feet) apart.

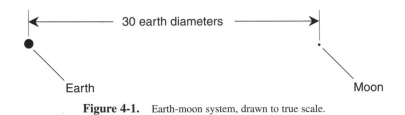

Figure 4-1.   Earth-moon system, drawn to true scale.

PERIOD, PERIGEE, AND APOGEE

With respect to the Sun, the Moon takes 29½ days to make one orbit around Earth. The exact *synodic* (sun-based) *lunar orbital period* varies slightly from one orbit to the next because the orbit of the Moon around Earth is not a perfect circle and the orbit of Earth around the Sun is not a perfect circle either. However, for most amateur astronomers (including us), 29½ days is a good enough figure. Relative to the stars, the Moon's orbit is faster; the *sidereal lunar orbital period* is about 27 days and 7 hours.

The synodic and sidereal lunar orbital periods differ for the same reason the synodic day is longer than the sidereal day. Every time the Moon makes one trip around Earth, our planet has moved approximately one-twelfth of the way around the Sun. The Moon has to travel further to come into line with the Sun from one orbit to the next than it must travel to come into line again with some distant star (Fig. 4-2).

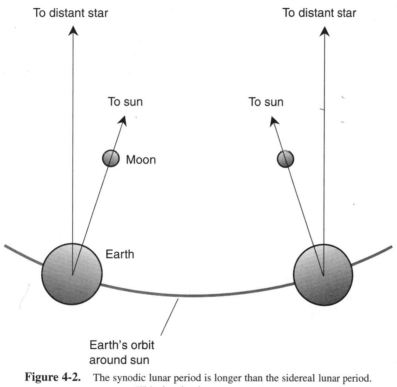

Figure 4-2. The synodic lunar period is longer than the sidereal lunar period. (This drawing is not to scale.)

Have you ever looked at the Moon, especially the full Moon, and imagined it to be closer than you remember previous full Moons to have been? Maybe it's not your imagination. The Moon orbits Earth in an elliptical path, with Earth at one focus. The Moon can get as close as 356,000 kilometers (221,000 miles) and as distant as 407,000 kilometers (253,000 miles) from Earth. This is a difference of 13.5 percent of the Moon's mean distance. Sometimes the Moon's disk appears 13.5 percent larger than at other times. This is enough to make a difference, especially when the Moon passes precisely between an observer and the Sun. The Moon's closest approach is the *lunar perigee*; this term also applies to the minimum-distance figure. The Moon's furthest retreat is the *lunar apogee*, a term that also is used in reference to the maximum-distance figure.

LIBRATION

As the Moon makes its way around Earth, it keeps the same face toward us, more or less. This is so because the Moon's mass is not uniformly distributed within the globe, and Earth's gravity has managed, over millions of centuries, to tug the Moon's rotation rate into near-perfect lockstep with its revolution. But the Moon still wobbles back and forth a little; it has not completely "settled down." We can see 59 percent of the Moon's surface from Earth if we make enough observations. The wobbling of the Moon's face relative to Earth is called *libration* (not to be confused with libation).

Libration can give rise to interesting phenomena. For example, when amateur radio operators bounce their signals off the Moon to communicate with their fellows on the opposite side of the world, libration produces multiple signal paths whose lengths vary constantly, making the radio waves add and cancel in a manner so complicated that precise analysis would challenge any computer. The resulting received signals sound like someone babbling or hooting underwater. The wavelengths of light are too short for this effect to be observed visually. If we could see at radio wavelengths, the Moon would seem to sparkle and scintillate as if fireworks were constantly being set off all over its surface.

PHASE

The appearance of the Moon is drastically affected by its orientation relative to the Sun. When the Earth, the Moon, and the Sun are in line or nearly in line, the Moon is said to be *new*, and its existence is not visually apparent unless there happens to be a solar eclipse. As the Moon orbits Earth, a journey that takes place in a counterclockwise direction as viewed from high above Earth's north pole (Fig. 4-3), it presents more and more and then less and less of its lit-up face to us. Three or four days after the new Moon, it is a *waxing crescent*. About a week after the new Moon, we see half its globe illuminated by the Sun; this is *first quarter*. Three or four

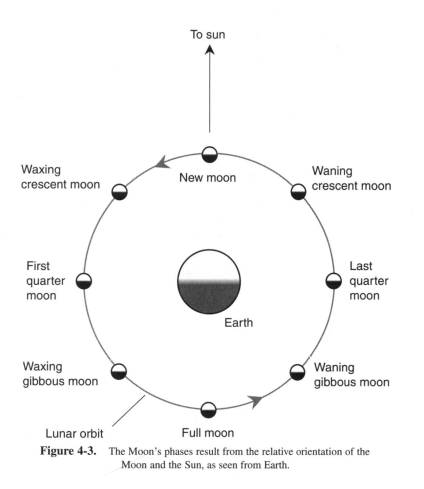

To sun

Waxing
crescent moon

New moon

Waning
crescent moon

First
quarter
moon

Last
quarter
moon

Earth

Waxing
gibbous moon

Waning
gibbous moon

Lunar orbit

Full moon

Figure 4-3.   The Moon's phases result from the relative orientation of the
Moon and the Sun, as seen from Earth.

days after that, most of the Moon is illuminated as we see it; this is *waxing gibbous*. Two weeks and 18 hours after the new Moon, it is entirely illuminated for us unless a lunar eclipse happens to be taking place. This is the *full Moon*. Phases proceed in timely fashion after the full Moon through *waning gibbous*, *last quarter*, *waning crescent*, and finally, back to new again.

Almost nobody lives in this world without getting to know the lunar phases before they get into kindergarten. The waxing crescent is visible just after sunset; the first-quarter Moon can be seen until midnight. The waxing gibbous Moon stays in the sky into the wee hours of the morning, and the full Moon is above the horizon all night, setting as the Sun rises. After the full phase, the waning gibbous Moon rises a couple of hours after sunset; the last-quarter Moon rises around midnight; the waning crescent Moon waits until the predawn hours to rise. Moonset in the waning phases takes place in the daytime, and some people say that there is "no moonset" during this half of the lunar cycle.

OF NORTH AND SOUTH

Most people envision the Moon's phase-to-phase progress and appearance as seen from the northern hemisphere. This is natural because there are more people living north of the equator than south of it. As far as Earth itself is concerned, however, this is only half the story.

Figure 4-4 shows the way the Moon looks at various stages in its orbit around Earth as seen from a midlatitude northern location such as Kansas City, Missouri, or Rome, Italy. (There is some variance in the tilt, depending on the season of the year; moonrise and moonset occur somewhat north

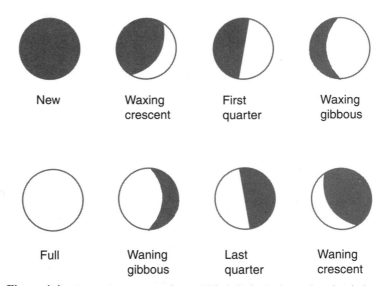

New   Waxing   First   Waxing
             crescent   quarter   gibbous

Full   Waning   Last   Waning
             gibbous   quarter   crescent

Figure 4-4.   Lunar phases as seen from middle latitudes in the northern hemisphere. The Moon's tilt varies somewhat, depending on the season and the time of night.

or south of due east or west.) The waxing crescent appears in the southwestern or western sky just after sunset and sets 2 to 4 hours after the Sun. The Moon at first quarter is in the southern sky at sunset, moves generally westward, and sets around midnight. The waxing gibbous Moon is in the southeast at sunset, moves generally westward, and sets in the predawn hours. The full Moon is opposite the Sun, rising at sunset and setting at or near sunrise. The waning gibbous Moon rises some time after sunset and sets after sunrise the next day. The last-quarter Moon rises around midnight and sets around noon. The waning crescent waits until the predawn hours to rise and sets in the afternoon.

Figure 4-5 illustrates the appearance of lunar phases as seen from a mid-latitude southern location such as Perth, Australia, or Napier, New Zealand.

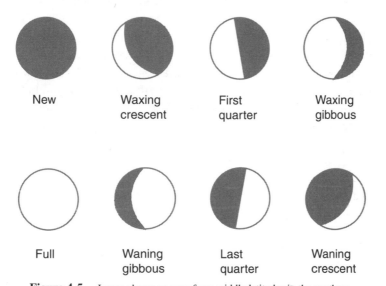

Figure 4-5. Lunar phases as seen from middle latitudes in the southern hemisphere. The Moon's tilt varies somewhat, depending on the season and the time of night.

(As with the northern-hemispheric situation, there is some variance in the Moon's tilt. Depending on the season of the year, moonrise and moonset occur somewhat north or south of due east or west.) The waxing crescent is in the northwestern or western sky just after sunset and sets 2 to 4 hours after the Sun. The Moon at first quarter is in the northern sky at sunset, moves generally westward, and sets around midnight. The waxing gibbous Moon is in the northeast at sunset, moves to the west, and sets a couple of hours before dawn. The full Moon rises around sunset and sets around sunrise, tracking across the northern half of the sky during the night. The waning gibbous Moon rises shortly after sunset and sets after sunrise the next day. The last-quarter Moon rises at about midnight and sets around noon. The waning crescent rises a couple of hours before dawn and sets in the afternoon.

THE FACE OF THE MOON

When you look at the Moon without the binoculars or a telescope, it's impossible to know much about the true nature of the surface. Before Galileo Galilei and other astronomers began looking at the heavens through "spy glasses" a few hundred years ago, no one could be certain that the terrain was dry, scarred, and lifeless. In fact, the true austerity of the Moon would surprise even the most pessimistic dreamers of old.

The naked-eye Moon, especially the full Moon, has light and dark features. In absolute terms, the whole Moon is a rather dark object; it reflects only a few percent of the solar light that strikes it. If the Moon were as white as snow or powdered sugar, it would shine several times more brightly. Even without the help of telescopes, people long ago surmised that the Moon's light areas represent irregular terrain and the dark regions are flat by comparison. Some people thought the light regions were clouds and the dark zones were areas of clear weather, but after observing the Moon for many nights and seeing that the "clouds" never moved, most people rejected that theory. However, these general ideas were as far as pretelescopic people got. Many people considered the dark areas to be liquid oceans made up of water. They were called *maria* (pronounced "MAH-ree-uh"), a word that means "seas." To this day, flat plains or plateaus on the Moon have names such as the *Sea of Tranquillity* and the *Sea of Crises*. The light areas were assumed to be land masses, but few people supposed they were strewn with mountain ranges and crater fields.

When Galileo and others began looking at the sky with telescopes in the seventeenth century, humanity's ideas did not change overnight, as they might have if our race had been driven more by hunger for knowledge and less by ego, fear, and superstition. People's imaginations were more active in Galileo's day than they had been a few centuries earlier, but they were not quite as daring as we are now. When Galileo announced that the Moon had craters and mountains, his fellow scientists became interested right away, but those who held power over people's lives had other notions. To them, Galileo was a troublemaker, and he was treated as one. He ended up spending his last years under house arrest. It was not a tyrannical government dictator that subjected him to this, but the Pope. Imagine the reaction the Pope would get today if he demanded that some scientist spend the rest of his life under confinement!

THE TIDES

The Earth-Moon system stays together because of gravitation. Earth pulls on the Moon, keeping it from flying off into interplanetary space. The

Moon also pulls on Earth, although we are not aware of it unless we make certain observations. The Moon's gravitational effects vary depending on where in the sky (or beneath the horizon) it happens to be at any given time.

The lunar day is about an hour longer than the synodic day—roughly 25 hours—but the Moon, like the Sun, appears to revolve around any stationary earthbound observer. The effects of gravity propagate through space at the speed of light, some 299,792 kilometers (186,282 miles) per second. This covers the Earth-Moon distance in only a little more than 1 second, so there is essentially no lag between the Moon's position and the direction in which its "tugging" takes place. The Moon's pull is extremely weak compared with the gravitation of Earth at the surface; it is nowhere near enough to affect the reading you get when you stand on a scale and weigh yourself, for example. But this does not mean that the gravitational effect of the Moon can be disregarded, as any ocean beach dweller will attest.

The Moon's gravitation, and to a lesser extent the Sun's gravitation too, cause Earth's oceans to be slightly distorted relative to the solid globe of the planet. When considered to scale, the oceans form a thin, viscous coating on Earth. Even the deepest undersea trenches reach less than 0.2 percent from Earth's surface down to the center, and most of the oceans are far shallower than this. Even so, the depth of the oceans is affected by the combined external gravitational effects of the Sun and the Moon. The effect is greatest when the Sun, the Moon, and Earth are all in line, that is, at the times of new and full Moon. When the Moon is at first or last quarter, its gravitational field acts at right angles with respect to that of the Sun, and the two almost cancel each other out, although the Moon's effect is a little greater than the Sun's.

As Earth rotates under its slightly distorted "coating" of ocean, the level of the sea rises and falls at specific geographic locations. In certain places, the rise and fall is dramatic, and people who live on the shore must take its effects seriously. In other places, these *tides* are much less extreme. There are two high tides and two low tides during the course of every lunar day; the reason for this can be envisioned by looking at Fig. 4-6. However, these drawings represent an oversimplification. The actual tides are delayed by the fact that on a planetary scale water behaves more like molasses than the freely running liquid with which we are familiar. Also, the contours of the ocean floor and the continental shelves have an effect. This is not all: Land masses break the planetary ocean up, so wave effects cannot propagate unimpeded around the world. The tides are waves, although they are very long, having two crests and two troughs with the passage of every lunar day. Actually, the tides consist of two waves of different frequencies.

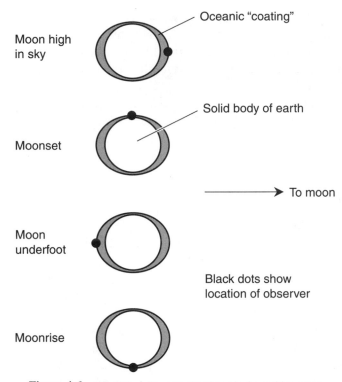

Figure 4-6.   Simplified diagrams showing why lunar tides occur.
These views are from high above Earth's north geographic pole.

Superimposed on the *lunar tidal waves*, which have a period of about 25 hours, are *solar tidal waves* (truly tidal in nature, unlike *tsunamis*, which are caused by undersea earthquakes, not by tides) with a period of 24 hours, but whose crests and troughs have smaller magnitude.

The Moon and Sun are not the only natural entities that affect sea level. Weather systems, especially ocean-going storms, have an effect, and in some places a storm surge can cause the sea to rise 10 times as much as the astronomical tide. A tsunami comes in and pounds away at the shoreline in a manner similar to that of a storm surge, except that the tsunami is caused by a jarring of the sea floor (or, occasionally, by a volcanic eruption) rather than by high winds piling the water up onto shore. Neither of these phenomena are true tides.

Tides don't occur to a significant extent in land-locked seas and lakes. This is so because in order for water to rise in one location, it must fall somewhere else where the lunar-solar gravitational composite is different.

This can't happen in a small body of water, on which, relative to every point, the Moon and Sun are in the same positions at any given time.

There has been some debate about the effects of the solar and lunar gravitational fields on the behavior of living cells. No one has yet come out with a respected scientific study that quantifies and defines exactly how such effects, if any, are manifest. For example, so-called Moon madness (lunacy) has not been explained on the basis of increased intracellular tidal effects during the full Moon. Because a similar loss of reasoning power does not seem to grip its perennial victims during the new Moon, it is almost certain that Moon madness (if it really exists) is not caused by gravitation.

HOW THE MOON WAS FORMED

Earth-Moon pair is unique in the Solar System; no other major planet has a satellite so large in proportion to itself. (Pluto and its moon Charon are sometimes called a double-planet system, but both of them are hardly larger than asteroids and might better be classified as such.) If the Moon were only a few hundred miles across, say the size of the asteroid Ceres, there would be little doubt that it was captured from solar orbit by Earth's gravitation. However, because the Moon practically qualifies as a planet, being almost as large as Mercury, there are competing theories about its origin.

One theory holds that the Moon was once a planet in its own right, orbiting the Sun rather than Earth, but something, such as a collision with a large asteroid, deflected it from its solar orbit and caused it to pass so close to Earth that Earth captured it permanently. Few astronomers believe this. If it were true, then the Moon's orbit likely would be an elongated ellipse sharply inclined to the plane of the ecliptic.

Another theory suggests that Earth originally had no Moon but that a Mars-sized object struck Earth a glancing blow and sent vast quantities of material into orbit. This would have produced rings around Earth, something like those around Saturn. These rings eventually would have congealed into the Moon. One problem with this theory is that an impact severe enough to blast that much matter off Earth might have shattered our planet. However, an increasing number of astronomers accept this theory, and it is lent support by the fact that the Moon lacks a metallic core; most or all of the blown-off material would have come from Earth's mantle and crust. It also might explain why Earth's equator is tilted significantly with respect to the plane of the ecliptic. A major impact could have jarred Earth and caused its axis to shift by 23.5 angular degrees.

Still another theory contends that Earth and the Moon formed as a double planetary system over the same period of time as the other planets, condensed from the disk of gas and dust orbiting the newborn Sun. This is a popular theory and is fairly consistent with what we observe about the Moon. It explains why the Moon orbits Earth in a nearly circular path and why the Moon's orbit is tilted only a little (about 5 angular degrees) with respect to Earth's orbit around the Sun.

The Sun

The Sun has been worshipped, despised, and feared by humans throughout history. Ideas about the Sun are as strange nowadays as they have ever been, but there are some statements about our parent star that we can make with confidence.

SIZE AND DISTANCE

The Sun is the largest nuclear reactor from which humanity has ever derived energy, and until we venture into other parts of the galaxy, this will remain the case. The Sun has a radius of about 695,000 kilometers (432,000 miles), more than 100 times the radius of Earth. If Earth were placed at the center of the Sun (assuming the planet would not vaporize), the orbit of the Moon would fit inside the Sun with room to spare (Fig. 4-7).

The commonly accepted mean distance from Earth to the Sun is 150,000,000 kilometers (93,000,000 miles) in round numbers. But the day-to-day distance varies up to a couple of million kilometers either way. Earth's orbit around the Sun, like the Moon's orbit around Earth, is not a perfect circle but is an ellipse with the Sun at one focus. Earth's closest approach to the Sun is called *perihelion*, and it occurs during the month of January. Earth is farthest away from the Sun—*aphelion*—in July. Surprisingly enough, for those of us in the northern hemisphere, the Sun is closest in the dead of the winter. It is not Earth-Sun distance that primarily affects our seasons but the tilt of Earth on its axis.

MEASURING THE SUN'S DISTANCE AND SIZE

Centuries ago, people did not know how large the Sun was, nor how far away it was. Estimates ranged from a few thousand miles (kilometers

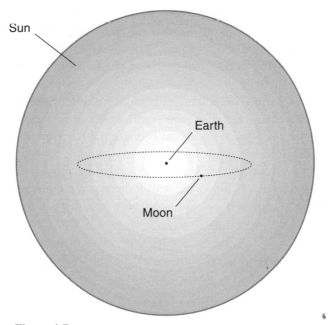

Figure 4-7.   Earth-moon system would easily fit inside the Sun.

hadn't been invented yet) to a few million miles. The distance to the Sun could not be measured by parallax relative to the background of stars because the Sun's brilliance obliterated the stars near it. The distance to the Moon had been measured by parallax, as well as the distances to Mars and Venus at various times, but the Sun defied attempts to measure its distance until someone thought of finding it by logical deduction. What follows is an example showing the sort of thought process that was used, and can still be used, to infer the distance to the Sun. Let's update the measurement techniques from those of our forebears and suppose that we have access to a powerful radar telescope, with which we can measure interplanetary distances by bouncing radio beams off distant planets and measuring the time it takes for the signals to come back to us.

Given a central body having a known, constant mass, such as the Sun, all its satellites obey certain physical laws with respect to their orbits. One of these principles, called *Kepler's third law*, states that the square of the orbital period of any satellite is proportional to the cube of its average distance from the central mass. This is true no matter what the mass of the orbiting object; a small meteoroid obeys the rule just as does Earth, Venus, Mars, and Jupiter. We know the length of Earth's year and the length of Venus's year; from this we can calculate the ratio (but not the actual values) of the two planets' mean orbital radii. Knowing this ratio is not enough, all by itself, to solve the riddle of Earth's mean distance from the Sun, but it solves half the problem.

The next step involves measuring the distance to Venus. If we could do this when Venus is exactly in line with the Sun, then we could figure out our own distance by simple mathematics. Unfortunately, the Sun produces powerful radio waves, and our radar telescope won't work when Venus is at *inferior conjunction* (between us and the Sun) because the Sun's radio noise drowns out the echoes. However, when Venus is at its maximum *elongation* (its angular separation from the Sun is greatest either eastward or westward), the radar works because the Sun is out of the way. At maximum elongation, note (Fig. 4-8) that Venus, Earth, and the Sun lie at the vertices of a right triangle, with

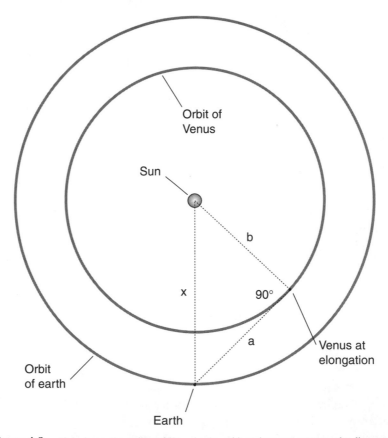

Figure 4-8.   If we know the radius of Venus' solar orbit and we can measure the distance to Venus at its maximum elongation, then we can calculate our own distance from the Sun.

the right angle at the vertex defined by Venus. One of the oldest laws of geometry, credited to a Greek named *Pythagoras*, states that the square of the length of the longest side of a right triangle is equal to the sum of the squares of the other two sides. In Fig. 4-8 this means that $a^2 + b^2 = x^2$, where x is the elusive thing we seek, the average distance of Earth from the Sun.

Now that we know the value of a in the equation (by direct measurement) and also the ratio of b to x (by Kepler's third law), we can calculate the values of both b and x because we have a set of two equations in two variables. Let's not drag ourselves through a detailed mathematical derivation here. If you've had high school algebra, you can do the derivation for yourself. It should suffice to say that this scheme can give us a fairly good idea of Earth's mean distance from the Sun if the measurement is repeated at several maximum elongations and the results averaged. However, even this will only give us an approximation because the orbits of Earth and Venus are not perfect circles. In recent decades, astronomers have made increasingly accurate measurements of the distance from Earth to the Sun using a variety of techniques.

Once the distance from Earth to the Sun was known, the Sun's actual radius was determined by measuring the angular radius of its disk and employing surveyors' triangulation in reverse (Fig. 4-9).

HOW IT "BURNS"

The Sun "shines" by constantly "burning" hydrogen, the most abundant element in the universe. You know, if you have taken chemistry classes, that hydrogen is flammable and that it burns clean and hot. (It might someday replace natural gas for heating if a method can be found to cheaply and abundantly produce it and safely distribute it.) However, the Sun "burns" hydrogen in a far more efficient and torrid fashion: by means of *nuclear fusion*. The enormous pressure deep in the Sun, caused by gravity, drives hydrogen atoms into one another. Hydrogen atoms combine to form helium atoms, and in the process, some of the original mass is converted directly into energy according to Einstein's famous equation $E = mc^2$ (energy equals mass times the speed of light squared).

The earliest theories concerning the Sun involved ordinary combustion, the only question being what, exactly, was burning. Coal was suggested as a fuel for the Sun, but if this were the case, the Sun would have burned out long ago. Besides this, there was the little problem of how all that coal got up there into space. Another idea involved the direct combination of matter with antimatter, resulting in total annihilation. However, if this theory were

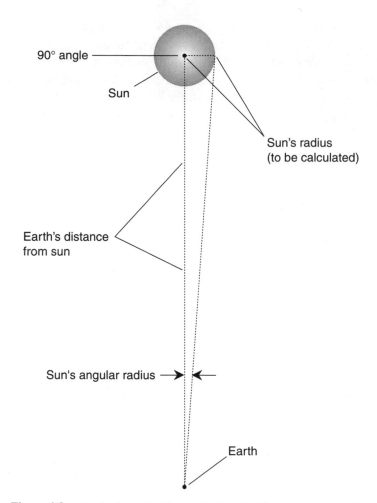

Figure 4-9.   The Sun's actual radius can be determined by reverse triangulation, on the basis of its apparent radius and its known distance from Earth.

true, the Sun would be far brighter and hotter than it is. The hydrogen-fusion theory accounts for what we see and is consistent with theories concerning the age of the Universe and the age of Earth.

ABOUT THE SUN'S LIFE

How long will the Sun's supply of hydrogen fuel last? Should we worry about the possibility that the supply will run out soon and Earth will cool off and freeze over?

Eventually, the Sun will burn out, but it will not happen for quite awhile. In fact, most scientists believe that the Sun will continue to shine for at least 1 billion more years at about the same level of brilliance as it does today. There are a lot of hydrogen atoms in that globe. It has a radius, remember, of 695,000 kilometers, or 69,500,000,000 centimeters. Scientists would write that as 6.95×10^{10} cm. Remember the formula for calculating the volume of a sphere from your middle school geometry class: $V = \frac{4}{3}\pi r^3$, where V is the volume and r is the radius. Using your calculator, you can figure out, using 3.14 as the value of π, that there are about 1.4×10^{33}, or 1.4 decillion, cubic centimeters of matter in the Sun. Written out in full, that number looks like this:

$$1,400,000,000,000,000,000,000,000,000,000,000$$

With this number in mind—or out of mind, because it's incomprehensibly large—you might be willing to accept practically any claim as to the Sun's longevity, except, of course, life everlasting. The Sun *will* perish. The symptoms of aging will begin in 1000 or 2000 million years.

As the supply of hydrogen runs out, the Sun will expand, and its surface will cool off. However, Earth will heat up because the bloated Sun will appear much larger in the sky and will send far more energy to Earth's surface than is the case now. The climate will become intolerably hot; the polar ice caps will melt; wildfires will reduce all plant life to ashes. Sometime during this process, any remaining humans and other mammals will die off. The oceans, lakes, and rivers will boil dry. All living things, even the hardiest bacteria and viruses, will die. The atmosphere will be blown off into space. Some astronomers think that the Sun's radius will grow until it exceeds the radius of Earth's orbit so that the Sun will swallow Earth up and vaporize it.

Don't let this scenario depress you. By then humanity will have colonized a couple of dozen other planets and will grieve no more about the fate of Earth than we do today about the buried houses of ancient cities. If our descendants remember us at all, it will be with fascination. Knowledge of our present society might be conveyed by legend, by stories told to children at bedtime, by tales about a place called *Terra* that sank beneath the surface of a stormy star after its inhabitants had fled, a place where people burned the decomposed by-products of dead plants and animals in order to propel surface transport vehicles. And the children will laugh and say that such a ridiculous place couldn't have existed.

After the *red-giant* phase, the Sun will fuse helium into carbon, iron, and other elements, and will shrink as gravity once again gains dominance over the pressure of nuclear heat. However, this process cannot continue forever. A point will be reached at which no further nuclear reactions can take place, and then gravitation will assert its ultimate power. The Sun will be crushed into an orb of planetary size and, as the last of its heat dissipates, will fade away and spend the rest of cosmic time as an incredibly dense, dark ball.

SPOTS AND STORMS

If you've never looked at the Sun through an appropriately filtered telescope, don't pass up the chance. Be sure that you use the proper type of filter (one that fits over the telescope's objective lens, not in the eyepiece or the star diagonal), and use it according to the manufacturer's instructions. Chances are good that you will see at least one dark spot on that bright disk.

Before the time of Galileo in the seventeenth century, the idea of sunspots did not cross the minds of people in Western civilizations—or if it did, no one ever voiced their thoughts aloud. The Sun was regarded as perfect, and if someone had suggested otherwise, they would have been disciplined or put to death. There is reason to believe that the ancient Chinese knew about sunspots and accepted their existence, having seen them, most likely, when the Sun was rising or setting in a hazy sky.

Sunspots enable astronomers to calculate the rotational period of the Sun because the blemishes seem to move across the solar disk, disappear over the edge, and then reappear a couple of weeks later on the opposite limb. The Sun rotates approximately once a month. Careful observation of spots has revealed that the Sun spins faster at lower latitudes than at higher latitudes. Using spectroscopy, this has been verified, and the solar "day" is in fact 25 Earth days long at the equator and 34 Earth days long at the poles. This fact has been used to prove that the Sun is not a solid body like Earth or the Moon.

Sunspots have an appearance that reminds some observers of biological cells (Fig. 4-10). The darkest part, at the center, is called the *umbra*; it is surrounded by a brighter region called the *penumbra*. Sunspot sizes vary, but they can be, and often are, much larger in diameter than Earth. The spots tend to form in groups and are believed to be depressions in the solar surface resulting from magnetic disturbances. They are, in a sense, storms on the Sun. The overall average number of sunspots rises and falls in a cycle of roughly 11 years. The most recent sunspot maximum took place in late 2000 and early 2001.

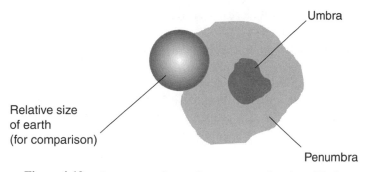

Figure 4-10. A sunspot can have a diameter greater than that of Earth.

If sunspots are like hurricanes on the Sun, then *solar flares* are like nuclear explosions: sudden, bright eruptions that send high-speed, charged subatomic particles flying off into space. Solar flares are more difficult to see with an ordinary telescope than are sunspots, but astronomers constantly watch the Sun for signs of these outbursts. A large solar flare is followed, in a day or so, by destabilization of Earth's magnetic field. As the charged particles come near our planet, they accelerate toward the north and south *geomagnetic poles*. This acceleration produces its own magnetic field, which interacts with that of Earth. As a consequence of this, the ionized layers of Earth's upper atmosphere change dramatically; this can create vast halos of light around the geomagnetic poles. If you happen to live at a high latitude, especially in North America, you are familiar with this glow as the *aurora borealis* (northern lights). Similar effects take place in the south polar regions, but the *aurora australis* (southern lights) are not spectacular, except as seen from Antarctica and from the far southern ocean on rare occasions when the sky over them is not socked in with grim overcast. The ionospheric disturbances affect radio communications and broadcast, especially on the so-called shortwave bands. In the extreme, even wire, cable, and satellite communications systems are disrupted by the powerful, erratic magnetic fields.

Eclipses

The Moon's orbit around Earth does not lie exactly in the plane of the ecliptic but instead is tilted about 5 degrees. The angular diameter of the Moon's disk, as seen from Earth, is only about half a degree; the same is true of the Sun's disk. Therefore, when the Moon passes between Earth and the Sun, the Moon's shadow almost always misses Earth, and when Earth passes between the Moon and the Sun, the Moon usually misses Earth's shadow. Once in a while, however, shadow effects occur and are observed by people. When the Moon's shadow falls on Earth, we have a *solar eclipse* or *eclipse of the Sun*. When Earth's shadow falls on the Moon, we have a *lunar eclipse* or *eclipse of the Moon*.

A FORTUNATE COINCIDENCE

The fact that the Sun and the Moon are almost exactly the same angular size as we see them from Earth is a convenient coincidence, especially when it comes to eclipses of the Sun. The Moon is just about exactly the right size to blot out the Sun's disk but little or none of the surrounding space, allowing us to see, during total solar eclipses, features of the Sun that would otherwise be invisible except from outer space. These include the *corona*, a pearly white mane of glowing, rarefied gases, and *solar prominences*, which are bright red or orange and look like flames leaping thousands of miles up from the solar surface.

PARTIAL SOLAR ECLIPSES

The Moon's shadow consists of two parts, called the *penumbra* and the *umbra* (the same names as are used with sunspots but with different meanings). Any object exposed to sunlight casts a shadow consisting of an umbra and a penumbra. From the point of view of an observer in the Moon's penumbral shadow, part of the Sun is obscured, but not all. Observers within the Moon's umbra see the entire solar disk covered up by the Moon.

The umbra of the Moon's shadow is a long cone extending approximately 395,000 kilometers (245,000 miles) into space opposite the Sun. The penumbra extends much further from the Moon and gets wider and wider with increasing distance, as shown in Fig. 4-11. The penumbra, while much longer than the umbra, does not have infinite length; it fades away as the Moon's apparent diameter shrinks with respect to that of the Sun. At a distance of several million miles, the penumbra dissipates as the Moon's disk as viewed against the Sun becomes comparable with the size of a sunspot.

When an observer is located within the penumbra of the Moon's shadow, a *partial solar eclipse* is observed. At the edge of the penumbra, the

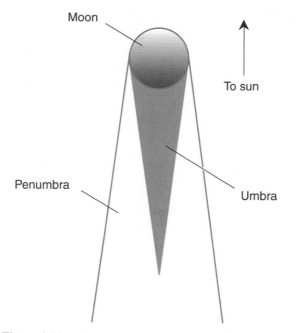

Figure 4-11.   The umbra and penumbra of the Moon's shadow.

Moon seems to "take a bite out of the Sun." Further within the penumbra, the Sun's disk takes the shape of a crescent. The closer to the umbra the observer gets, the slimmer the crescent Sun becomes. However, even a narrow crescent Sun is bright and has the same observed brilliance per unit area as the full Sun. Even with three-quarters of the Sun covered by the Moon, daylight still looks quite ordinary. This is where careless observers get into trouble. The exposed portion of the Sun is likely to cause permanent eye damage if a partial eclipse is viewed directly. This is true even if 99.9 percent of the Sun is obscured. The only safe way to look at an eclipse is the hole-projection method. Punch a 1- or 2-millimeter hole in a piece of cardboard, let the Sun shine through the hole, and look at the Sun's image on a piece of white paper held in the shadow of the cardboard. *Never look at the Sun directly during an eclipse.*

TOTAL AND ANNULAR SOLAR ECLIPSES

The length of the Moon's umbral shadow is almost exactly the same as the mean distance between Earth and the Moon. Thus, when the Moon passes between Earth and the Sun, the tip of the umbra sometimes reaches Earth's surface, but not always. If the Moon is at perigee and Earth is at aphelion when a solar eclipse takes place, the Moon is at its largest possible angular diameter, whereas the Sun is at its smallest. This results in a spectacular *total solar eclipse*, and, under ideal conditions, it can last about 7 minutes.

The worst conditions for eclipses of the Sun occur when the Moon is at apogee and Earth is at perihelion. Then the Moon is at its smallest possible apparent size, and the Sun is at its largest. In this case, a total eclipse does not occur anywhere on Earth. For observers fortunate enough to have the tip of the Moon's umbral shadow pass overhead, an *annular eclipse* takes place. The term *annular* means "ring-shaped," a good description of the appearance of the Sun during such an event. During an annular eclipse, the landscape and sky appear as they would in ordinary daylight through dark sunglasses.

LUNAR ECLIPSES

Earth, like the Moon, casts a shadow into space, but Earth's shadow is longer and wider. The full Moon usually misses Earth's shadow, but it sometimes enters the penumbra, and once in while it makes it into the umbra. If the Moon passes into Earth's penumbral shadow, we usually don't notice anything; for a few hours the full Moon might shine a little less brightly than it ought, but that is all. However, if part or all of the Moon moves into Earth's umbral shadow, a lunar eclipse occurs, and anyone on Earth who can view the Moon will see the eclipse.

When the Moon first begins to enter Earth's umbral shadow, the darkness "takes a bite out of the Moon" in much the same way as the Moon obscures the Sun during the beginning of a solar eclipse. The Moon might pass beneath or above the shadow core, so darkness never covers the Moon completely; in these cases we see a *partial lunar eclipse*. At mideclipse, the Moon seems to either "smile" or "frown" depending on which side of the umbra it passes.

The diameter of Earth's umbral shadow at the Moon's distance (381,000 kilometers, or 237,000 miles, on average) is several times the diameter of the Moon itself. For this reason, there is a fair chance that the Moon will plunge entirely into the umbra for a time, causing a *total lunar eclipse* (Fig. 4-12). These happen more often than total solar eclipses occur on Earth. The Moon rarely goes black during totality but exhibits a dim coppery or rusty glow caused by sunlight passing through Earth's atmosphere; the redness occurs for the same reason that some sunrises or sunsets appear red. An observer on the Moon would see a total solar eclipse of a truly alien sort. Imagine it: a thin red, orange, and yellow ring hanging in a black sky filled with unblinking stars, the Moonscape aglow as if with energy of its own.

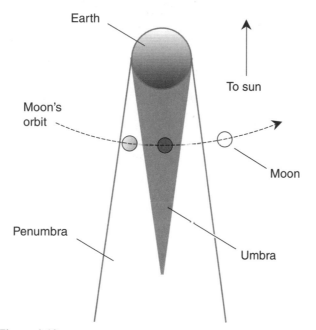

Figure 4-12.   Progress of a lunar eclipse. This drawing is not to scale.

Quiz

Refer to the text if necessary. A good score is 8 correct. Answers are in the back of the book.

1. The rotational period of the Sun
   (a) is synchronized with the orbit of Earth.
   (b) is shorter at the equator than at the poles.
   (c) varies in an 11-year cycle.
   (d) is not defined; the Sun does not rotate at all.

2. Which of the following theories concerning the Moon's formation is most popular?
    (a) The Moon condensed from the ejecta of thousands of earthly volcanoes many millions of years ago.
    (b) The Moon spun off Earth because of centrifugal force when Earth was in its early, molten state.
    (c) The Moon and Earth formed as a double planet.
    (d) All three of the preceding theories have been disproven conclusively.

3. Eclipses would be more common if the Moon's orbit
    (a) were more tilted with respect to the plane of the ecliptic.
    (b) were in the plane of the ecliptic.
    (c) had a longer period.
    (d) were more elongated.

4. The northern lights owe their existence to
    (a) Earth's magnetic field.
    (b) charged particles ejected from the Sun.
    (c) solar flares.
    (d) more than one of the above.

5. On a particular day, the Moon sets around high noon. What phase is the Moon in or near?
    (a) New
    (b) First quarter
    (c) Full
    (d) Last quarter

6. The composite tidal pull of the Sun and Moon is greatest when
    (a) Earth, the Sun, and the Moon are at the vertices of an equilateral triangle.
    (b) Earth, the Sun, and the Moon all lie along the same straight line.
    (c) Earth, the Sun, and the Moon are at the vertices of a right triangle.
    (d) This question is irrelevant; the composite tidal pull of the Sun and the Moon never varies.

7. Earth is
    (a) about 81 times more massive than the Moon.
    (b) about 10 times the diameter of the Moon.
    (c) always in full phase as seen from the Moon.
    (d) the same angular diameter as the Sun, as seen from the Moon.

8. In an annular eclipse of the Sun,
    (a) the path of totality is narrow.
    (b) the Sun's disk is totally covered for only a few minutes.
    (c) the Sun's disk is never totally covered.
    (d) the Moon takes on a copper-colored glow.

9. Suppose that a small stone with a mass of 10 grams and a large boulder with a mass of 10,000 kilograms are both put into circular orbits 15,000 kilometers above Earth's surface. Which of the following statements is true?
   (a) The stone will take longer to orbit Earth than the boulder.
   (b) The boulder will take longer to orbit Earth than the stone.
   (c) The stone and the boulder will take the same amount of time to orbit Earth.
   (d) The relative orbital periods will depend on how close together the stone and the boulder are placed.

10. The sidereal lunar orbital period (around Earth) is
    (a) longer than the synodic lunar orbital period.
    (b) the same as the synodic lunar orbital period.
    (c) shorter than the synodic lunar orbital period.
    (d) sometimes longer than and sometimes shorter than the synodic lunar orbital period.

Test: Part One

Do not refer to the text when taking this test. A good score is at least 30 correct. Answers are in the back of the book. It's best to have a friend check your score the first time, so you won't memorize the answers if you want to take the test again.

1. A solar flare can produce
   (a) disturbances in the Earth's magnetic field.
   (b) lunar libration.
   (c) a total solar eclipse.
   (d) a dark spot on the Sun.
   (e) A reddish glow on the Moon.

2. Longitudes are not assigned values greater than 180 degrees east or 180 degrees west because
   (a) doing so would result in a redundant set of coordinates.
   (b) the King of England forbade it in the seventeenth century, and no one has contradicted him since.
   (c) Galileo saw that such a thing could not possibly occur.
   (d) 180 degrees represents a full circle.
   (e) Oh, but they are! Latitudes are commonly assigned values greater than 180 degrees east or 180 degrees west.

3. The Sun's declination is 0 degrees on
   (a) January 21.
   (b) July 21.
   (c) October 21.
   (d) December 21.
   (e) none of the above dates.

4. The Moon's diameter is
   (a) about 1/81 that of the Earth.
   (b) about 1/30 that of the Earth.

    (c) about 1/4 that of the Earth.

    (d) about the same as that of the Earth.

    (e) variable, depending on its phase.

5. Which of the following constellations consists of a group of stars early in their lifespans, and still shrouded in the gas and dust from which they formed?

    (a) The Pleiades

    (b) Orion

    (c) Ursa Minor

    (d) Pollux

    (e) Bootes

6. In the southern hemisphere, the Sun's right ascension is 12 hours on or around the twenty-first day of

    (a) September.

    (b) December.

    (c) March.

    (d) June.

    (e) no month; the Sun never reaches a right ascension of 12 hours in the southern hemisphere.

7. The reddish color of the Moon during a total lunar eclipse is caused by

    (a) sunlight passing through the Earth's atmosphere.

    (b) sunlight reflected from the Earth back to the Moon.

    (c) sunlight scattered by particles in interplanetary space.

    (d) solar flares.

    (e) solar prominences.

8. If you were at the south geographic pole, the elevation of Polaris, the North Star, would be approximately

    (a) $-90°$.

    (b) $0°$.

    (c) $+90°$.

    (d) 180°W.

    (e) 180°E.

9. Scattered artificial light, such as that produced by the lights of a large city,

    (a) obscures many of the dimmer stars and constellations, which can be seen easily from locations in the outback.

    (b) causes the Moon to appear larger than it really is.

    (c) affects star visibility near the zenith more than star visibility near the horizon.

    (d) has no effect on stargazing whatsoever.

    (e) renders the dimmest stars more visible than they would be in the outback.

10. If Orion, the hunter, appears to be standing upright in the southern sky on a January evening as seen from New Hampshire, then at the same time, to viewers in Santiago, Chile, the hunter is

    (a) standing upright in the northern sky.

    (b) lying on his side in the northern sky.

    (c) standing on his head in the northern sky.

    (d) invisible because the seasons in the southern hemisphere are inverted with respect to the seasons in the northern hemisphere.

    (e) invisible because it would never rise above the horizon.

11. The constellation Octans, near the south celestial pole, can be seen rising in the east on evenings in the month of

    (a) April.

    (b) July.

    (c) October.

    (d) January.

    (e) None of the above

12. The relatively dark, central part of a sunspot is called the

    (a) core.

    (b) ecliptic.

    (c) pole.

    (d) umbra.

    (e) synodic.

13. Assuming that interstellar travel is possible and that humans will do it someday, the constellations will not be usable by the captains of interstellar space ships because

    (a) stars are invisible at warp speeds.

    (b) the constellations have their characteristic shapes only from the vantage point of our solar system and its vicinity.

    (c) all the stars will have moved to new positions and the present constellations will no longer exist.

    (d) they do not radiate electromagnetic fields of the proper type.

    (e) Wrong assumption! The constellations will be perfectly good navigational tools for long-distance interstellar travel.

14. Meridians on the Earth

    (a) are circles centered at the equator.

    (b) are circles centered at the south pole.

    (c) are half circles connecting the north and south poles.

    (d) are half circles parallel to the equator.

    (e) are straight lines passing through the Earth's physical center.

15. At which of the following times of year would the number of hours of daylight change the least rapidly from one day to the next?

    (a) Late June

    (b) Late August

    (c) Early October

    (d) Early March

    (e) Early April

16. The angular diameter of the Moon, considered as a whole and as viewed from the Earth, is
    (a) about ¹⁄₆ the angular diameter of the Earth.
    (b) about ¹⁄₂ the angular diameter of the Sun.
    (c) largest at apogee, and smallest at perigee.
    (d) tilted about 5 degrees with respect to the ecliptic.
    (e) about the same as the angular diameter of the Sun.

17. How many hours of right ascension are there in 30 degrees of arc, measured along the ecliptic?
    (a) One hour
    (b) Two hours
    (c) Three hours
    (d) Four hours
    (e) Six hours

18. Which constellation, easily visible at temperate latitudes in both the northern and southern hemispheres, is also known as the hunter?
    (a) Hercules
    (b) Libra
    (c) Orion
    (d) Pollux
    (e) Ursa Major

19. An effect of *libration* is:
    (a) that the Moon's orbit is not a perfect circle.
    (b) to accelerate solar particles as they encounter the Earth's magnetic field.
    (c) to cause the tides to lag the gravitational effects that produce them.
    (d) to let us see slightly more than half the Moon's surface.
    (e) to allow viewing of the solar corona during a total solar eclipse.

20. The summer solstice in London, England, occurs within a day of
    (a) March 21.
    (b) June 21.
    (c) September 21.
    (d) December 21.
    (e) None of the above

21. At the time of the first-quarter moon,
    (a) the Earth is directly between the Sun and the Moon.
    (b) the Moon is directly between the Sun and Earth.
    (c) the Sun is directly between Earth and the Moon.
    (d) the Moon is 90 angular degrees from the Sun in the sky.
    (e) the Moon is 180 angular degrees from the Sun in the sky.

22. Which of the following constellations lies in the same direction as the center of our galaxy?
    (a) Cancer

(b) Ursa Minor
(c) Gemini
(d) Orion
(e) Sagittarius

23. From which vantage point on the Earth would the star Sirius rise in the west and set in the east?
    (a) The far northern hemisphere
    (b) The north pole
    (c) The equator
    (d) The far southern hemisphere
    (e) Nowhere

24. What do the two stars Epsilon Eridani and Tau Ceti have in common?
    (a) They are both red giants.
    (b) They are members of a double-star system.
    (c) They have both been suggested as possibly having Earthlike planets.
    (d) They are both in the process of formation.
    (e) Nothing; these are the names of constellations, not stars.

25. As seen from a midlatitude location in the southern hemisphere at around sunset, the first-quarter moon
    (a) would be in the northern sky.
    (b) would be in the southern sky.
    (c) would be rising in the east.
    (d) would be beneath the horizon and therefore not visible.
    (e) might or might not be visible depending on the season of the year.

26. The Sun's declination is approximately +90° on or around the twenty-first day of
    (a) September.
    (b) December.
    (c) March.
    (d) June.
    (e) no month; the Sun never reaches a declination of +90°.

27. The brightest star in the sky, other than the Sun, is
    (a) Procyon.
    (b) Deneb.
    (c) Rigel.
    (d) Sirius.
    (e) Aldebaran.

28. Suppose that we make a scale model of the Earth-Moon system. The Earth's is represented by a beach ball 1 meter in diameter. The Moon would best be represented as:
    (a) a basketball 30 meters away.
    (b) a baseball 5 meters away.

    (c) another beach ball 100 meters away.

    (d) a marble 3 meters away.

    (e) another ball, but we need more data to know how big it should be and how far from the beach ball to place it.

29. Azimuth is essentially the same thing as
    (a) celestial longitude.
    (b) right ascension.
    (c) compass bearing.
    (d) elevation.
    (e) declination.

30. In the southern hemisphere, azimuth $0°$ is sometimes considered to be
    (a) east.
    (b) west.
    (c) south.
    (d) at the zenith.
    (e) at the nadir.

31. The Sun's rotational period, averaged between the poles and the equator, is roughly
    (a) the same as the period of the Moon's orbit around the Earth.
    (b) the same as the Earth's rotational period.
    (c) one synodic day.
    (d) synchronized with the tides.
    (e) synchronized with the equinoxes.

32. At the north celestial pole,
    (a) none of the observed constellations are circumpolar.
    (b) the stars all stay above the horizon for 11 hours and 58 minutes a day and stay below the horizon for the other 11 hours and 58 minutes.
    (c) half the constellations are circumpolar.
    (d) all the observed constellations are circumpolar.
    (e) Polaris is exactly on the northern horizon.

33. Suppose that you are in Calcutta, India, on March 21. For how long is the Sun above the horizon that day, measured with respect to the center of its disk?
    (a) Much longer than 12 hours
    (b) A little longer than 12 hours
    (c) Exactly 12 hours
    (d) A little less than 12 hours
    (e) Much less than 12 hours

34. The constellations as we know them today would no longer exist if we were to travel in time
    (a) 100 years into the future.
    (b) 200 years into the future.
    (c) 300 years into the future.

    (d) 400 years into the future.

    (e) The constellations would appear the same as seen from all the above time-journey destinations.

35. Magellanic Clouds are

    (a) high-altitude weather phenomena visible long after sunset.

    (b) vast tracts of interstellar dust.

    (c) a part of the Sun's corona.

    (d) caused by solar flares.

    (e) none of the above.

36. The Sun derives its energy primarily from

    (a) hydrogen combustion.

    (b) hydrogen fusion.

    (c) matter-antimatter reactions.

    (d) nuclear fission.

    (e) Gravitational pressure.

37. From April 1 to July 1, as viewed at 10:00 P.M. from a midlatitude northern place such as Colorado, the circumpolar constellations appear to rotate

    (a) 90 degrees clockwise.

    (b) 90 degrees counterclockwise.

    (c) 180 degrees.

    (d) 0 degrees; they are in the same positions.

    (e) by some amount, but we need more information to quantitatively answer this question.

38. In a few minutes on a clear, starry night, a star at the zenith will move toward the

    (a) north celestial pole.

    (b) south celestial pole.

    (c) vernal equinox.

    (d) celestial equator.

    (e) none of the above.

39. As seen from the Earth's equator, Polaris is approximately how many angular degrees from the zenith?

    (a) 0°

    (b) 30°

    (c) 60°

    (d) 90°

    (e) There is not enough information given here to answer this.

40. The circumference of the ninetieth parallel in the northern hemisphere is

    (a) the same as that of the equator.

    (b) half that of the equator.

    (c) one-third that of the equator.

    (d) zero because it is the north geographic pole.

    (e) impossible to determine without more information.

PART TWO

The Planets

CHAPTER 5

Mercury and Venus

Earth is the third planet in the Solar System. The two closer-in planets are *Mercury*, named after the Roman messenger god, and *Venus*, named after the goddess of love and beauty. Both of these *inferior planets*, as they are called, suffer temperature extremes that make them forbidding places for life to evolve or exist. Conditions on both planets are so severe that future space travelers will dread the thought of spending time on either of them. The harsh environments might, however, serve as testing grounds for various machines and survival gear, using robots, not humans, as subjects.

Twilight Stars

Neither Mercury nor Venus ever strays far from the Sun in the sky because both orbits lie entirely inside Earth's orbit. Mercury rarely shows itself to casual observers; you have to know when and where to look for it, and it helps if you have a good pair of binoculars or a small telescope. Venus, in contrast, is at times the third brightest object in the sky, surpassed only by the Sun and the Moon.

OBSERVATION

The best time to look at Mercury is when it is at or near maximum elongation and when the ecliptic is most nearly vertical with respect to Earth's horizon. Maximum elongations happen quite often with Mercury because it travels around the Sun so fast. But ideal observing conditions are rare.

Suppose that you live at temperate latitudes in the northern hemisphere. If Mercury is at maximum eastern elongation (the planet is as far east of the Sun as it ever gets) near the March equinox, it can be spotted with the unaided eye low in the western sky about a half hour after sunset. If Mercury is at maximum western elongation near the September equinox, look for the planet low in the eastern sky about a half hour before sunrise.

If you live in the southern hemisphere, the situation is reversed. If Mercury is at maximum eastern elongation near the September equinox, it can be spotted with the unaided eye low in the western sky about a half hour after sunset. If Mercury is at maximum western elongation near the March equinox, look for the planet low in the eastern sky about a half hour before sunrise.

Venus shows itself plainly much more often than does Mercury. In fact, this planet has sometimes been mistaken for an unidentified flying object (UFO) because it tends to "hover" in the sky and can be as bright as an aircraft on an approach path several miles away. With a light haze or with high-altitude, thin cirrostratus clouds, the planet can be blurred and the effect exaggerated. Nevertheless, ideal observing conditions occur under the same circumstances as with Mercury; Venus can be seen for several hours after sunset when it is at maximum eastern elongation and for several hours before sunrise when it is at maximum western elongation.

PHASES

The inferior planets go through phases like the Moon. This fact was not known to astronomers until Galileo and his contemporaries first turned "spy glasses" to the heavens. The phases occur for the same reason the Moon goes through phases, and they can range all the way from a thin sliver of a crescent to completely full.

Figure 5-1 shows the mechanism by which an inferior planet attains its phases. The half-illuminated phases occur, in theory, at the points of maximum elongation, that is, when the angle between the planet and the Sun is greatest as seen from Earth. (In the case of Venus, this is not quite true because the thick atmosphere of that planet has a slight effect on the position of the twilight line.)

The full phase of an inferior planet takes place at and near *superior conjunction*. When it is exactly at superior conjunction, the planet is obscured

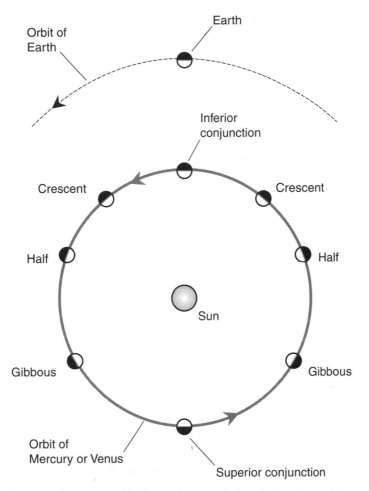

Figure 5-1.   Phases of the inferior planets result from their relative positions
with respect to Earth and the Sun.

by the brilliance of the Sun and might even be eclipsed by our parent star. The new phase, which takes place at *inferior conjunction*, is usually invisible too, but not always. Sometimes Mercury passes so nearly in line between Earth and the Sun that it can be seen against the Sun's disk when observed through a filtered telescope. On rare occasions, the same thing happens with Venus. When this happens, Mercury or Venus is said to *transit* the Sun.

Profile of Mercury

Let's take a close look at the innermost known planet in our solar system. This is one of six planets, other than Earth, that has been known since antiquity. It got its name from the fact that it moves fast, not only as we see it, but literally, because of its proximity to the Sun.

THE YEAR AND THE DAY

Mercury's mean orbital radius is 38.7 percent that of Earth, or about 58 million kilometers (36 million miles). The planet's distance from the Sun varies considerably; the orbit is far from a perfect circle. At perihelion, Mercury is 46 million kilometers (29 million miles) from the Sun; at aphelion, it is 70 million kilometers (44 million miles) distant. This, together with the fact that Mercury's year is 88 Earth days long while its day (measured relative to the distant stars) is 59 Earth days long, gives the planet "seasons" of a sort.

Even though the equatorial plane of Mercury is tilted by only 2 degrees relative to the plane of its orbit around the Sun, the planet's changing distance from the Sun produces a maximum-to-minimum solar brilliance ratio of more than 2:1. That is, at perihelion, Mercury gets more than twice as much energy from the Sun as it does at aphelion. This does not matter in any practical sense; from a human point of view, the surface broils in daylight and freezes at night. However, if Earth's orbit were as eccentric as that of Mercury, life on this planet would have evolved along a much different course. The weather would be more violent, the tides would be greater, and things in general would be rougher than they are. Earth's 23.5-degree axial tilt causes seasonal variability, but nothing of the sort that would take place if the irradiation from the Sun were twice as great at some times compared with other times.

COMPOSITION

Mercury is a lone planet; it has no moons, except perhaps for a few tiny, captive asteroids or degenerate comet nuclei. Mercury is more dense than any other planet in the solar system except Earth and is considerably smaller, only about 4,880 kilometers (3,030 miles) in diameter. Figure 5-2 shows the relative sizes of Mercury and Earth.

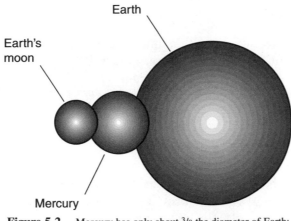

Figure 5-2.   Mercury has only about ³/₈ the diameter of Earth;
it is only a little larger than Earth's moon.

Mercury has a surface pockmarked with craters. The planet is even more rugged than our own Moon. If Mercury were suddenly put in place of our Moon, we would notice only a slight difference in size and brightness, but the "facial features" would be markedly different. Moreover, the tides on our planet would be much more variable, for Mercury has a core of iron that is unusually large in proportion to the planet's overall size. The dense core is overlaid with a crust of silicate rock.

There is no evidence of geologic activity on Mercury. Because it is so small, it has cooled off to the point that future volcanism will not occur. There is plenty of evidence that volcanoes spilled lava out onto the landscape in the distant past, but today the planet's surface heat is attributable entirely to the Sun. This keeps the daytime temperature hot enough to melt lead. In one region, known as the *Caloris Basin* because the Sun is near the zenith there when Mercury is at perihelion, the midday temperature rises above 400°C (750°F).

The surface of Mercury is crisscrossed by a network of cliffs 2 to 4 kilometers (3 to 6 miles) high. This suggests that as the interior of the planet cooled off, it shrank enough to cause the crust to fall inward in places. Imagine the violence of such an event if Earth were to endure it! When a piece of our own *terra firma* collapses or slides a few meters along a fault line, it makes news headlines all around the world. Suppose that the San Andreas fault were rocked by a quake that sent the coast of California plunging 3 kilometers in a couple of minutes? Imagine the same thing happening along all the major fault lines of the world at once!

ATMOSPHERE

Surprisingly, Mercury has a thin atmosphere. It is only a tiny fraction as dense as that of Earth at the surface; in fact, the "air" on Mercury would constitute an excellent laboratory vacuum. Argon and helium gases have been detected. The Sun continually blows Mercury's atmosphere off into space, but at the same time, the heat of the Sun drives new atmospheric atoms out of the surface.

There is no weather, as such, on Mercury. However, there is evidence that water ice might exist in some of the craters near the pole. The *Mariner 10* mission in the mid-1970s analyzed the surface of Mercury extensively, photographing most of it, and there is apparently some sort of frost that has settled in the depths of polar craters where the Sun never shines. The minimal tilt of the planet's axis keeps certain small regions of the polar surface in shadows at all times. If humans ever attempt to land on Mercury, these would be the logical places to touch down.

Profile of Venus

The second planet from the Sun is named after the Roman goddess of love and beauty. This doubtless arose from its striking appearance in the evening and morning skies, especially near the times of maximum elongation. However, as the literal facts about Venus emerged from flybys and landings that took place in the mid-twentieth century, it became clear that the surface of Venus is neither a lovely nor a beautiful place. In fact, conditions there resemble medieval humanity's conception of hell.

THE YEAR AND THE DAY

Mercury's mean orbital radius is 72 percent that of Earth—roughly 108 million kilometers (67 million miles). The perihelion and aphelion of Venus differ by less than one percent, so the orbit of the planet is nearly a perfect circle. At times, Venus comes within 38 million kilometers (24 million miles) of Earth. Venus takes 225 Earth days to orbit the Sun.

The day on Venus is strange because it is longer than the year. Venus rotates once on its axis, relative to the distant stars, every 243 Earth days.

The peculiarity goes further: Venus spins from east to west rather than from west to east.

The tilt of Venus' equatorial plane, relative to that of its orbit around the Sun, can be interpreted in either of two ways. If we consider the rotation of the planet to be retrograde (that is, from east to west), then the tilt is only about 3 degrees. This is the conventional view. However, we also can imagine the tilt as being 177 degrees (3 degrees short of a complete flipflop), so the north pole points in the opposite direction from the north poles of all the other planets. No one on the surface would care about details like this. They would be preoccupied with surviving the heat, comparable with the temperature of Mercury's Caloris Basin at midday, and the pressure, which would necessitate that housing be designed as if it were for the deep sea. The daylight on Venus is comparable with that of a gloomy winter afternoon in London or Seattle.

COMPOSITION

Venus, like Mercury, is moonless. The planet is almost the same size as Earth (12,100 kilometers or 7,500 miles in diameter), and the internal composition is similar to that of our own planet. However, from the surface upward, Venus and Earth could hardly be more different.

The surface of Venus remained a mystery until it was mapped by radar. The microwave signals of radar equipment can penetrate Venus's thick clouds, which maintain an unbroken overcast at an altitude several times higher than the highest clouds on Earth. These clouds reflect light very well, and this accounts for the brilliance of the planet as we see it in our evening or morning sky. When the surface was finally observed using radar and computer-graphics programs, huge volcanoes were discovered. The largest of these would make Hawaii's Big Island seem tiny in comparison.

In the 1970s, Russian *Venera* probes landed on Venus and survived long enough to take a few photographs. The surface appeared strewn with rocks that cast no shadows on account of the overcast. The landscape and sky were bathed in rusty light, produced by filtration of sunlight through the sulfurous clouds. There was no sign of life, and it was surmised that life as we know it could not exist there. Along with the searing heat, the pressure at the surface of Venus is 90 times as great as the pressure at Earth's surface.

The surface of Venus is not cratered in the same way as the Moon or Mercury because the atmosphere of Venus has eroded all minor impact marks that might have been formed long ago. Small meteors are burned up by the atmosphere; Venus's shroud of heavy carbon dioxide provides better protection than does Earth's atmosphere. Gigantic meteors and, of course, asteroids can survive even the thickest atmosphere, and it is certain that Venus, like all the other planets, has been struck many times. However, the more notable source of cratering on Venus is volcanic activity.

ATMOSPHERE

When astronomers used spectroscopes to analyze the atmosphere of Venus by looking at the light reflected from the planet's clouds, they knew that the air on that world consists largely of carbon dioxide. This led them to suspect that Venus must be much hotter than Earth because carbon dioxide is well known for its ability to trap heat. It was not until space probes gave the planet close examinations that the truly Hades-like nature of the atmosphere was determined. While the air itself contains mainly carbon dioxide, especially near the surface, the clouds are laden with sulfuric acid.

The clouds that surround Venus are different from those in Earth's atmosphere in almost every possible way. Much of Earth's surface is obscured by overcast, but there are always clear spots through which the surface can be seen from space and through which direct sunlight can pass. This is not the case on Venus, at least not at the visible wavelengths. As seen through the most powerful Earth-based telescopes, Venus looks like a cue ball. At ultraviolet wavelengths, details in the clouds can be seen, and from analyzing such views it has been determined that the upper equatorial clouds race around the planet approximately once in every 100 Earth hours.

The clouds on Venus exist in a single layer several kilometers thick and much higher above the surface than any clouds on Earth (except the *noctilucent clouds* sometimes seen in Earth's stratosphere). No place on the surface of Venus is ever subjected to direct sunlight. The sky appears angry because of the orange color produced by the sulfur compounds. The winds at the surface of Venus are light, at least in an absolute sense, never moving more than a few meters per second. However, because the air at the surface is thick, a breeze that would be a whisper on our planet carries considerable force on Venus. Perhaps the landscape on Venus is subjected to an ongoing severe thunderstorm with sulfuric acid "rain" that evaporates before it can get down to the surface. Figure 5-3 provides a comparison between Earth's atmosphere and that of Venus.

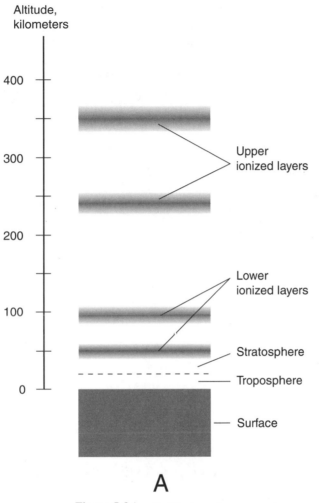

Altitude, kilometers

400

300 — Upper ionized layers

200

Lower ionized layers

100 —

Stratosphere

Troposphere

0 —

Surface

A

Figure 5-3A.   Earth's atmosphere.

From Earth to Mercury

Imagine that you get on board an imaginary space vessel, the *Valiant*, for a trip that will take you all the way to the planet Mercury to begin an inside-to-outside tour of the Solar System. This vessel was built in Earth orbit and is large enough to keep a crew of 25 people comfortable. Artificial gravity is provided by a spinning, ring-shaped residential deck. The rate of spin can be adjusted during the journey from one planet to the next so that when (or

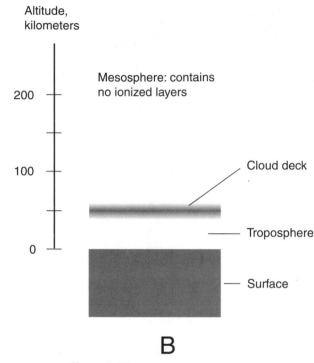

Figure 5-3*B*.   The atmosphere of Venus.

if) you step off the vessel and onto the surface of another world, you body will be spared gravitational shock.

GETTING THERE IS HALF THE TROUBLE

Mercury lies much closer to the Sun than does Earth, and the little planet races around the Sun in only about one-quarter of an Earth year. At first thought it might seem like an easy task to hurl a space object inward toward the Sun, but it requires considerable energy. In order for an object to fall in toward the Sun from Earth orbit, that object must decelerate, and this takes as much energy as the acceleration necessary to achieve orbits more distant from the Sun than that of Earth. When you apply the brakes on a vehicle for a long time, they get hot; energy is expended slowing you down. Powerful retrorockets must be employed to begin our journey inward toward Mercury.

Along the way to Mercury, you cross the orbit of Venus. If you time things just right, you can use the gravitational field of Venus to help the ship attain a course for a rendezvous with Mercury (Fig. 5-4). This is known as a *gravitational assist* and has been a useful maneuver ever since the first interplanetary probes were launched by humankind back in the 1970s. Gravitational assists can either accelerate or decelerate a space ship and can help interstellar space vessels get underway out of the Solar System. In your case, a Venus nearmiss sends the ship plunging inward toward the Sun. Timing is critical; the slightest miscalculation might send you into an eccentric orbit between Mercury and Venus, missing Mercury by tens of thousands of kilometers. Or worse, it could put you on an irrevocable course into the Sun.

ATTAINING ORBIT

Even if you don't fall into the Sun, there is danger in a trip to the inner planets. The Sun occasionally has "tantrums" called *solar flares* that eject large quantities of high-speed subatomic particles. These particles affect living tissue in much the same way as the high-intensity gamma rays produced by

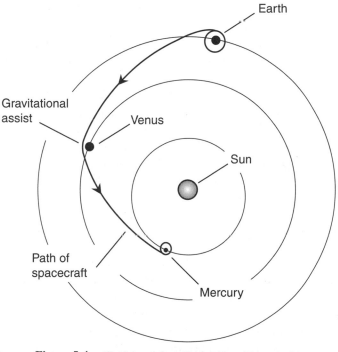

Figure 5-4.   The journey from Earth orbit to Mercury orbit
employs a gravitational assist from Venus.

nuclear bombs. Earth's atmosphere protects you from these particles when you're on your home planet's surface, but in Earth orbit or when traveling between Earth and the Moon, there is some risk. The peril increases according to the *inverse-square law* as you approach the Sun. If a solar flare were to take place while you were in orbit around Mercury, you would be subjected to almost 10 times the radiation from these particles as you would get if you were in orbit around Earth. For this reason, your itinerary planners saw to it that you should make your journey near the time of sunspot minimum, when solar flares are least likely to occur.

As you attain orbit around Mercury, you see the surface up close. It looks similar to Earth's moon, although there are more cratered areas and less *maria*, or flat regions. The *escarpments*, or cliffs, produced when the crust collapsed around the cooling core long ago are vividly apparent, especially near the twilight line where the Sun shines down on the surface at a sharp angle, producing long shadows. The captain has selected a polar orbit around Mercury—one that passes over the north and south poles—so that you can land easily in one of the craters near the south pole. Your trip planners left nothing to chance. They found a crater and a specific spot within that crater where the Sun never shines. Thus, while you are on the surface, you'll be safe even if a solar storm happens to take place. If humankind ever sees fit to put a permanent base on Mercury, it will be important that it be in a spot such as this. The temperature is far below zero; this will require that the landing craft's heat generator work hard.

"There it is," says the captain, and you look down to see a ring of mountain peaks lit up by the Sun, with shadows so black that the crater looks like a bottomless well. "We will land right there. But not on this pass; on the next one." You don't have long to wait. Mercury is a small world, and you are in a low orbit, having no worries about atmospheric drag. In less than an hour you'll be on your way down.

DOWN AND BACK

The landing shuttle is named *Eagle*, after the *Apollo 11* lunar landing module, and that's a good name for this contraption. The Mercury lander looks very much like the *Apollo* mission landing craft. "I know what you're thinking," says the captain as you get into the little chamber. "But if it worked in 1969, it will work now. The biggest difference is that this one can deal with greater temperature extremes. And the communications and navigation equipment are a lot better than they were in 1969."

The first officer will be our guide for the landing mission. The captain must stay with the *Valiant*. The first officer also will serve as our teacher during long periods of interplanetary travel. "One hour of general astronomy and cosmology training every day will make cosmic gurus out of you by the time this journey is over," he says.

As you break free of the main vessel, you get a feeling like that of a first free dive when learning SCUBA diving. It's one thing to practice SCUBA diving in a swimming pool or a small pond; it is entirely another to dive in the open ocean, miles from land. If something goes wrong while you're isolated in this little landing craft, only the skill of the first officer and a good measure of luck will stand between you and disaster.

As you near the crater, the mountains loom. "That looks like water ice," you say, noticing a grayish sheen inside the crater. "It might be exactly that," says the first officer with a hint of a smile. The craft slows; the landing lights come on. Then suddenly you are in pitch darkness. The Sun has dipped below the rim of the crater. Stars flash into view as the Sun no longer dominates the sky. Your eyes adjust to the darkness; the first officer scans the surface below with high-resolution submillimetric and infrared radar. You hover. The craft seems to move a little. The glint of the Sun-scorched peaks on the far side of the crater lends some perspective. "Hmm," says the first officer. Then again, "Hmm." His eyes are glued to the screen and to a group of lights and dials. "Is this good or bad?" you ask. "It depends," says the first officer. "If you are worried about liftoff, it is good. If you are looking forward to touchdown, it is bad."

You begin the ascent back up to the *Valiant*. You will not land.

"I couldn't find a suitable spot," apologizes the first officer. "We have limited fuel, and that means we have limited time. The captain didn't get the *Valiant* into the orbit she had hoped for. The orbit is a little too high; we had to travel a little farther to get down here than I was expecting."

"Is that good or bad?" you ask.

"It is not good," says the first officer. "But it could be worse. If our orbit had been a few kilometers higher or slanted by another degree with respect to the poles, I would not have attempted this trip at all."

AN ORBITAL INTERLUDE

After you return to the *Valiant*, the captain has little to say. "This is the nature of life in outer space," she says; "you take what you can get. Now settle in and get comfortable. We have to stay in orbit around Mercury for awhile."

"Why?" you ask. But the captain has already left the room; she has more important business.

The first officer explains. "The captain selected a polar orbit when we made our rendezvous with Mercury. It was relatively easy to get into that sort of orbit. We drifted in and swooped under Mercury's south pole. The planet's gravitational field took care of the rest. But now we find ourselves in an orbit that is inclined 90 degrees to the plane in which the planets orbit the Sun."

"So?" you ask.

"If we shoot out of this polar orbit in the planetary plane right now," explains the first officer, "we won't be heading for Venus. But we have to stay in the planetary plane. If we did not shoot out of Mercury orbit in the plane of the planets, we would find ourselves bound for interstellar oblivion. Our fuel would never be sufficient to get us back to the Solar System. Our ship must leave this orbit along essentially the same line from which it approached (Fig. 5-5). This greatly restricts our options."

"So what do we do?" you ask.

"We wait," says the first officer, "until Mercury is in the proper position with respect to Venus. Then we will fire our engines, shoot back out from under the south pole, and coast to Venus, where we will enter an equatorial orbit. That will be much less inconvenient to get out of properly."

"How long do we have to wait here?"

"Eight days."

"It is a good thing we have broadband Internet access on this ship," you say.

"Yes," says the first officer. "But the latency is horrible. We're several light-minutes away from the hub in Dallas, Texas. I suggest that you watch old videos or listen to old music albums. We have a huge selection. And don't forget to work your body out for two solid hours a day. Otherwise you will lose calcium from your bones."

"Won't the artificial gravity prevent calcium loss from bones?"

"To some extent. But remember that it's less than one Earth gravity and will increase slowly because the geometry of our trajectory means that we will be en route to Venus for a long time."

"How long?" you ask.

"You don't want to know," says the first officer.

From Mercury to Venus

Life in space can become tedious if you are the sort of person who likes genuine outdoor air, real rain, and real snow. Maybe some day virtual real-

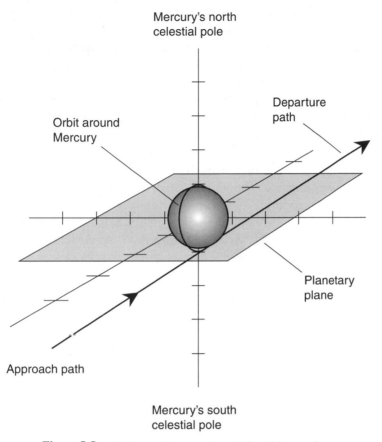

Mercury's north
celestial pole

Departure
path

Orbit around
Mercury

Planetary
plane

Approach path

Mercury's south
celestial pole

Figure 5-5.   A ship must leave a polar orbit from Mercury along
essentially the same line from which it approached.

ity will help alleviate some of this tedium, but it will always take a certain
type of person to deal with the rigors of long-distance space travel.

BOOSTING SPEED

The rotation rate of the artificial-gravity wheel increases gradually, beginning immediately after your return from the aborted Mercury landing. Over
the next several weeks the spin rate will be controlled by a computer program that will optimize your adaptation to Venus gravity just as you enter
orbit around that planet.

The first officer wasn't joking when he said there is a good selection of
videos on the spacecraft. In total, there are more than 2 million hours'

worth of audiovisual entertainment available. You are, after all, just beginning your journey; you will eventually be going all the way to Mars. There are music albums, virtual-reality games, and ironically, video games in which you get to pretend that you are a starship captain. However, the holographic environmental simulators (the rooms you have seen in science-fiction shows and movies) are something you'll have to wait a few decades to experience. "The present government administration," explains the captain, "does not believe the budget should allow for such frivolities."

Finally, one happy morning (according to clocks in Texas) the first officer announces that you are leaving Mercury orbit. "We will notice a shift in the axis of gravitation," he says.

"What does that mean?" you ask.

"When this vessel is not accelerating, the artificial gravity pulls straight outward at exactly a right angle with respect to the ship's course through space (Fig. 5-6)," he says. "But when the ship is accelerating or decelerat-

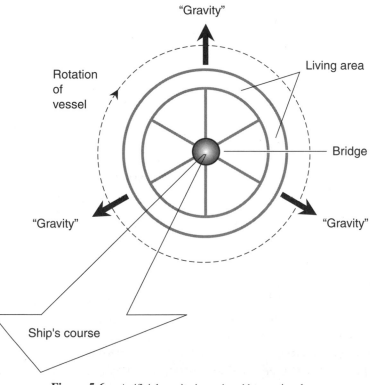

Figure 5-6.   Artificial gravity is produced by rotating the living space on the interplanetary vessel.

ing, there is an additional vector either backward or forward, and this adds to the outward pull of the artificial gravity. The result is the peculiar illusion that gravity does not pull you straight down. If you were in the living space during acceleration, it would seem as if the floor were slanted. But you will have to come back to the bridge area and strap yourselves in while we leave Mercury orbit and accelerate on our way to Venus. Fortunately, this process won't take long. In order to get on course to Venus, we need only a tiny bit more speed than was necessary to escape Mercury."

"Good," you say. "I can hardly wait to watch some more videos."

"You work out," says the captain. "Then you watch videos."

The acceleration process takes about an hour. This is enough time to digest breakfast before you use the stationary bicycle, the all-in-one universal gym, the punching bag, and that strange running track that runs completely around the outer circumference of the living space and that seems to forever curve uphill.

ATTAINING ORBIT

In space there are no days and nights, except those that are produced artificially by the environmental systems on board the ship. In the living quarters of the *Valiant* there are no windows; the spinning of the gravity wheel would give you vertigo if you had windows to gaze from. However, there are pictures on the walls that change on a 24-hour cycle: landscapes with illumination that mimics the spectrum of the Sun on Earth's surface. Your room has scenes of the rural Midwest. It is winter there right now, and the ship's computer has determined that there has been plenty of snow in the last couple of weeks.

With realistic views out the "windows," plenty of videos, music albums, a big library of books (the real, bound paper kind), an increasing workout schedule, good food, and a friendly crew, time goes fast. You wake up one morning and notice that all your "window" scenes have been reprogrammed to show the clouds of Venus swirling beneath the ship, whitish yellow, and a ruddy horizon that quickly fades away to black higher up.

"We are in an equatorial orbit," explains the first officer over breakfast. "Our descent and flight will be along the equator, where the clouds race around the planet in four Earth days. You might notice, if you're astute, that our orbit is retrograde."

Venus spins contrary to all the other planets, that is, from east to west. The clouds, too, move in that direction. But what does the first officer mean by "flight"?

SHUTTLE OR AIRPLANE?

The first officer will not attempt to land on Venus. "Landing craft have been devised that can survive conditions at the surface," says the first officer, "but they are prohibitively expensive. We could never venture outside. All we could do would be sit around and peer out through thick, reinforced little portholes, like those in an undersea vessel."

You will fly just beneath the clouds, in a hybrid shuttle/aircraft contraption reminiscent of the X-15 high-altitude craft used by the United States in the middle of the twentieth century. The upper-level winds rush along in the Venusian atmosphere at 400 kilometers (250 miles) per hour. "Venus has a huge, continuous, retrograde jet stream," says the first officer, "and we are going to ride it once around the planet at an airspeed of 800 kilometers (500 miles) per hour."

You do some quick calculating. Venus is about 40,000 kilometers in circumference; you will be moving at 800 + 400, or 1,200, kilometers per hour with respect to the surface. You decide to use your wrist calculator. "Calculator," you say, "divide 40,000 kilometers by 1,200 kilometers per hour." The little thing speaks up in its synthesized voice: "Solution: 33 hours and 20 minutes."

You look quizzically at the first officer, who has now been joined by the captain who will wish you a happy trip.

"Don't worry," says the captain. "The planetary atmospheric reconnaissance vehicle (PARV) has a bathroom, a refrigerator, and a stove. Just like home."

"There will be gravity for most of the trip," says the first officer. During the ride beneath the clouds of Venus, during which you hope to get a good view of the surface on the sunlit side and some spectacular lightning shows on the nighttime side, you will be riding in an aircraft in a gravitational field almost of exactly the same strength as that at the surface of Earth.

"There are plenty of airsick bags," says the first officer. "Venus can be a stormy place. I hope you didn't eat too much for breakfast."

A WILD RIDE

The descent to the cloud tops goes smoothly, and the yellowish white barrier rises up to meet the shuttle just as you cross the twilight line into darkness. The rumble of the rocket engines fades as the first officer allows atmospheric drag to slow the aircraft down. "We won't come through the bottom of the cloud layer until halfway through the night," says the first officer as he sips on a glass of lemonade. "We are letting the atmosphere do all the work of slowing us down. The jet engines will start after we get under the clouds."

The view out the windows turns from cream-colored to yellow, then orange, then rusty, then brown, and finally black. You imagine that you can hear the hiss of the sulfuric acid droplets as they eat away at the exterior of the craft, but you know this cannot actually be taking place because of the protective coatings that keep heat, radiation, and corrosion from affecting the shell of the vessel. Just as the windows have completely blacked over, you feel the first jolt.

For the next 8 hours you lie flat on your back, your seat all the way down to horizontal, strapped in tight, and try to sleep. You don't get airsick, but you worry that something will go wrong, the craft will shake apart, and you will be sent tumbling down into hell. Then finally the violent ride becomes smooth, and you hear the whooshing sound of the jet engines, which are now propelling the craft beneath the clouds of Venus at midnight.

All around the ship, lightning flashes: yellow, blue-white, brilliant white. With each flash, you can see the ceiling of cloud deck above. You cannot yet see the planet's surface.

"We're 40 kilometers (25 miles) above the landscape down there," says the first officer. "You'll be able to see it as soon as we come back into daylight. That will be in about 8 hours."

"How high are we going to fly?" you ask.

"We have leveled off now," says the first officer. "This is as low as we dare go. If we went a little lower, the atmosphere would slow us down too much. Then we would lose altitude, and the drag would increase further, slowing us down still more, and we would plunge to the surface and crash."

"That is not what I want to hear," you say.

"Don't worry," says the first officer. "I am well trained."

THE LANDSCAPE OF HADES

Finally, you fall asleep. As you awaken, it is starting to grow light outside. This is a morning like none you have seen before. Above you are orange clouds, and beneath you stretches a rolling, sullen, reddish plateau advertising the fact that life cannot exist there. You know that the air pressure at the surface of Venus is as great as the pressure 1,000 meters (3,300 feet) beneath the surface of the sea, deeper than SCUBA divers ever venture. A

Venus lander would have to endure pressure like a submarine, be able to fly like an airplane, take off like a rocket, and all the while endure temperatures that rise higher than the hottest day in the Caloris Basin of Mercury.

On the windows you see what looks like rain. But this is no ordinary rain; it is *sulfuric acid virga*, liquid that precipitates from the noxious clouds and evaporates long before it can reach the ground.

Highlands loom. "That's the continent called *Aphrodite*," says the first officer. "Of course, it is not a continent in the same sense as those on our home planet because there is no ocean on Venus. We just call it a continent because it is one of two large regions of high land. The other is near the pole and is known as *Ishtar*."

Suddenly you see a bright orange glow and then what looks like a dull flash coming from a ragged, mountainous area.

"As you can see, that is an active volcano. We are lucky to be here while it's putting on a show. The volcanoes on Venus are something like those on the Big Island of Hawaii, with one important difference. As far as we know, the crust of Venus does not move over the volcanic hot spots the way the crust of Earth does, or if there is movement, it is much slower. The chain of the Hawaiian Islands was formed as Earth's crust moved over a hot spot welling up from Earth's mantle (Fig. 5-7A). On Venus, however, a single hot spot gets a chance to produce a volcano for a longer time (Fig. 5-7B).

BACK TO THE MAIN SHIP

No sooner have you passed over the volcano, which is too far below to be cause for concern, than the first officer orders you to strap down again. "It's time to begin our ascent," he says. "This ought to be a smoother ride than the one on the way down; we will be using rockets all the way." All the food and drink must be put away; seat backs and tray tables are returned to their upright and locked positions. Then the thunder of the rocket engines drowns out all other sound, and the force of acceleration pushes you back into your seat, harder and harder, with force you have not felt since you blasted off from the space center at the beginning of this voyage.

"You know," says the first officer, "if Earth had formed a few million kilometers closer to the Sun, it would have turned out like this place. There would have been a little more carbon dioxide (CO_2) in the atmosphere. That would have retained a little more solar heat, which would have increased the CO_2 level. The result would have been a vicious circle like the one that took place here on Venus. Carbon dioxide makes up almost all

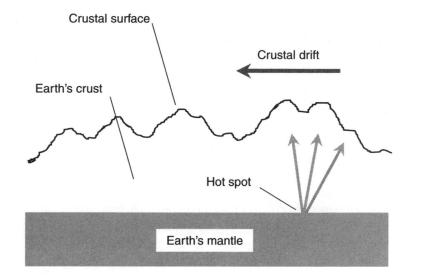

A

Figure 5-7A.   On Earth, the crust drifts over a hot spot, as in this rendition of the Hawaiian Island chain.

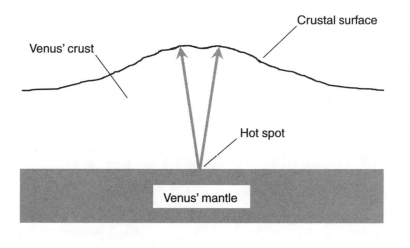

B

Figure 5-7B.   On Venus, volcanoes tend to stay put over hot spots for a longer time than on Earth.

the atmosphere of Venus below its clouds. As you know, CO_2 is called a *greenhouse gas* because it tends to trap heat. The principle is much the same as that which keeps a greenhouse warm in winter. Short-wave infrared and visible-light rays penetrate the glass or the CO_2 and heat up the surface. The surface, in turn, radiates long-wave infrared, but the glass or the CO_2 is opaque to that, so instead of being reradiated into space, the long-wave infrared is absorbed and turned into heat. If this had happened on Earth, the oceans would have boiled away, the oxygen would all have been bound up with carbon, and our planet would be almost as hot and every bit as dry as this place. Of course, we would not care. We would not exist."

"I've heard some people say that a runaway greenhouse disaster could occur on Earth if humans keep generating CO_2 and cutting down forests without replenishing them," you say. "But I don't know what to believe. Is that really true? How critical is the balance? Are we on the verge of tipping it the wrong way?"

"No one knows exactly," says the first officer. "There's only one way we will ever find out for sure whether or not the danger is real, and that is for the worst to happen. I would rather not learn the truth that way. Would you?"

The landscape below melts away into a ruddy blur, the cloud ceiling seems to rush down on the craft and swallow it up, and then, having seen enough for one day, you pull down the window blind, close your eyes, and try to imagine yourself strolling through a grassy field or along a windy beach on Earth, a place that you are now beginning to realize is special indeed.

 Quiz

Refer to the text if necessary. A good score is 8 correct. Answers are in the back of the book.

1. In one day on Earth, the upper equatorial clouds of Venus
    (a) travel about one-quarter the way around the planet.
    (b) travel about one-half the way around the planet.
    (c) travel all the way around the planet.

(d) travel twice around the planet.

2. The "seasons" on Mercury are caused mainly by
    (a) the tilt of the planet's axis.
    (b) the clouds that cover the planet most of the time.
    (c) the greenhouse effect.
    (d) the difference between perihelion and aphelion.

3. Excellent conditions for observing Mercury occur when the planet
    (a) is at inferior conjunction.
    (b) is at superior conjunction.
    (c) is at opposition.
    (d) None of the above

4. When Mercury or Venus is at its greatest elongation either east or west, approximately how much of the surface do we see illuminated?
    (a) None of it
    (b) Half of it
    (c) Three-quarters of it
    (d) All of it

5. Mercury is
    (a) smaller than the Moon.
    (b) the same size as the Moon.
    (c) larger than the Moon but smaller than Earth.
    (d) the same size as Earth.

6. Mercury's core is almost certainly made up primarily of
    (a) silicate rock.
    (b) iron.
    (c) volcanic lava.
    (d) uranium.

7. High noon on Venus would be just about as bright as
    (a) the brightest midday on Mercury.
    (b) a typical day on the Moon.
    (c) a sunny summer afternoon on Earth.
    (d) a gloomy winter day on Earth.

8. The greenhouse effect
    (a) increases a planet's surface temperature.
    (b) reduces a planet's surface temperature.
    (c) increases the radiation that reaches a planet's surface from the Sun.
    (d) keeps heat energy from reaching a planet's surface.

9. If either Mercury or Venus were to transit the Sun, to which lunar phase would its appearance most nearly correspond?
    (a) Full
    (b) First quarter

(c) New

(d) Last quarter

10. Which gas is the most abundant in the atmosphere beneath the clouds of Mercury?

(a) Oxygen

(b) Nitrogen

(c) Carbon dioxide

(d) This is an improper question; Mercury has no clouds.

CHAPTER 6

Mars

Mars is the fourth known planet in order outward from the Sun, and is in many ways the most Earthlike. Science-fiction writers have used Mars more than any other extraterrestrial place as the setting for civilizations, outposts, and evolution. It is interesting to suppose that the first Martians will come from Earth. Will you live to see the invasion of Mars by Earthlings?

The Red Planet

Mars is also known as the *Red Planet*, although its true color varies from rusty orange to gray to white. Some casual Earthbound observers mistake it for a red-giant star. However, because of its significant apparent diameter, it does not twinkle as does a star. In ancient mythology, Mars was the god of war. The planet has two moons, named *Phobos* (Greek for "fear") and *Deimos* (Greek for "terror" or "panic").

CONJUNCTIONS AND OPPOSITIONS

Mars never passes between Earth and the Sun because the orbit of Mars lies entirely outside that of Earth. Mars occasionally lines up with the Sun when it is exactly opposite the Sun from us. This is called *conjunction*. There is no need to use the word *superior* because there is only one kind of Martian conjunction. (With Venus and Mercury, there are two kinds, inferior and superior, as you know from the preceding chapter.)

Mars does not pass through significant phases as do Mercury, Venus, and the Moon. We always see Mars with most of its face lit up by the Sun. However, the brightness of Mars in the sky does vary greatly. It is dimmest when it is at and near conjunction. When it is very close to conjunction, Mars is invisible because its wan glow is washed out by sunlight. Earth travels more rapidly around the Sun than does Mars, so this unfavorable condition never lasts for long. After a conjunction, Mars begins to show itself in the eastern sky before dawn. As time passes and Earth begins to catch up with Mars, we get closer and closer to the Red Planet. As this happens, Mars appears earlier and earlier in the predawn hours, and its brilliance increases. Eventually, Mars reaches a position opposite the Sun so that it rises when the Sun sets, is visible all night long, and sets at sunrise. Then the Red Planet is at *opposition*, and it rivals Jupiter in brightness. As seen with the unaided eye, Mars at opposition is a more attention-getting sight than any other planet except Venus.

Opposition is the best time to view Mars, but some oppositions are better than others. This is so because the orbit of Mars is far from a perfect circle around the Sun. While Earth's orbit only varies a percentage point or so either way from perfect circularity, Mars follows a decidedly elliptical path with the Sun at one focus. The best oppositions, in terms of viewing Mars from our planet, occur when three things happen at the same time:

- Mars is at opposition.
- Mars is at perihelion (closest to the Sun).
- Earth is at aphelion (farthest from the Sun).

This can only take place during the northern hemispheric Earth summer, especially during the month of July, because that is when Earth is at aphelion (Fig. 6-1). The ideal state of affairs happens only about once every 15, 16, or 17 years.

YEAR, DAY, AND SEASONS

Mars's mean orbital radius is half again that of Earth, or about 228 million kilometers (142 million miles). As a result, Mars has a longer year, in terms of Earth days, than does Earth; in fact, it is 1.88 times the length of an Earth year. The Martian day is about 2.5 percent longer than that of Earth.

When and if we humans set up bases on Mars, we will be perfectly comfortable with our 24-hour time system, although we will have to lengthen the

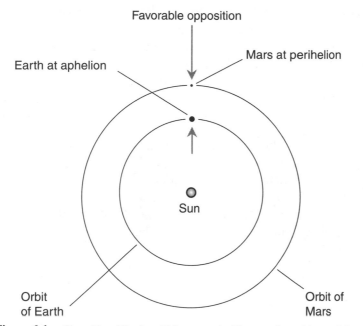

Figure 6-1.   The orbits of Earth and Mars, to scale. The most favorable conditions
for viewing Mars occur when Earth is at aphelion and Mars is at perihelion.

second, minute, and hour by about 2.5 percent. Most of us would not notice
such a difference in the face of the other adaptational problems inherent in a
relocation to the Red Planet, such as the reduced gravitation, the bitterly cold
temperatures, the lack of a breathable atmosphere, the lack of protection from
solar ultraviolet radiation, and the infrequent but brutal dust storms.

The equator of Mars is tilted about 24 degrees relative to the plane of its
orbit around the Sun. This is almost identical to Earth's axial tilt and results
in Martian seasons whose extremes are similar (in terms of proportionality)
to those on our planet. If you lived at 40 degrees north latitude on Mars, you
would see the Sun behave in a manner similar to the way it behaves on Earth
as the seasons pass, except, of course, that the seasonal progression would
take many more days to go full circle. The summer Sun in the northern
hemisphere would rise in the northeast, take a high course across the sky,
and set in the northwest, and daylight would last about two-thirds of the
solar day. The winter Sun in the northern hemisphere would rise in the
southeast, take a low course across the sky, and set in the southwest, and
daylight would last only about one-third of the solar day. People living in
the Martian southern hemisphere, hailing from such places as Sydney,

Cape Town, or Buenos Aires, also would find their seasons familiar, at least in terms of the Sun.

Anatomy of Mars

The Red Planet is about 53 percent the diameter of Earth, roughly 6,800 kilometers (4,200 miles). This would put Mars neatly between Earth and the Moon in size if the three orbs could be lined up next to one another (Fig. 6-2). If you were to stand on Mars and look toward the horizon, you would see a strangely foreshortened vista. Similarly, standing on top of one of the highest mountains, you would be able to perceive the curvature of the planet.

THE INTERIOR

The smaller size of Mars, combined with a density somewhat lower than that of Earth, produces a less intense gravitational field than the one we know. Your weight on Mars would be 37 percent of your weight on Earth. If you weigh 160 pounds here, you would weigh 59 pounds there. You would be able to throw a baseball much farther on Mars than you can on Earth. While golf-loving astronauts might not be able to drive a ball as far

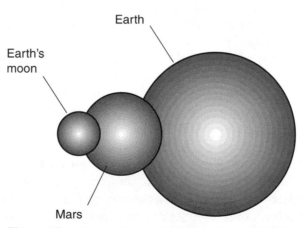

Figure 6-2.   Mars is a bit more than half the diameter of Earth.

on Mars as they did on the Moon, they would do better than they can on Earth. A golf course on Mars would have to be much larger than one on Earth, and to make things more interesting, there would be no shortage of boulders and sand traps.

There is evidence that the crust of Mars is thicker than that of Earth. There is also evidence that the Martian mantle historically has been less active than that of Earth.

The surface area of a sphere is proportional to the square of the diameter, whereas the volume is proportional to the cube of the diameter. Mars, being approximately half the diameter of Earth, has one-quarter the surface area but only one-eighth the volume. This means that the surface-area-to-volume ratio of Mars is twice as great as that of Earth, causing Mars to cool off faster after its formation along with the rest of the Solar System. All these factors have combined to create a world where the crust does not move very much.

On Earth, crustal plates float around over the mantle, so volcanoes move gradually away from the hot spots underneath. On Mars, however, little or no such movement has occurred, so some volcanoes have built themselves up to enormous proportions. The crowning glory of the Martian volcanoes is *Olympus Mons* (Mount Olympus), which is 24 kilometers (15 miles) tall and measures 600 kilometers (370 miles) in diameter at its base.

THE SURFACE

Mars is pitted in some places with impact craters. In fact, when *Mariner 4* took the first close-up photographs of Mars in 1965, coming within 9,800 kilometers (6,100 miles) of the surface, craters seemed to dominate the landscape. This led astronomers to believe that Mars might be as desolate as our own Moon. It was up to later missions to demonstrate otherwise. An entire planet cannot be characterized by looking at only one spot. Suppose an alien civilization were to send a probe past Earth and happened to obtain photographs of only the Sahara Desert?

The southern hemisphere of Mars consists of highlands, and this is where most of the impact craters are found. The northern hemisphere, in contrast, is several kilometers lower in elevation and appears flooded over by the lava from volcanic eruptions. Various parts of the surface have a dusky gray, almost green appearance. This greenish cast was seen by the first people who looked at Mars through "spy glasses." The dark regions, along with illusory straight dark lines that seemed to lead from them toward the polar ice caps, led some respected scientists to believe that Mars must be home to an intelligent civilization.

Numerous probes followed *Mariner 4*, and with each new set of photographs and data, the hope for finding life on Mars diminished. Many scientists believe there is an abundance of water locked up on Mars in the form of permafrost beneath the surface. Some water ice also appears in the polar caps. In addition to water ice, during the polar winters there is frozen carbon dioxide (dry ice) in the polar caps, particularly the southern cap, which endures a longer winter.

Dried-up riverbeds are the most interesting features seen on the surface of Mars. After careful analysis, most geologists have agreed that these marks could only have been made by running water. One theory is that Mars was at one time covered to some extent by a shallow sea or perhaps by large glaciers that melted during a series of volcanic eruptions.

ATMOSPHERE AND WEATHER

Mars has oxygen, but almost all of it is bound up with elements in the surface and with carbon in the atmosphere. The result is a rusty world, with an atmosphere consisting almost entirely of carbon dioxide (CO_2). The barometric pressure at the surface of Mars is less than 1 percent of the pressure on Earth. Were it not for the fact that CO_2 is a heavy gas, the atmosphere of Mars would be even thinner than this.

Despite the thin air on Mars, weather occurs, and it can be extreme. Winds aloft can reach speeds of around 400 kilometers per hour (250 miles per hour); near the surface, they commonly rise to 120 kilometers per hour (75 miles per hour). It would be a mistake to say that such winds are of "hurricane force" because the thin air on Mars produces far less wind pressure for a given wind speed than the air on Earth. However, dust particles from the surface are picked up and travel right along with the wind, blowing high up into the atmosphere, where they at times shroud the planet completely. During these massive dust storms, which can be accompanied by lightning, the surface features of the planet practically disappear. As seen from the surface, such a storm would produce a dark red sky, obscuring the sun and casting an evil gloom over the landscape.

One of these planetwide dust storms was indirectly responsible for the discovery of the four largest Martian volcanoes. These include Olympus Mons (already mentioned as the largest mountain on the planet),

Ascraeus Mons, *Pavonis Mons*, and *Arsia Mons*. As the storm abated, the dust gradually settled. The peaks of the volcanoes were seen first; more and more of them appeared as the Martian sky regained its characteristic clarity.

High clouds, similar to cirrostratus and cirrus clouds on Earth, are sometimes observed on Mars. In addition, Olympus Mons is occasionally shrouded in a thin veil of cloud, in much the same way as high mountains are cloud-covered on Earth. These Martian clouds are far less substantial than their Earthly counterparts, and scientists doubt that they produce much, if any, precipitation. However, they do produce a sort of fog at the top of Olympus Mons. Standing inside the caldera of this monstrous mountain on a foggy morning, you might for a moment imagine yourself in the Namib Desert on the southwestern African coast.

Life on Mars

The temperature on Mars never rises above the freezing point as we know it on Earth. At night in the winter, Mars would make Antarctica seem inviting by comparison. As if this were not bad enough, the Sun blasts the surface with ultraviolet radiation because the air is not thick enough to shield against it. Some high-speed solar particles also might reach the surface following solar flares.

WHAT MIGHT LIVE THERE?

If there is any sort of life remaining on Mars, it must be a primitive sort of bacteria or virus or some hardy "germ" similar to the toughest organisms on Earth. Even these life forms would not be found on the surface but underground.

After the invention and deployment of the first telescopes in the seventeenth century, some observers of Mars claimed to see straight lines connecting the dark areas near the equator with the polar caps. Percival Lowell, one of the most noted astronomers of all time, theorized late in the nineteenth century that these canals logically would have been constructed by a civilization intent on surviving a planet whose climate was becoming ever-more hostile. Numerous *canals*, as they were called, were mapped by some observers. These were optical illusions; the orbiter probes showed no such canals (although the dried-up river beds they did see were every bit as interesting and were no illusion).

No sign of life has ever been found on Mars. There is no indication that intelligent life has ever set foot (or appendage of any other sort) on its surface.

Science-fiction writers have taken advantage of the fact that Mars, while not a hospitable place by Earthly standards, at least presents an environment where life might survive with the proper equipment. Thus H. G. Wells' novel *The War of the Worlds*, published around the year 1900, created a cult of people who believed in the existence of native Martians. Ironically, it was our own Earthly disease bacteria that prevented us, in this horrifying tale, from being annihilated by the gigantic, slimy aliens whose ships came streaking down like meteorites and who stalked our planet in armored contraptions resembling nothing humanity had ever seen before.

TRUE ALIENS

Mars has been suggested as a possible colony for pioneers from Earth. Perhaps the water ice in the permafrost can be released, plants can be introduced to provide oxygen for the atmosphere, and other large-scale operations can be launched in an attempt to make Mars into an Earthlike place. However, the obstacles to such a project are formidable indeed. The low surface gravity, the lack of a substantial magnetic field to protect against the solar wind, and the possibility that the undertaking could create some horrible, incurable new disease strains must all be taken into consideration. Arguably, it will be far easier to control the population explosion on our own planet so that it never becomes necessary to colonize Mars.

Despite all the naysayers, we Earth dwellers undoubtedly will try to go to Mars. Why? Because it is there, and we have the technology to get there. Who knows? Maybe we will find primitive life there. Maybe Mars bases will be built. Maybe people will learn to think of the Red Planet as their home, being born, educated, and employed there. We will then, by all rights, be entitled to call ourselves colonizers of space! However, it will take a special sort of human being to endure the rigors of a life spent on Mars.

From Venus to Mars

Imagine that you go on a mind journey and, for a few moments, become one of those privileged few who get to walk around on Mars, taking precautions, of course, to ensure that you do not suffer the fate of H. G. Wells' fictitious Martians and perish from some unknown disease for which your body has no defense.

SPEED, FUEL, AND PLANTS

As you accelerate away from Venus, the primary problem will be one of fuel. It will be necessary to accelerate considerably to hurl the vessel out to the orbit of Mars. Here you encounter one of the bugaboos of long-distance space travel. The more you accelerate, the more fuel you need at the outset, and the more fuel you tank up with, the harder it becomes to accelerate in the first place. Fortunately, there is a way around this problem on the way from Venus to Mars. You can refuel by making a rendezvous with one of the space stations in orbit around Earth (Fig. 6-3).

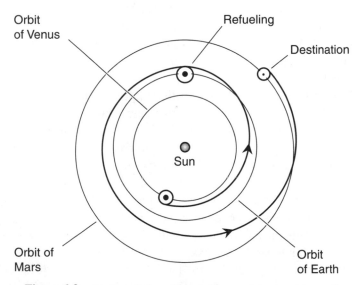

Figure 6-3.   On the way from Venus to Mars, an interplanetary vessel
can refuel at an Earth-orbiting space station.

The trip from Venus to Earth is uneventful, and you enter an equatorial orbit high above the surface.

"We'll be staying here for a couple of days," says the first officer. The ship needs to be checked over, and the Venus craft will be taken off our hands. We'll use the Mercury lander to set down on Mars, but we also need to get the MUV."

"What is the MUV?" you ask.

"That's short for 'Mars utility vehicle.'"

"I should have known."

"The MUV is a like the SUVs (sport utility vehicles) that were popular when cars used to burn fossil fuels to get around. Of course, this vehicle, like most modern Earth surface transport vehicles, is powered by compressed hydrogen. The only difference is that the MUV needs to take along its oxygen, too," says the first officer.

"Can't the oxygen be extracted from carbon dioxide in the Martian atmosphere?" you ask.

"If that were possible," says the first officer, "there would be hundreds of robotic MUVs roving Mars right now. It might someday be possible to get oxygen from subsurface water ice on Mars by melting it and electrolyzing it using solar energy, but that is not a convenient way to obtain oxygen for a moving vehicle run by a combustion engine."

The first officer has just been informed, by means of his digital communicator, that there is a problem with the *Valiant*, your main ship. Apparently, the life-support systems need some further work before you can embark on the journey to Mars.

"What is the problem?" you ask.

The first officer explains how the life-support system works. "It makes use of the Sun's ultraviolet radiation to manufacture oxygen by a sort of super-plant photosynthesis. Specially bred plants, a result of genetic engineering research, recycle the carbon dioxide from our breath and produce oxygen from it. The result is, ideally, a self-sustaining system that could, if it were possible to overcome other problems, work long enough for humans to go all the way to Saturn and back. (Beyond Saturn, solar radiation is not intense enough for the system to work.) The problem at the moment appears to be that the plants have come down with some sort of ailment," he says.

"What does this mean?" you ask.

"Replacement photoplants," says the first officer. "And a few new decorative plants as well."

"You mean all those potted plants in the residential areas are real?" you ask.

"Of course they're real," says the first officer. "You didn't think they're plastic, did you? They serve at least two important functions. They assist with the oxygenation of the air, and they help make the ship look less institutional."

THE RIGORS OF SPACE TRAVEL

The trip to Mars will take several months; this delay only adds to the tedium of interplanetary travel.

"Is there time enough for me to go down and visit my family for a while?" you ask.

"Yes," says the first officer. "But not enough money. The new space shuttles are smaller, faster, more efficient, and less expensive to operate than the gigantic rocket-boosted ships of the late twentieth and early twenty-first centuries, but they aren't free, and this mission was difficult to get approved. Taking civilians such as yourself into space has always been unpopular with certain people in the establishment."

You will watch more videos, read more books, and work out with ever-increasing devotion. The exercise is vital, and not just for staying in physical shape. An attitude problem can take hold of space travelers if they don't get enough exercise. The captain had explained it once, when she was in one of her rare talkative moods.

"It's called, logically enough, 'space-travelers' depression,'" she said. "It is like the old problem they used to call 'cabin fever,' except worse. Fortunately, there is a simple cure. It involves careful attention to nutrition, plenty of visible light at the same wavelengths as those from the Sun, and a great deal of aerobic exercise."

So you take your vitamins. You make sure you eat the right foods, in the right amounts, and on the right schedule. You drink plenty of water. And you increase your workouts to twice a day, for an hour and a half each session. The last thing you need is to get depressed 50 million kilometers (30 million miles) from your home planet.

ENTERING MARS ORBIT

By the time the ship nears the Red Planet, you are more video-literate, audio-literate, and aerobically fit than ever before in your life. "Most civilians," the first officer explains, "are mistaken for California natives when they return to Earth from one of these journeys."

"Why is that?" you ask.

"They are thin, they are fit, and they know every character in every movie produced during the last 100 years."

The orbit around Mars, just as the trip from Earth to Mars, will be exactly in the Earth's ecliptic plane (not that of Mars). There's a good reason for this: fuel economy. Altering the plane of travel, even minutely, is a fuel-guzzling business. Because of this efficiency, the round trip between Earth and Mars requires less fuel than did the journey from Earth to Mercury, then to Venus, and then back to Earth.

The Moons of Mars

Some astronomers think Phobos and Deimos are ex-asteroids that ventured too close to the planet and were captured by its gravitation, although there is reason to believe that one or both of them congealed from ejected material as the result of large asteroids or small protoplanets striking Mars long ago. Both moons are tiny compared with their parent planet, and both moons require large telescopes to be seen by Earthbound observers.

DEIMOS

"We'll be passing Deimos and then looking at Phobos from a distance as part of this tour," says the first officer. "Deimos is the smaller of the two. It orbits the planet in about 30 Earth hours. There has been some talk of putting several large communications satellites on Deimos so that it can serve as a repeater for maintaining contact among exploration crews."

"Landing on Deimos would be a problem, wouldn't it, because of the low gravity?" you ask.

"Deimos is too small to have any gravitation to speak of, at least from a practical point of view. It is a chunk of rock only about 13 kilometers (8 miles) in diameter. Anyone who wants to rendezvous with Deimos will have to dock with it instead. One suggested scheme has been to harpoon it. A small rocket would be fired at Deimos, would crash-land there, and then burrow into the surface. However, no one has been able to figure out how to make sure the harpoon wouldn't get pulled out and send construction workers scattering into Mars orbit. The escape velocity is less than 6 meters (about 10 feet) per second."

"Aren't regular communications satellites good enough?"

"Generally speaking, yes. But there would be room for gigantic storage batteries on a piece of rock like Deimos or Phobos, and these could serve several satellites. They could be charged with massive solar panels," says the first officer.

"There's something," you say, pointing to an irregular object, half lit by the Sun, the other half eerily glowing with Mars-shine. Deimos is only about 20,000 kilometers (12,500 miles) above the Martian surface, and the Red Planet looms large.

"That is Deimos," says the first officer. "It orbits Mars in a nearly perfect circle. That casts some doubt on the theory that Deimos is an asteroid that was thrown out of its original solar orbit by the gravitation of Jupiter. But that is a popular theory."

"How much extra fuel did it cost us to see this piece of rock?" you ask.

"Not much," says the first officer. "But it would cost too much to go right up to Phobos, the inner moon; we will have to be content to look at it through our telescopic cameras. Come with me."

PHOBOS

The first officer leads you into a dimly lit room with a huge screen on one wall. There is something strange about that screen; it is obviously there, but you can't ascertain how far away it is. "Is that a holographic projection system?" you ask.

"Yes," says the first officer. "And the first projection we'll see is an animated rendition of Mars, Phobos, and Deimos (Fig. 6-4). The piece of rock you just saw is the smaller, higher, and slower of Mars' two moons. Phobos orbits much closer to the planet. This illustration is to scale. Note that Phobos revolves around Mars much faster than Deimos. In fact, Phobos is inside what is called the *synchronous-orbital radius*. Phobos revolves around Mars faster than the planet rotates on its own axis. This means that an observer on Mars will see Phobos move across the sky from west to east."

"It almost looks like an artificial satellite in this picture."

"Phobos eventually will suffer the fate of most artificial satellites. It likely will spiral into Mars and crash. This will produce a significant impact. Phobos is not huge, but I wouldn't want to be on Mars when it hits.

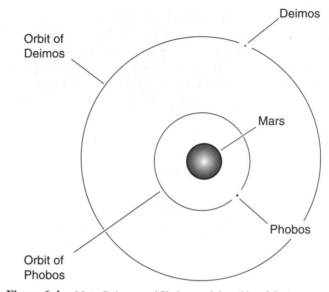

Figure 6-4. Mars, Deimos, and Phobos, and the orbits of the two moons, drawn to scale and viewed from high above the Martian north pole.

The gravitation of Mars may break Phobos up before it can crash, and then Mars will have a ring system."

The next image is of a dark, stonelike object that seems to be falling out from underneath a curved, inverted Martian horizon. "That is Phobos rising right now," says the first officer. "Actually we are looking back at it. We are still at a higher altitude than Phobos and have slowed down in preparation for Mars orbit. Phobos is catching up to us. We will be passing its orbital level while it is safely on the opposite side of Mars."

"It looks like a lump of coal," you say.

"Phobos is made of material called *carbonacious chondrite*, similar to that of many meteoroids and asteroids," says the first officer. "Its *albedo* is only 0.06. This means that it reflects only 6 percent of the light that strikes it."

"How large is Phobos?"

"Slightly bigger than Deimos, about 20 kilometers (12.5 miles) in diameter, but elongated."

"Right now it looks like a hand grenade with the top part taken off so that there's a hole in the top," you say.

"Wait a while and it'll change," says the first officer. He's right; a short while later it looks almost spherical.

"Is the hole an impact crater?" you ask.

"Yes. It is called *Stickney*," says the first officer. "If Phobos was originally an asteroid, it struck a lot of other asteroids before it attained orbit around Mars. The impact that made Stickney might be the one that knocked Phobos out of the asteroid belt, if, that is, Phobos was indeed an asteroid at one time. I am not sure that this is true."

HOW THE MOONS OF MARS FORMED

"This might sound like a stupid question," you say, "but—"

"There are no stupid questions."

"Okay. You showed me an image of Deimos and Phobos in orbit a while ago."

The image reappears on the holographic screen.

"All right," you continue. "The orbits of both Phobos and Deimos look like perfect circles."

"The orbit of Deimos is essentially a perfect circle. Phobos has an elliptical orbit, but the eccentricity is small, so the orbit is almost a perfect circle too," says the first officer.

"Can we look at the orbits as seen from the plane of Mars' equator?" you ask.

"Certainly," says the first officer. He smiles. "We'll look at the situation from just outside the equatorial plane so that we can get a little perspective." The view changes. The moons now appear to be orbiting Mars as seen from slightly above their orbital plane. As you suspected, they both orbit almost exactly above the equator of Mars (Fig. 6-5).

"If those moons were originally asteroids and were captured by the gravitation of Mars, why are their orbits both so nearly circular, and why are they both so nearly in the plane of Mars's equator? That's quite a coincidence, isn't it?"

"That," says the first officer, "is not a stupid question. In fact, it may answer the riddle of how these moons came into existence. I believe that both of these moons formed as the result of one or two major impacts in Mars's distant past, just as the Earth's moon is believed, by many astronomers, to have formed. It would have taken only a modest-sized object to blast that much material into orbit."

"But both moons are made of asteroid-like stuff," you say.

"Yes," says the first officer. "I think that one or two large asteroids—much bigger than either Phobos or Deimos—crashed into Mars. This melted the big asteroid, and most of it was absorbed by Mars. However, some

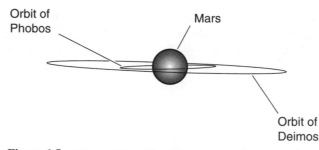

Figure 6-5. Mars and the orbits of its two moons, drawn to scale and viewed from nearly in the equatorial plane.

of this asteroid was cast into Mars orbit, along with some 'molten Mars,' and from that stuff, the moons formed. This is my theory," says the first officer.

We may never know exactly how these moons were created.

The Landing

The Mars lander, the *Eagle*, awaits. You're not eager to get inside its cramped cabin and endure weightlessness, even for a short while. The artificial-gravity wheel on the *Valiant* has been slowing down gradually since you left Venus, where the gravitational pull is nearly equal to that of Earth. The shock of weightlessness nevertheless will be unpleasant; you have never tolerated it well.

FINDING A SPOT

Your first question, naturally enough, is "Where will we land?"

The first officer responds, "The best places, in my experience, are the calderas (craters) of old Martian volcanoes. While some of the volcanoes on Mars might still erupt from time to time, none are active at the moment. If a volcanic eruption were imminent, there would be signs, just as there are on Earth."

"What kinds of signs?" you wonder.

"We have seismometers in all the landing-site calderas," says the first officer. "The one we will be visiting today is called *Pavonis Mons*. This means "Peacock Mountain." It lies almost exactly on the Martian equator. It happens to be only a few days past the Martian vernal equinox, so the Sun will rise directly in the east, follow a course right up to the zenith, and then set in the west, 12 Mars hours later."

"A Mars hour is . . . "

"About 62 Earth minutes," says the first officer. "We have decided to divide the Mars day up into 24 hours according to the Sun, just as is done on Earth. You won't notice any difference between Mars time and Earth time. We have special wristwatches with quartz oscillators aligned so that they function according to Mars time. Here." He hands you a watch. You strap it on over your pressure suit.

The descent proceeds smoothly enough. A huge crater yawns beneath the *Eagle*. "I don't like this," you say. You have visions of an impromptu Vesuvius or Krakatoa eruption replay, with the *Eagle* as part of the volcano's ejecta. "No need to worry," says the first officer. "If there is any sign of trouble, which is less likely than getting hit by a bolt of lightning on Earth, we'll be out of here. We have rehearsed all kinds of emergency evacuation scenarios."

It is almost sunset as the *Eagle* touches down. At the last moment before touchdown, the Sun vanishes beneath the rim of the crater. The sky above is pink where the Sun was, magenta all around, fading to deep purple and finally to black at the zenith. You think that you see a tiny white dot moving down toward the eastern horizon. "Is that Phobos?" you ask.

"No," the first officer says, "That is our main ship."

THE MARTIAN NIGHT

The outside temperature is −40°C, which happens also to be −40°F, at sunset. The thermometer plummets fast. It will drop down to −90°C, or −130°F, in the predawn hours.

"That's colder than it ever gets in Antarctica," you say.

"And the thin air, if you could stand outside and not die from the lack of pressure, would make it seem even colder than that."

"I can't imagine −130°F, no matter what the pressure," you say.

"Think of the worst possible arctic blizzard, with the temperature far below zero and the wind roaring like a hurricane. Then imagine getting into a swim suit and going outside and just standing there."

"I get the idea."

"At the poles during the Martian winter, it can get quite a lot colder even than that," says the first officer.

"It is beyond my comprehension."

"Now we need to get some sleep," says the first officer.

"In this cramped little vessel?" you ask.

"Well, not out there on the Martian desert sand. If you want to stay awake all night, go ahead, but don't keep me up." He nods his head and begins to doze off. All you can do is peer out the window and try to see if you recognize any constellations. You think you see Orion, tilted nearly on its side, hovering low in the eastern sky.

Then you, too, fall asleep. You wake up to sunshine on your face after what seemed like only a few seconds.

"The dreamless sleep of space explorers," says the first officer. "And now we will perform the little test for which we came."

AN EXPERIMENT

The Mars utility vehicle (MUV) reminds you of pictures you saw of the very first Moon rovers in the *Apollo* missions of the mid-twentieth century. And in fact, the two are quite similar.

"What's that?" you ask as the first officer unfolds a huge, gossamer-thin, butterfly-like sheet of material.

"A kite," he says.

"A kite! How will that fly here?"

"It is a windy day, or hadn't you noticed?"

You get into the MUV with the first officer. Then you feel a tug on your pressure suit and hear a whisper against the side of your helmet. "That's a little bit of breeze."

"A 20-meter-per-second breeze," says the first officer. "Or, in old-fashioned terms, a good 45-mile-an-hour gale. Look over that way." He points toward the southern rim of the caldera. Then you see plumes of pink dust rushing along from east to west.

"Is this wind enough to fly that kite?" you ask.

"More than enough," says the first officer. "We'll get away from the *Eagle* and then try to communicate with some other explorers that happen to be on the far side of the planet right now. This kite will support an antenna. A long-wire antenna, just like the first radio experimenters used around the year 1900 to see if they could send their signals across the Earth's Atlantic Ocean."

"Why can't we do this experiment from the *Eagle*? I feel nervous out here with nothing but a pressure suit for protection."

"That feeling is normal," says the first officer. "All astronauts, or nearly all, get the same feeling when they go on their first roves away from a space vehicle. It's like free diving in the middle of a big lake or in the ocean. That's not the same thing as paddling around in a swimming tank."

"But why do we have to be all the way out here in the middle of nowhere just to test a radio?" you ask.

"There would be too much radio noise near the *Eagle*. Electromagnetic interference. All the *Eagle's* computers and instruments generate electromagnetic noise. This is a sensitive little radio. It operates at a very low frequency, just 2 kHz, where the waves travel in contact with the surface of the planet," says the first officer. "Here. You drive the MUV."

"Two kilohertz! That's audible sound!"

"It would be if we connected a speaker to the transmitter output rather than an antenna," says the first officer.

You drive the MUV along toward the great plumes of pink. It's like riding in a golf cart, except faster and with a slower but more exaggerated rolling motion. Red Martian rocks and boulders litter the floor of the caldera, stretching away in all directions as far as you can see. After about 20 minutes, the first officer says, "We stop here."

He unreels the antenna line, a thin aluminum wire, and the delta-wing kite sails upward. "Don't try this at home," says the first officer.

"Why not?"

"Static electricity can build up, even on a clear day, and reach dangerous levels. I've got a couple of scars to show you exactly what it can do."

"Can't the same thing happen here?" you ask.

"Yes," says the first officer. "But our pressure suits are metal-coated to protect against the solar wind and the ultraviolet. That also will discharge any . . . "

At that moment a spark jumps from the kite line to the first officer's sleeve and from his ankle to the ground. You can't hear it because of your protective headgear and because the Martian atmosphere is so thin, but you can imagine the "Pop!" it would make back home on Earth.

"Why must you use such low frequencies?"

"Higher frequencies require an ionosphere, or else artificial satellites, to propagate over the horizon. However, very low frequency (VLF) radio waves do not, at least not on a planet that can conduct electricity to any significant extent," says the first officer. "Mars, according to our data, should conduct well enough to allow VLF waves to travel all the way around the planet." He pulls out a sheet of paper from the pocket of his pressure suit and hands it to you. "Please see Fig. 6-6."

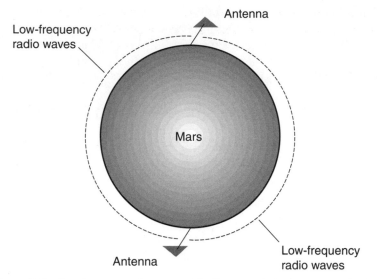

Figure 6-6.   Low-frequency radio waves might allow over-the-horizon communications between exploration parties on planets that have no ionosphere, such as Mars.

"Interesting," you say. "Primitive but interesting."

"This MUV rolls on wheels, and they are more primitive than this antenna."

"That's a good point," you say.

"That's high enough," says the first officer. The kite is now a tiny triangle against the sky, almost straight overhead. "Two and a half kilometers up."

The radio tests are conducted. The radio itself is a small, battery-powered box with an old-fashioned telegraph key. The first officer taps on the key, then listens, then taps some more, then listens some more.

"Well?" you ask.

"Negative," he says."

"Is it supposed to work?"

"In theory, yes, if we have enough transmitter power and a long enough antenna."

"Has anyone ever done this before?" you ask.

"Not successfully," says the first officer. "Not from such a vast distance."

"Why can't you use communications satellites? Why this old-fashioned stuff?"

"We can use satellites once they are up and working. This is only an experiment. If we can ever get this type of communications system to work, explorers to the moons of the outer planets and someday to worlds beyond our Solar System might use it for communication before any satellites are launched."

A HASTY RETREAT

The first officer spends the next 2 hours verifying that the people on the opposite side of the planet actually have been testing their radio, then testing some more, and even trying a couple of different frequencies. All the results are negative.

The Sun is near the zenith in a sky that has become a uniform pinkish orange when the first officer says, "Time to pack up and head on back to the *Eagle*. We'll be taking off early. There's no time to lose."

"What's the hurry?" you ask.

"Do you see all the dust in the sky?"

"Yes. Isn't that normal?"

"No. We have reports from the *Valiant*, as well as from general observation stations, that a planetwide dust storm might be brewing. We must get off the surface soon, before the winds aloft get so strong that we can't get back to the *Valiant* at all," says the first officer.

"I was just starting to feel safe down here," you say.

"My friend, we are on the planet Mars. We are millions of kilometers from Earth and within real-time communications range of only a few other human beings in the entire Universe. The pressure outside your suit is so low that you wouldn't stay conscious for 3 minutes without it. The temperature right now is $-40°C$, which happens also to be $-40°F$. However, the wind chill is much colder. It is hard to say whether you would die of suffocation or exposure if your pressure suit failed. There are several weak links in the chain that is keeping us alive. We must be certain that not a single one of those links is allowed to break. A full-Mars storm can last for months. By the end of it, links would not only be missing, but the whole chain would be gone."

"So this means . . . "

"It means a total change of plan. I am going to make sure we get off of this planet as soon as possible. Premature termination of mission," says the first officer.

You ask, "Didn't you know about the impending storm before you decided to bring us down here?"

"No. There were no signs of a storm when we left the *Valiant*. At least, none that we yet have the ability to detect. This storm developed suddenly."

"It's as if a hurricane formed out of a clear blue sky in a single day," you say. "I've never heard of such a thing."

"You speak of Earth," says the first officer. "This is Mars."

You ride the MUV back to the *Eagle*. Mars, which smiled in the morning, scowls now. The horizon has become brown as the *Eagle* rises from the surface. The winds begin to buffet the craft.

"Don't crash," you say.

"Don't worry," says the first officer. "I am well trained."

Within seconds the *Eagle* has cleared the dust, which for now is confined to the first 200 or 300 meters above the floor of the caldera. The crater rim and the slopes of Pavonis Mons are in the clear, but the crater floor is an obscure mass, as if bathed in smog. "If the storm becomes intense enough, these dust clouds will be kicked up high into the atmosphere, possibly covering the entire mountain below us. In the extreme, the dust might obliterate all surface features, ascending several kilometers into the sky," says the first officer.

"You say you cannot forecast the severity of the storm?" you ask.

"Weather forecasting on Mars is an inexact science, to say the least," says the first officer. "Have you ever heard of the *butterfly effect*?"

"Yes, that's the principle that deals with large, long-term consequences arising from small causes."

"If the butterfly takes off from Olympus Mons, the storm will cover the southern hemisphere. If the butterfly takes off from Pavonis Mons, the storm will cover the northern hemisphere. However, if the butterfly takes off from the Huygens Crater, the storm will envelop the entire planet."

"You are joking, of course," you say.

"Of course," says the first officer. "But the principle is clear, isn't it?"

"Yes, but there is one flaw in that theory."

"There are no butterflies on Mars."

"I know what you mean anyway."

"The butterfly effect is the reason we have no way of knowing for certain how extensive this particular storm will be. Martian weather, it seems, is more sensitive than Earth weather."

"Maybe the whole Martian ecosystem, such as it is, is more sensitive than that of the Earth," you say.

"We don't know until we test it. There are people back home who want to try to change the climate of Mars. Make a new world out of it. Try to get plants, or at least some sort of lichens, to grow. Maybe even try mold spores, bacteria, viruses. Anything. Anything that might change this planet into a world that humans can exploit," says the first officer.

"Do you think humanity will ever make a livable planet out of Mars?"

"I don't know."

"That's not a scientific answer," you say.

"I prefer to let certain mysteries remain mysteries," says the first officer. "I don't think we humans are ready to make a planet of our own."

"Time will tell," you say.

"Time always tells," says the first officer.

Quiz

Refer to the text if necessary. A good score is 8 correct. Answers are in the back of the book.

1. If the Martian day were divided into 24 hours of equal length, then one Mars hour would be approximately how long?
   (a) A little shorter than an Earth hour
   (b) Exactly the same as an Earth hour
   (c) A little longer than an Earth hour
   (d) Variable, depending on the time of year

2. The mean orbital radius of Mars is
   (a) about two-thirds that of Earth.
   (b) about the same as that of Earth.
   (c) about 1.5 times that of Earth.
   (d) about twice that of Earth.

3. Suppose that an object has a weight of 50 pounds on Mars. On Earth it would weigh approximately
   (a) 18 pounds.
   (b) 37 pounds.
   (c) 74 pounds.
   (d) 135 pounds.

4. Phobos appears to traverse the Martian sky from west to east because
   (a) Phobos' orbital period is less than Mars' rotational period.
   (b) Phobos' orbital period is greater than Mars' rotational period.
   (c) Phobos' orbit is retrograde.
   (d) That's not true! Phobos traverses the Martian sky from east to west.

5. In terms of size, Mars is
   (a) larger than the Moon but smaller than Mercury.
   (b) larger than Mercury but smaller than Earth.

(c) larger than Venus but smaller than Earth.
(d) larger than Earth.

6. With respect to its orbit around the Sun, the equatorial plane of Mars is
   (a) on a level.
   (b) tilted about 24 degrees.
   (c) tilted about 45 degrees.
   (d) tilted about 90 degrees.

7. The smaller of the two Martian moons is called
   (a) Deimos.
   (b) Phobos.
   (c) Olympus Mons.
   (d) Pavonis Mons.

8. Suppose that you stand at the Martian equator when the planet's north pole is at its maximum tilt away from the Sun (that is, the Sun's declination is the most negative). The Sun will rise
   (a) directly in the east.
   (b) somewhat south of east.
   (c) somewhat north of east.
   (d) in the west.

9. Volcanoes can build up to larger size on Mars than they can on Earth because
   (a) the Martian crust does not float around in plates on the mantle the way Earth's crust does.
   (b) Mars has a more intense gravitational field than does Earth.
   (c) the atmosphere of Mars is thinner than that of Earth.
   (d) This statement is not true! Volcanoes on Mars never get as big as the volcanoes on Earth.

10. Mars appears as a crescent through a small telescope as viewed from Earth
    (a) when it is near inferior conjunction.
    (b) when it is near superior conjunction.
    (c) when it is near opposition.
    (d) at no time.

CHAPTER 7

The Outer Planets

Beyond Mars lies a vast gap in the Solar System that is occupied only by a swarm of relatively small rocks. These rocks, more appropriately called *planetoids* or *asteroids*, will be discussed later in this book. For now, we will turn our attention to the five known outer planets: *Jupiter*, *Saturn*, *Uranus*, *Neptune*, and *Pluto*.

Jupiter

Jupiter is the Roman name for the Greek god *Zeus*, who was, according to legend, the most powerful of all the gods. If size and mass translate into power, then Jupiter is the most powerful of the planets.

Jupiter has power in tangible ways. Its gravitational pull is 2.5 times as strong as that of Earth. You would have to be in good physical condition to stand up for long on Jupiter without fainting, if it had a surface and if you could get down to it. However, you would never make it down. The atmosphere would blow your landing vessel out of control and ultimately crush it, but you would be dead or near death from radiation sickness before then. The barrage of high-speed subatomic particles commanded by Jupiter's immense magnetic field would be lethal to astronauts who ventured near the giant planet.

THE YEAR AND THE DAY

Jupiter orbits the Sun at a distance of 5.2 *astronomical units* (AU). This means that its orbital radius is 520 percent that of Earth (Fig. 7-1). This is about 778 million kilometers (484 million miles). The best viewing of Jupiter is done when the planet is at opposition. Jupiter only receives 3.7 percent as much sunlight per unit area as Earth. However, Jupiter reflects sunlight well. This fact and its immense size mean that Jupiter is usually the second brightest planet in the sky after Venus. Only Mars, when at a favorable opposition, outshines Jupiter at its opposition.

Jupiter does not pass through phases; it always appears full or almost full. Its brilliance in the sky, as we see it, changes because its distance from us varies. In general, the greater the angle between Jupiter and the Sun, the brighter Jupiter appears as seen from Earth. Jupiter orbits the Sun much more slowly than Earth; it takes nearly 12 of our years to revolve once around the Sun with respect to the distant stars. Jupiter reaches an opposition approximately once every 13 Earth months.

Jupiter rotates rapidly on its axis. The complete Jovian day, midnight to midnight, lasts for 9 hours and 51 minutes Earth time. We would need to adopt a different method of time measurement if we were to visit this planet.

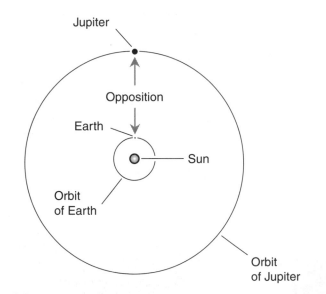

Figure 7-1.   The orbits of Earth and Jupiter, to scale. The most favorable conditions for viewing Jupiter occur when Earth is directly between Jupiter and the Sun.

Most likely we would divide the day into 10 hours and sleep every other night, bedding down before sunset and getting up after sunrise. However, as we have already seen, a colony on Jupiter will never exist; it is doubtful that there will even be permanent space stations in near orbit.

COMPOSITION

The volume of Jupiter is 1,300 times that of Earth and is greater than the combined volumes of all the other planets. At the equator, Jupiter measures 143,000 kilometers (89,000 miles) in diameter. This is more than 11 times the diameter of Earth (Fig. 7-2). The pole-to-pole diameter of Jupiter is somewhat less than the equatorial diameter. This difference is easy to see, even in a small telescope. The *oblateness* (flattening) of the planet is the result of its rapid spin and its largely liquid and gaseous composition.

Early astronomers were able to deduce the mass of Jupiter. First, they had to know how far away the planet was at the time of observation. This was determined by parallax among the distant stars and double-checked by measuring the time it takes Jupiter to make one complete revolution around the Sun. Once the distance to Jupiter was known, the orbital radius of one of its moons was determined by reverse triangulation. Then the period of revolution of that moon was measured; from this, the mass of Jupiter was calculated according to straightforward physics laws. The result was surprising: Jupiter is much less dense than the Earth. Whatever Jupiter is made

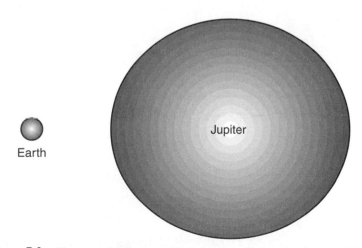

Figure 7-2.   The equatorial diameter of Jupiter is about 11 times the diameter of Earth.

of, it is nothing like our planet. The only elements light enough to explain Jupiter's low density are hydrogen and helium. These two elements make up almost all the planet.

Nowadays astronomers believe that Jupiter has a rocky, molten core several times the volume of Earth. This is surrounded by hydrogen under so much pressure that it acts like a metallic liquid; if we could get a sample of it, it would look like elemental mercury. This metallic hydrogen is a good conductor of electricity. Because of this, and because the planet is spinning rapidly, enormous electric currents are induced. A powerful magnetic field is generated by the current, and this field, called the *magnetosphere* of Jupiter, extends millions of kilometers beyond the visible sphere. High-speed solar particles, called the *solar wind*, squash the magnetic lines of flux on the Sunward side of Jupiter and stretch the lines of flux on the side of Jupiter opposite the Sun (Fig. 7-3). The magnetic field has effects that extend far beyond Jupiter's orbit.

Above the metallic hydrogen layer there is syrupy hydrogen-helium mixture that gradually becomes a liquid and finally thins out until it is gaseous. It is believed that there is no defined surface. In the upper 1,000 kilometers (about 600 miles) of Jupiter's visible atmosphere, various elements in small amounts produce the clouds we see through our telescopes and that were so vividly rendered by the *Pioneer* and *Voyager* space probes.

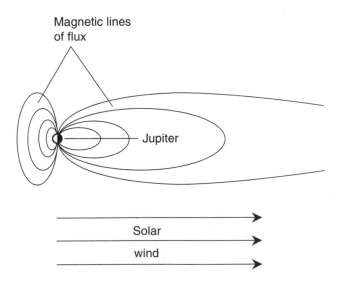

Figure 7-3.   The solar wind blows the Jovian magnetosphere into an elongated shape.

ATMOSPHERE AND WEATHER

The plane of Jupiter's equator is tilted 3.1 degrees relative to the plane of its orbit, so there are in effect no seasons on Jupiter. However, this is more than made up for by the violent and turbulent winds that roar around the planet in light-colored *zones* and dark-colored *belts*. At the boundaries between the zones and belts, the largest of which can be seen though a small telescope, eddies occur, similar to the high- and low-pressure systems we have here on Earth. The typical wind speeds on Jupiter, however, are many times greater than those we consider normal on our planet. In the extreme, Jovian winds exceed 400 kilometers (250 miles) per hour, comparable with the gales inside the most severe Earthly tornadoes.

The light-colored zones on Jupiter are the tops of the highest clouds, reflecting sunlight in the same way as the tops of thunderstorms or hurricanes reflect sunlight back into space from Earth's atmosphere. The dark belts are gaps in the high clouds, but further down, Jupiter is cloudy everywhere.

Imagine what it would be like to ride a hardy hybrid space/air shuttlecraft down into Jupiter in the middle of one of the dark belts. You would see massive white clouds above and on either side, rising like walls, boiling and shredding as the winds blew them around and ripped wisps of cloud off into the clear. You would see lightning strokes hundreds of kilometers long, and you would hear interminable peals of thunder. As you continued your descent, the sky would get red, then brown, and finally would deepen to black, punctuated by flashes of lightning, each flash followed by thunderclaps so loud that the whole vessel would shudder. Then your instruments would go crazy, indicating that the ship was tumbling, swooping, and diving as updrafts and downdrafts tossed it around like a falling snowflake. At that moment you would decide that it was time to return to the main ship!

THE GREAT RED SPOT

Jupiter's face is pockmarked by oval-shaped disturbances. The most prominent of these is known as the *Great Red Spot* and has been watched by astronomers for centuries. It spins outward, which happens to be counterclockwise because the spot is in the southern hemisphere of Jupiter. (If it were in the northern hemisphere, it would spin clockwise).

Until the Great Red Spot was observed by space probes close up, astronomers were not sure what it was. Some people thought the spot was a solid object, floating on a liquid sea and poking up through the clouds. Others thought it was caused by a volcano beneath the clouds; this theory was lent support because of the reddish hue. (Astronomers are still not certain why the spot is red.) It has been known to fade to dusky gray from time to time, and once in a while it seems almost to vanish, although the irregularities in the adjacent cloud bands betray that it is still there. It always returns to its full red glory sooner or later. The *Voyager* photographs convinced almost everyone that the Great Red Spot is a revolving high-pressure weather system. It is not alone. Smaller systems dot the face of Jupiter.

How can a weather disturbance stay active for so long? To answer this, we have only to look at our own Earth. The *Azores-Bermuda high*, which dominates the weather over the North Atlantic Ocean, is present almost all the time. It is not visually obvious, as is Jupiter's spot, but the high is no less permanent. In fact, the Azores-Bermuda high, which carries tropical storms into the Caribbean in summer and temperate storms into Europe in autumn, has been around longer than the spot. There are other semipermanent systems on Earth too. A low-pressure center exists in the far North Pacific, just south of the Aleutian Islands. It grows powerful every fall and winter, hurling storm after storm at the North American coastline. Sometimes it seems almost to disappear, although motions of nearby clouds and the jet streams betray that it is still there. It always swells to full force once the brief Arctic summer ends.

Certain climate phenomena persist because they feed on heat from the Sun (and in the case of Jupiter, from inside the planet as well); once they get going and get large enough, they form positive-feedback systems. Unless some tremendous outside force intervenes, such systems just keep on swirling around. If there were no land masses on Earth, hurricanes might persist for years, decades, or even centuries, traveling around and around the planet, because there would be nothing to break them up.

Saturn

Saturn is one of the most familiar planets because of its appearance. It is surrounded by a system of rings that can be seen though a small amateur telescope. It is the only planet with rings substantial enough to be seen easily. However, this planet is unique in other ways. It is the least dense of the planets. Its *specific gravity* (density relative to the density of water) is only 0.7. This means that Saturn is only 70 percent as massive as it would be if it were made entirely of water. The planet would float if it could be placed in a large enough, deep enough lake. It has the greatest amount of oblateness of any planet. The axis of Saturn's magnetic field corresponds almost precisely with its rotational axis, a fact that has befuddled scientists in their attempts to explain the dynamics of planetary magnetism.

In mythology, Saturn is the Roman god of agriculture. The name also refers to the father of the Greek god Zeus. Because Zeus and Jupiter are the same entity, Saturn might well be attached in mythology with an importance equal to or greater than that of Jupiter. Jupiter is Saturn's mythical son; without Saturn, Jupiter would never have been born, or would have turned out much different. (Of course, this is only according to the ancient myths; we know better than to believe that those tales are true.)

Before telescopes revealed the ring system, the name *Saturn* was associated with old age and dullness. If it were not for the rings, Saturn would indeed be a somewhat less interesting version of Jupiter, at least from an observational point of view.

THE YEAR AND THE DAY

Saturn orbits the Sun at a distance of 9.54 AU (Fig. 7-4). It orbital radius is about 1,430 million kilometers (888 million miles). The best viewing of Saturn is done when the planet is at opposition. Saturn is almost twice as far away from the Sun as is Jupiter, and the ringed planet receives only 1.1 percent as much sunlight per unit area as Earth. Saturn reflects sunlight well, and this is enhanced by the ring system. Saturn looks similar to Jupiter with the unaided eye but is somewhat dimmer, comparing favorably with Mars most of the time. Saturn is easy to relocate once you have found it on any given night.

Saturn, like Jupiter, does not pass through phases; it always appears full or almost full. Its brilliance in the sky, as we see it, changes because its distance from us varies. In general, the greater the angle between Saturn and the Sun, the brighter Saturn appears as seen from Earth. The brilliance of Saturn is also affected by the angle at which the rings are presented to us. If the rings are edge-on, the planet looks dimmer at a given distance from us than if the rings are seen from above or below. Saturn takes $29^1/_2$ Earth years to make a complete revolution around the Sun with respect to the distant

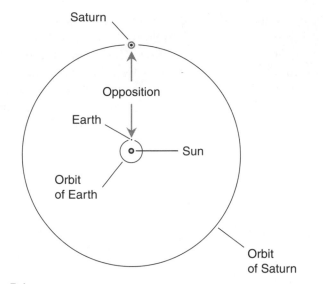

Figure 7-4. The orbits of Earth and Saturn, to scale. The most favorable conditions for viewing Saturn occur when Earth is directly between Saturn and the Sun.

stars. Thus Saturn reaches an opposition approximately once every $12\frac{1}{2}$ Earth months.

Saturn, like Jupiter, rotates rapidly on its axis. The complete day, midnight to midnight, lasts for about 10 hours and 40 minutes Earth time, as determined by observations of the magnetic field. The planet's upper clouds rotate slightly faster than this at latitudes near the equator. Near the poles, the atmosphere appears to rotate at about the same speed as the planet's magnetic field.

COMPOSITION, ATMOSPHERE, AND WEATHER

Saturn might be considered a little brother (rather than the father) of Jupiter on casual observation, were it not for the ring system. Saturn is almost as large as Jupiter. At the equator, Saturn's diameter is 121,500 kilometers (75,500 miles), more than nine times that of Earth (Fig. 7-5).

Saturn is comprised of about three-quarters hydrogen and one-quarter helium, with trace amounts of ice, methane, ammonia, and silicate molten rock. The inner core is where this mineral matter is found; if all the hydrogen and helium on Saturn were blown away, the remaining body would be a planet similar to Earth but several times more massive. As with Jupiter, the inner core is surrounded by liquid metallic hydrogen mixed with heli-

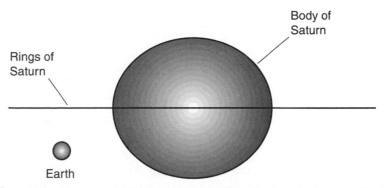

Figure 7-5.   The equatorial diameter of Saturn is about nine times the diameter of Earth.

um. As we progress further and further from the center of the globe, the liquid hydrogen becomes nonmetallic; then it becomes a dense gas, thinning out and topped with the yellowish clouds we see from a distance.

The rather bland appearance of Saturn's cloud bands, compared with those of Jupiter, belie the violent winds that continuously blow around Saturn. At their strongest, these winds are several times hurricane force on Earth. Vortices (eddies) occur on Saturn, but they are less visible than those on Jupiter because the upper cloud layer is more uniform and makes it difficult to see what is going on further down.

Saturn's equator is slanted by 26.7 degrees relative to the plane of its orbit around the Sun. Thus Saturn has seasons of a sort, although the deep and windy atmosphere tends to equalize the temperatures between the equator and the poles. An observer on Saturn, supposing that it were possible to exist there, would notice changes in the amount of daylight for each rotational cycle; "winter days" would be much shorter than "summer days." However, no one will ever go to Saturn and experience these seasonal fluctuations. It is believed that Saturn has no definable surface on which to land, and merely contending with the violent winds would totally preoccupy anyone venturing below the cloud tops. If Saturn did have a surface—liquid hydrogen, say, like an ocean—any ship that set sail there would be plucked up instantly and whisked away into the darkness like a toy boat in a tornado.

The outer atmosphere of Saturn is about 90 percent hydrogen. This means that the interior must contain relatively more helium. Some scientists believe that helium is constantly precipitating down toward the center of Saturn, in much the same way as the heavier components in a vinegar-and-oil salad dressing settle out. This process apparently has been going on ever since the birth of the Solar System and contributes to Saturn's internal heat.

THE RINGS

The ring system of Saturn, as we see it from Earth through telescopes, is about 250,000 kilometers (155,000 miles) in diameter. The rings are extraordinarily thin in proportion to their width. The drawing of Fig. 7-5 greatly exaggerates the thickness of the rings. If the drawing were true to scale in this respect, with the rings viewed edge-on, they would not be visible without a magnifying glass. Estimates of the rings' thickness range between 100 meters (about 300 feet) and 1 kilometer (about 3,000 feet). The only reason we can see them from Earth is that they are excellent reflectors of light.

Even before the rings of Saturn were photographed at close range by space probes, scientists knew they were comprised of countless chunks of icy material, ranging in size from grains of dust to boulders bigger than a house. They could figure this out because of what the rings do to electromagnetic waves from distant stars, galaxies, and other sources passing through the ring system. The number of particles is inversely proportional to their size; that is, there are far more tiny grains than medium-sized rocks and far more medium-sized rocks than large boulders.

The ring system presents several mysteries. Astronomers think they have solved some of these, but others remain inexplicable. One theory holds that the rings formed from the breakup of an icy moon that ventured too close to Saturn and was torn apart by gravitational forces. Every planet's gravitational field, even that of the Earth, has a minimum orbital radius within which large natural satellites cannot stay in one piece. This is known as the *Roche limit*. For Saturn, the Roche limit is roughly $2^1/2$ times the radius of the planet. Boulder-sized rocks and even a few small asteroids continue to orbit Saturn in one piece within this limit; the maximum limiting size depends on what the particle is made of. There are a few especially large boulders that orbit Saturn inside the ring system, and these are believed to be responsible for the *gaps*, also called *divisions*, that appear in the ring system.

Uranus

Uranus (pronounced "YOU-run-us") is approximately twice as far from the Sun as is Saturn: 2,871 million kilometers (1,784 million miles). This is 19.22 AU, so Uranus receives only about $(1/19.22)^2$, or 1/369, as much sunlight per unit area as does our planet.

FINDING AND OBSERVING URANUS

Uranus is so far from the Sun that its brightness does not change very much as the Earth revolves. However, observation is still the best when Uranus is at or near opposition (Fig. 7-6). If you were to travel to Uranus, the Sun would be as dim as it is during the darkest part of an annular solar eclipse as seen from Earth.

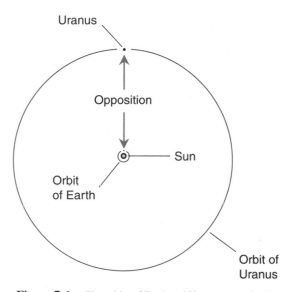

Figure 7-6.   The orbits of Earth and Uranus, to scale. As with the other outer planets, the best viewing is at opposition.

Uranus can be seen with the unaided eye, but just barely. It is better if you have a pair of strong binoculars or a small telescope. You must know exactly where to look if you are to find it. A good Web-based astronomical map can be found at the Weather Underground Web site:

http://www.wunderground.com

Click on the "Astronomy" link and proceed according to the instructions to get a sky map for your area. Celestron also publishes a CD-ROM called *The Sky* that works in a similar way. When you find Uranus, don't expect much. If you have a large telescope, you should be able to resolve it into a bluish green disk, and you also might see one or two of its moons.

YEAR, DAY, SEASONS

In several respects, Uranus is unique. It is tilted on its axis so much that the axis is only 8 degrees away from lying in the planet's orbital plane. Some texts say the axis is tilted by 98 degrees; however, because the north pole of a planet is generally defined as the pole that lies "above" the planet's orbital plane (generally toward the star Polaris), it is more precise to say that Uranus' axis is tilted by 82 degrees and its rotation is retrograde.

Uranus has dramatic, exaggerated seasons of variable daylight and darkness. When *Voyager 2* visited the planet in 1986, one pole was facing almost directly toward the Sun. Despite the fact that half the planet remains in total darkness for many Earth years at a time, however, the temperature on the dark side is just about the same as that on the daylight side.

The Uranian year is 84 Earth years long. The period of rotation is a little less than 18 hours, about three-quarters of an Earth day. We must be careful about how we define a "day" on Uranus. The best scheme is to use the planet's rotational period. If someone lived on Uranus (not likely for the same reasons as Jupiter and Saturn are unlivable), they would want to use a 24-hour system where each Uranian "hour" lasts about 45 Earth minutes.

Because the axis of Uranus is almost parallel to the plane of the planet's orbit around the Sun, daylight over much of the planet lasts for many Earth years, followed by an interval of Earthlike daylight and darkness hours, followed by many Earth years of continuous darkness, followed by another interval of Earthlike daylight and darkness hours, followed by continuous daylight for many years, and so on. Only in the immediate equatorial region do Earthlike daylight and darkness occur in sync with the planet's rotation all the Uranian year round.

A LIFETIME ON URANUS

Imagine what it would be like to live your whole life on Uranus, born on January 1, 1986. (Suspend your disbelief for a moment and pretend that Uranus has a surface on which humans can live; astronomers believe that it does not.) Let us say you dwelt at 45 degrees latitude, the equivalent of Minneapolis, Minnesota.

When *Voyager 2* made its visit in 1986, you would have just been born into the middle of a long night. You would never have seen daylight. The Sun would never come anywhere near rising above the horizon. It would not be until you were in grade school that, every 18 Earth hours, you would see a glimmer of twilight. As winter's dark grip (though no colder than any other season) came to an end, the Sun would give you a peek. The days would grow longer, and the Sun's course across the sky would get higher.

At the Uranian vernal equinox, when you were 21 Earth years old, the Sun would travel in an Earthlike way across the sky, behaving like it does in Minneapolis around the end of March or September, but with daylight lasting only 9 Earth hours and darkness another 9 Earth hours. Perhaps you would read in books (or on computer screens) about the changes of seasons in places like Minnesota and be thankful that no such fluctuations in temperature occurred on Uranus. The season would evolve, the Sun would move further toward the celestial pole, and the hours of daylight would grow long. For a while, the Sun would follow a path similar to that in midsummer in Minneapolis, but it would still be early spring on Uranus. The daylight hours would grow longer yet and the darkness hours shorter. One night darkness would never fall. After that, you would have a full 18 hours of daylight every day, and as the Earth years continued to progress, the Sun would describe a smaller and smaller circle in the sky around the celestial pole.

On the day of the summer solstice when you were 42 Earth years old, the Sun would follow a circle only 8 angular degrees in radius around the celestial pole. You would have become used to continuous daylight. The Sun would shine down through the Uranian haze at a favorable angle, casting a shadow just about as long as you were tall. As summer progressed, the circle would widen; your shadows would change length and orientation more noticeably with the time of day. After a few more Earth years, the Sun would rise to near the zenith at noon and dip to near the horizon at midnight. Then one midnight, the Sun would plunge below the horizon beneath the celestial pole. Twilight would once again become a familiar sight. The twilights would become longer and darker until one midnight darkness would become total for a while. Imagine the emotional effects of the first total darkness you had seen in a quarter of a lifetime!

At the autumnal equinox, when you were 63 Earth years old, the daylight and darkness hours would be of equal length. The days would grow shorter. You might remember the first time you saw the Sun as a child. As an elderly adult, you would see the Sun for the last time. In the days that followed, the noon twilight would grow dimmer and shorter until once again continuous darkness would reign. You would never see the Sun again, unless, of course, you moved to a southerly latitude.

COMPOSITION, ATMOSPHERE, AND WEATHER

Uranus is about 51,000 kilometers (32,000 miles) in diameter, which is almost exactly four times the diameter of Earth (Fig. 7-7). When the *Voyager* space probe approached this planet in 1986, Earth-based scientists were disappointed. The face of Uranus is featureless, light aquamarine in color, and reminiscent of a greenish blue–tinted, enlarged version of Venus. Beneath that bland face, however, Uranus is nothing like Venus. Uranus consists of hydrogen, helium, methane, ammonia, and perhaps water ice mixed in with rocky material.

Uranus is cold in its upper layers but is believed to have a hot interior. Just how hot is a mystery, though, because Uranus, unlike the other three "gas giants" Jupiter, Saturn, and Neptune, does not radiate more energy toward the Sun than it receives. Perhaps Uranus is cooler inside than the other "gas giants," or maybe Uranus is better insulated. By examining the extent to which Uranus is flattened by its rotation, and by measuring the relative proportions of hydrogen, helium, methane, and other substances in its visible outer layer, astronomers have begun to suspect that Uranus might be a gigantic ball of dirty slush. Such a planet would have no defined boundaries in its

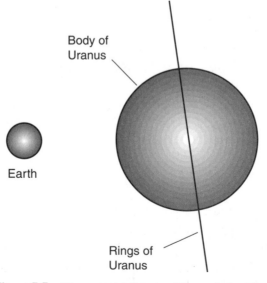

Body of
Uranus

Earth

Rings of
Uranus

Figure 7-7. The equatorial diameter of Uranus is about four
times the diameter of Earth.

depths. It would be gaseous on the outside, then liquid, then slush, and then goo at the core.

Further missions to Uranus will be necessary to get a better idea of what, exactly, this planet is made of. One thing is certain, however: It manages very well to maintain a constant temperature in its outer layers despite the exaggerated seasonal changes in solar irradiation.

The climatic system that governs the temperatures on Uranus might be called the *great thermal equalizer*. If Earth were tilted on its axis as much as is Uranus, the weather on our planet would be incredibly severe. Winters would be brutal everywhere except at the very lowest latitudes. The prevailing winds would be fierce as they attempted to equalize the radical annual seesaw of solar energy received at most points on the surface. Hurricanes of unimaginable size would prowl the seas and slam into land masses. The whole course of the evolution of life on our planet would be much different from what it was. We cannot be certain that intelligent life would have evolved at all.

THE RINGS

When the *Voyager* probe visited Uranus in January of 1986, the rings, which exist in the plane of the planet's equator, were seen in detail for the first time. Astronomers were not surprised to find them; their existence was known already because, as Uranus passed near distant stars, those stars seemed to blink several times. The only possible cause for such blinking was the existence of thin, nearly opaque rings around the planet. The fact that Uranus is tipped on its side so that the rings sometimes appear as pronounced ellipses (almost circles when either pole is nearly facing us) was an assist in their discovery prior to the *Voyager* "grand tour."

The rings of Uranus are much different than those of Saturn; they more resemble the faint rings around Jupiter. The *albedo* (proportion of light reflected) of the rings is only about 1 percent, similar to that of charcoal. If you were to visit the Uranian system in a space ship, you would have a hard time seeing the rings even if you passed right through their plane. Only if you actually struck one would you notice it easily; this would not be likely because the rings are exceedingly narrow. However, the rings consist of good-sized rocks, generally on the order of 70 centimeters (28 inches) or larger in diameter. You would not want to navigate your ship through them.

The narrowness of the Uranian rings seems to run contrary to dynamics. The natural tendency, over time, is for the rings to spread out and become more flattened, like those of Saturn (although much less prominent). However, small moons orbit near the rings, and the gravitational fields from these moons tends to force and keep the rings into narrow circles. Moons of this type have been observed inside the system of Saturn, too, accounting for some of the narrow rings there. Because these moons act to confine the ring particles and hold them within specific orbits, they have been termed *shepherd moons*.

Neptune

Neptune, named after the mythical god of the sea, is more than half again as far from the Sun as is Uranus: 4,504 million kilometers (2,799 million miles). This is 30.06 AU. Neptune receives only about 1/900 as much sunlight per unit area as does the Earth. If you're into electronics, acoustics, or physics, you might get some idea of the difference by noting that a ratio of 1:900 is equivalent to approximately 30 decibels (dB). If light were sound and the Sun shining on the Earth were like a loud vacuum cleaner, then the solar illumination on Neptune would be like a small fan running at low speed.

FINDING AND OBSERVING NEPTUNE

As is the case with the other outer planets, we see Neptune best when it is at or near opposition (Fig. 7-8), although its absolute brightness does not vary much as the Earth revolves around the Sun. Neptune cannot be seen with the unaided eye; powerful binoculars or, better yet, a good telescope is necessary to observe it. You need to know exactly where to look; the Weather Underground or Celestron CD-ROM maps can be used to locate it. Even when viewed through a large amateur telescope, Neptune will only look like a blue star.

THE YEAR AND THE DAY

Neptune is tilted on its axis by $29\frac{1}{2}$ degrees. This is to say, the equatorial plane of Neptune intersects its orbital plane at an angle of $29\frac{1}{2}$ degrees. This compares with $23\frac{1}{2}$ degrees for the Earth. The seasonal variations in

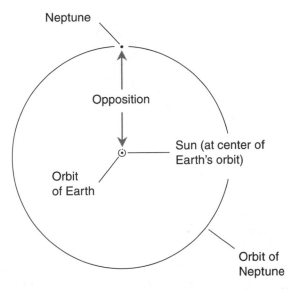

Figure 7-8.   The orbits of Earth and Neptune, to scale. As with
the other outer planets, the best viewing is at opposition.

the Sun's path across the sky on Neptune would be somewhat familiar to
Earthlings, except for one fact: No human would ever live long enough to
see all four Neptunian seasons go by. As a matter of fact, no human likely
would be able to survive long enough beneath the sapphire-blue haze and
snow-white clouds of Neptune to eat a decent supper, let alone carry out a
lifetime's research.

Neptune takes 165 Earth years to make one complete journey around the
Sun. Its orbit is almost a perfect circle, so any seasonal effects on Neptune's
climate must be caused entirely by the tilt of its axis and not by variations
in the amount of sunlight it receives.

COMPOSITION, ATMOSPHERE, AND WEATHER

In composition, Neptune is thought to be similar to Uranus, but it is more
dense. Neptune is about 49,500 kilometers (30,800 miles) in diameter; this
is a little less than four times the diameter of the Earth (Fig. 7-9). Neptune
generates more internal heat than Uranus; in this respect it more nearly
resembles Jupiter and Saturn. Neptune is more blue in color than Uranus,
and astronomers are not quite certain what is responsible for this vivid
sapphire hue.

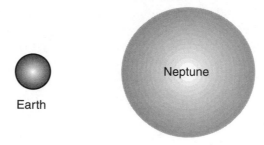

Figure 7-9. The equatorial diameter of Neptune is about the
same as that of Uranus, four times the diameter of Earth.

When *Voyager* passed by Neptune in 1989, it proved to have a more
interesting atmosphere, at least visually, than its aquamarine cousin. There
were dark spots and bright clouds; one of the clouds raced around Neptune
independently of other weather phenomena and was named *Scooter* for this
reason. There was an oval-shaped spot of deep indigo, similar in shape to
but about half the diameter of Jupiter's Great Red Spot. This system had
winds that blow faster than those on any other planet in the Solar System,
approximately three times the speed of the wind in a maxitornado on Earth.
The *Great Dark Spot*, which was as large in diameter as the Earth, disap-
peared after a few years. Apparently the self-sustaining forces that keep
storms alive on Jupiter are not as effective on Neptune.

1989N1R

Neptune has a ring system similar to those of Uranus and Jupiter but
fainter. These rings were revealed by the *Voyager* probe. The rings of
Neptune are unique in the Solar System because the outermost one, called
1989N1R, is nonuniform. This name derives from the year of discovery
(1989), the planet (Neptune), the ring number (1 means "outermost"), and
R (which stands for "ring").

Astronomers are not sure why 1989N1R is "clumped." One theory holds
that it was created only a little while ago, on a cosmic scale, possibly within the last few centuries. If an asteroid or comet ventured too close to the
planet, it would break up because of gravitation. Eventually the particles
would spread all the way around Neptune and form a uniform ring, but this
process would require some time. Maybe it hasn't had time to do this yet.

Another theory for the clumping of 1989N1R involves gravitational
interaction between the ring particles and a tiny moon, *Galatea*. It is possible that certain gravitational resonances could cause the clumping.

The particles that make up the rings around Neptune are in general less than an inch in diameter. This was revealed by analyzing radio waves passing through or reflected from the particles. Large chunks of rock affect radio waves at medium and long wavelengths, and as the rocks become smaller, they affect radio waves at shorter wavelengths. Neptune's rings were not detected when observations were made at radio wavelengths.

Pluto and Charon

Pluto, named after the god of the underworld, follows an eccentric orbit that ranges from slightly inside that of Neptune (29.7 AU) to about 50 AU at aphelion. Pluto and its moon, *Charon*, named after the ferry boatman who took dead souls to Pluto for judgment, receive 1/900 as much sunlight per unit area as Earth when the two are at perihelion. At aphelion, the system receives only 1/2500 as much sunlight per unit area as Earth.

Figure 7-10 illustrates the orbits of Pluto-Charon, Neptune, and Earth to scale. You might wonder if either Pluto or Charon will ever crash into Neptune. The answer is no because of a phenomenon called

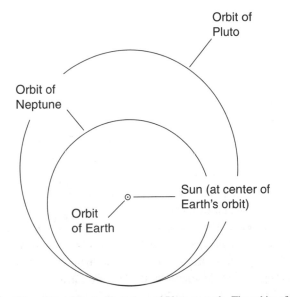

Figure 7-10.   The orbits of Earth, Neptune, and Pluto, to scale. The orbits of Pluto and Neptune overlap, but orbital resonances ensure that the two planets will never collide.

orbital resonance. The Pluto-Charon system makes exactly two solar orbits for every three orbits of Neptune; as a result, the two systems can never get any closer than 17 AU to each other. Unless some other celestial object intervenes and gravitationally upsets the orbit of Neptune or the orbit of Pluto-Charon, a cosmic collision will never take place.

THE YEAR AND THE DAY

The equatorial plane of Pluto intersects its orbital plane at an angle of 58 degrees, but the rotation of Pluto is retrograde. There are pronounced seasonal changes in the path of the Sun across the sky at any particular location; the Sun always appears to rise in the west and set in the east.

Pluto takes about 6 days and $9^{1}/_{2}$ Earth hours to rotate once on its axis. Charon follows a prograde orbit (in the same direction as Pluto rotates) over Pluto's equator and completes one orbit every Plutonian day, so Charon always stays over the same spot on Pluto. An observer on Pluto would see Charon hanging almost perfectly still in the sky. In addition, Charon, like most planetary moons, keeps the same side toward Pluto constantly.

The Pluto-Charon system takes 248 Earth years to make one complete journey around the Sun. Its orbit is a pronounced ellipse. Thus the variations in this system's distance from the Sun, as well as its extreme axial tilt, affect the seasons. The maximum-to-minimum ratio of solar irradiation is about 2.8:1.

COMPOSITION

Pluto is approximately 2400 kilometers (1500 miles) in diameter; this is smaller than the Earth's moon. Charon is about half the diameter of Pluto. The centers of the two objects are about 20,000 kilometers (12,500 miles) apart. If the Pluto-Charon system could be brought close to Earth for size comparison, the result would look like Fig. 7-11.

Both Pluto and Charon have about twice the density of water. This implies that they consist of a combination of ices and rocky materials. They may in fact be huge "dirty snowballs," consisting of primordial matter that never accreted into an object large enough to properly be called a planet. Controversy has arisen here; an outspoken group of astronomers has expressed their belief that Pluto would not be called a planet if it were discovered today.

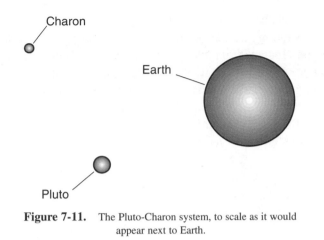

Figure 7-11.   The Pluto-Charon system, to scale as it would
appear next to Earth.

ATMOSPHERE

Pluto has a thin atmosphere consisting largely of nitrogen. However, astronomers think that this atmosphere, which is on the order of one-millionth the density of Earth's atmosphere at the surface, exists only when Pluto is near perihelion. When the system moves farther from the Sun, the atmosphere is believed to freeze onto the surface. Because Pluto's gravitation is weak, the atmosphere extends to considerable distances from the planet, enveloping Charon. The atmosphere might be blown into a teardrop shape by the solar wind, in much the same way as a comet's tail is blown away from the Sun.

Although no probe has yet flown near Pluto, images of the planet have been obtained through the Hubble Space Telescope. The surface is pinkish red; this is thought to be caused by the presence of methane ice. There are bright and dark regions, with the south polar region being especially reflective. High-resolution images of Pluto and Charon will be obtained when and if a close flyby is made. Some astronomers believe that when this happens, Pluto and Charon might be reclassified as a double comet.

What Makes a Planet?

Astronomers have been searching for a large planet beyond Neptune ever since Neptune itself was discovered. Pluto is not massive enough to account for observed aberrations in the orbits of Uranus and Neptune. Perhaps such a "Planet X" does not exist, and the so-called perturbations in the orbits of Uranus and Neptune are caused by some unseen (or unseeable) object or effect. Maybe "Planet X" is a large, massive object with albedo (reflectivity) so low that we cannot see it even with the largest Earth-based telescopes.

Astronomers are almost certain that there are thousands or millions of asteroids and dormant comets in solar orbits beyond the orbit of Neptune. This *Kuiper Belt* is a disk-shaped swarm of primordial rocks and "dirty snowballs"; the *Oort Cloud* is a larger, spherical congregation of such objects that encloses the Solar System like a bubble. Every once in a while, an object from one of these swarms undergoes a gravitational interaction or collision with another object and is hurled into the main part of the Solar System. If the object passes near Neptune or Uranus, the gravitation of the large planet can send it diving toward the Sun. A few decades later, we on Earth discover a new asteroid or comet.

We might say that in order to be a planet, a celestial object must be spherical, must orbit the Sun (and not some other planet), and must be larger than a certain diameter (say, 500 kilometers) or have more than a certain amount of gravitation (say, 5 percent that of the Earth). However, no official standard yet exists. Depending on the set of criteria adopted, assuming scientists ever agree on one, Pluto-Charon may be "demoted" to the status of a double comet or else hundreds, maybe thousands, of objects now considered primordial matter will be reclassified as planets.

Quiz

Refer to the text if necessary. A good score is 8 correct. Answers are in the back of the book.

1. Which of the following planets generates the least amount of internal heat?
   (a) Jupiter
   (b) Saturn
   (c) Uranus
   (d) Neptune

2. The Pluto-Charon system is unique in that
   (a) they always keep the same sides facing each other.
   (b) they are the smallest of the gas giants.

(c) they have atmospheres consisting entirely of helium.

(d) they actually orbit Neptune, not the Sun.

3. The dark side of Uranus is much colder than the sunlit side because
(a) the axis of Uranus is so greatly tilted.
(b) the atmosphere is so thin.
(c) there are no winds on Uranus.
(d) No! The dark side of Uranus is just as warm as the sunlit side.

4. Most astronomers believe that the surfaces of the gas giant planets
(a) are liquid water.
(b) are liquid methane.
(c) do not exist as definable boundaries.
(d) are peppered with craters.

5. The tops of the highest clouds on Jupiter
(a) reflect sunlight very well.
(b) are red or brown.
(c) spin counterclockwise because they are high-pressure systems.
(d) are actually smoke from volcanic eruptions.

6. An Earthly analog of Jupiter's Great Red Spot might be
(a) a tornado.
(b) a hurricane.
(c) a high-pressure system.
(d) a volcano.

7. Saturn appears in its crescent phase, as seen from Earth, when it is at
(a) conjunction.
(b) quadrature.
(c) opposition.
(d) No! Saturn never appears as a crescent to Earth-bound observers.

8. The magnetosphere of Jupiter is distorted by
(a) the solar wind.
(b) Jupiter's gravitation.
(c) Jupiter's rings.
(d) Jupiter's moons.

9. The most oblate planet is
(a) Jupiter.
(b) Saturn.
(c) Uranus.
(d) Neptune.

10. An astronomical unit is
(a) the mean distance of Earth from the Sun.
(b) 299,792 kilometers (one light-second).
(c) the mean distance of the Moon from Earth.
(d) the radius of the Solar System.

CHAPTER 8

An Extraterrestrial Visitor's Analysis of Earth

Suppose that you were an extraterrestrial being visiting Earth for the first time. What would you see? How would you interpret your observations? Imagine yourself in the role of the explorer/reporter assigned the task of visiting Earth and writing a report about the planet for a magazine article back home. This chapter is written from the point of view of an imaginary explorer/reporter from the fictitious planet called *Epsilon Eridani 2*.

Until now in this course, measurement units such as kilometers, miles, and degrees usually have been written out in full. However, starting with this chapter, I will use more abbreviated symbology. This is the way scientists usually write such expressions, so you should get used to it too.

This chapter is written as a fictitious story, but all the scientific information is based on well-documented knowledge. While the explorers from Epsilon Eridani 2 are make-believe characters, the things they see are real, although viewed from perspectives we don't normally consider. It has been said that it is difficult to see a big picture when you are inside the frame. Let's step outside the frame for awhile. Here is the log of the first officer of the fictitious Epsilon Eridanian exploration vessel, the *Dragon*.

The Blue and White Planet

As our ship approaches the planet *Sol 3*, the third planet in orbit around the star that we call *Sol*, we are struck by the amazing blue and white colors. Previous probes have shown us that the white regions are clouds of water vapor and ice ranging in altitude from zero (at the surface) up to about 16 kilometers (km) or 10 miles (mi). The surface is well-defined, and more than half of it is liquid water. Some of the surface is frozen water; other regions show an amazing variety of features, the details of which it is part of this mission to catalogue.

PRELIMINARY OBSERVATIONS

We settle into a circular, polar orbit at an altitude of 500 km (300 mi). From this vantage point, over time we will be able to map in detail the entire surface of the planet using radar and optical equipment. We all look forward to the landings. We will go down two-by-two, and because there are 20 of us and we each are to be allowed only one trip, we will make 10 different landings in 10 different places on the surface.

There are regions on Sol 3 where plant life abounds, sometimes only a millimeter (less than 1/16 inch) tall and in other places upwards of 100 m (330 ft) in height. Plants also live in and around bodies of water. The largest water zones are tainted with sodium chloride (salt) and other minerals. At low latitudes, but never exactly at the equator, revolving storms occasionally occur; from our initial orbit we count five of these. The largest has a diameter of more than 1000 km (600 mi). Animal life also has been observed both on the solid surface and beneath the liquid surface. These animals exist in a range of sizes similar to those of the plants.

PRELIMINARY MYSTERIES

The most interesting structures on the planet will require extensive investigation. Their geometry suggests life forms having great intelligence some of the time and amazing stupidity at other times.

Some of these structures appear as monoliths in great congregations, perforated by holes covered over with glass. Narrow, solid strips serve as pathways for objects resembling gigantic rolling insects that go from place to place in an orderly fashion but without apparent overall purpose. Two-legged life forms have been seen entering and exiting these rolling insects. Apparently the insects are not life forms themselves but rather are vehicles designed by the life forms that enter and leave them and are intended to transport those life forms from place to place within and between the great congregations of monoliths, which, for lack of any other name at this point, I will henceforth call *anthills*.

I can hardly wait to land right in the middle of one of these anthills and see what happens close-up. I have already picked the one I want to check out. It is located at approximately 41°N and 73°W. According to electromagnetic signals from this anthill, it calls itself *New York*. There is an open, green area in the middle of this congregation of monoliths that appears ideal for a landing.

Other large insect-like vehicles have been seen flying through the air at altitudes approaching those of the highest icy clouds. When these flying vehicles are on the ground, the two-legged life forms have been seen entering and leaving them in large groups. Apparently these vehicles are designed for the purpose of transporting the life forms between anthills separated by great distances. The two-legged life forms also have been seen entering and leaving objects that slowly float on and across bodies of water, avoiding, of course, the revolving storms but nevertheless sometimes enduring wave action that would challenge the stomachs of our hardiest space travelers. Sometimes it seems as if these two-legged life forms use their vehicles for the sole purpose of having a violent ride!

YEAR, DAY, AND SEASONS

The equatorial plane of Sol 3, which the two-legged inhabitants call *Earth*, is tilted by approximately 23.5 degrees with respect to the plane of its orbit around the parent star Sol, which they call the *Sun*. This results in considerable seasonal variations in the weather that become increasingly dramatic as the latitude increases. The hours of daylight and darkness are always equal at the equator, but fluctuations become greater and greater as one goes nearer to the poles. North of 66.5°N and south of 66.5°S, there are periods when the Sun stays above the horizon for days at a stretch. At the poles themselves, the daylight period lasts for fully half the year, and the darkness period lasts for the other half.

There are about 365.25 solar days in each Earth year. It is difficult for me to describe the length of the Earth day except to say that the two-legged creatures divide each day into 24 equal units called *hours*. Each hour is divided into 60 minutes, and each minute is divided into 60 seconds. Fractions of a second are expressed in decimal form. For some reason, most Earth inhabitants divide the solar day into two 12-hour segments called A.M. and P.M. Some of their scientists use an undivided hour system that runs from 0000 (zero hours, zero minutes) to 2359 (23 hours and 59 minutes) and then starts over again at 0000 in the middle of the dark period.

The Earth is farthest from the Sun (that is, it is at aphelion) in the month called *July* and is closest to the Sun (that is, at perihelion) in the month called *January*. The mean distance of the Earth from the Sun is 149.6 million km (93 million mi). The variation in orbital radius is only about ±1 percent. The Earth's greater distance from the Sun in the northern-hemispheric summer results in less solar irradiation over the planet's greatest land masses at that time. However, this effect is balanced by the fact that the season is lengthened; Earth moves more slowly around the Sun at that time (Fig. 8-1). Conversely, the Earth's lesser distance from the Sun during the northern-hemispheric winter produces more solar irradiation, but the season is shorter because the Earth moves faster around the Sun.

The earth's axis *precesses*, or wobbles, slowly like the axis of a spinning top. Every 25,800 Earth years, the axis describes a complete circle whose angular radius is 23.5 angular degrees on the celestial sphere. This means that in 12,900 years, Earth will be closest to the Sun in July and farthest from the Sun in January (Fig. 8-2). It is difficult to say what effect this might have on the overall climate of the planet. There is much more land mass in the northern hemisphere than in the southern; land masses heat up and cool off more rapidly than the oceans. This could have a tremendous

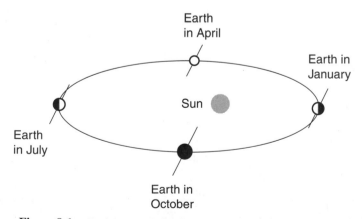

Figure 8-1.   Earth is at perihelion in January and at aphelion in July. (In this drawing, the eccentricity of Earth's orbit is exaggerated for clarity.)

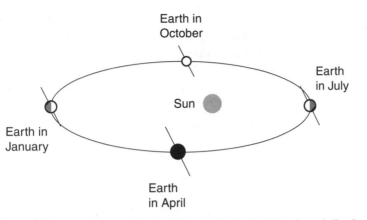

Figure 8-2.   In 12,900 years, Earth will be at perihelion in July and at aphelion in January. (In this drawing, the eccentricity of Earth's orbit is exaggerated for clarity.)

cumulative effect when the northern-hemispheric summer is shortened and the winter is lengthened. It is known that the repeated cycles of glaciation that take place on our own planet, Epsilon Eridani 2 (the second planet in orbit around the star Earth inhabitants call *Epsilon* in the constellation Eridanus), also have taken place on Earth; axial precession might be a contributing factor to these so-called ice ages.

Anatomy of Earth

The Earth has a slightly larger diameter at the equator than at the poles. This *oblateness*, or flattening, is caused by the rotation of the planet. The effect is too small to be visually apparent as the planet is seen from space; the outer planets, especially Jupiter and Saturn, are much more oblate than is the Earth and actually look that way. The Earth's diameter is 12,756 km (7,926 mi) in the plane of the equator and 12,714 km (7,900 mi) as measured along the rotational axis from pole to pole.

THE POWER OF TIME

Time is one of the most powerful forces in the Cosmos. The Earth-dwelling two-legged creatures who call themselves *Homo sapiens* have not developed a mature concept of this power and how it can be harnessed. If they had a better understanding of time and how things happen in the long term—millions upon millions of their years—many of the mysteries that befuddle them would become clear and simple in their minds.

Suppose it were possible to look at time so that a year seemed to pass in a fraction of a second? How would Earth look when beheld from such a perspective? The precession of the axis would be apparent; Earth would look like a furiously spinning top. The continents would drift around like ice floes on an Arctic lake during the springtime thaw. In some places large chunks of land would break away from continents. In other places islands would bump into continents and join up with them. Crumpling of the crust, caused by the drifting of land masses, would create mountain ranges. The Hawaiian Islands would drift toward the northwest, eroding down into the ocean at the northwestern end and being born anew in continuous volcanic eruptions at the southeastern end. The sea level would rise and fall periodically; glaciers would advance and retreat. The Earth, which seems like a stable place on a day-to-day scale, would be revealed as dynamic, fluid orb. One might be tempted to suppose that Earth has a life of its own, that it is a gigantic biological cell. An idea of this sort has been posed by some respected Earth scientists. It is called the *Gaia hypothesis*. However, this notion has not been proven true, and many academics have dismissed it as unscientific.

Given sufficient time, rivers cut canyons hundreds of meters deep. One of the best-known examples is the Grand Canyon in the southwestern United States. It is hard to imagine, on an hour-by-hour or day-by-day scale, how the little Colorado River could have gouged out such a ravine, but time is patient beyond all human understanding. Time has unlimited endurance. It works day and night; it never rests. It carves and chips and grinds, builds new structures atom by atom or cell by cell, and keeps on doing its work for human lifetime after human lifetime, generation after generation, age after age. Time has been at work on Earth for more than 4 billion human-defined years and will continue to mold and change the planet for at least that many years yet to come.

THE INTERIOR

The Earth can be considered to have four distinct layers. The central portion, called the *inner core*, is believed to be solid. It is extremely hot and consists mainly of iron and nickel. These metals are *ferromagnetic*, meaning that they can be magnetized. Surrounding the inner core is a liquid iron and nickel layer called the *outer core*. This liquid flows in huge eddies that are thought to be responsible for the magnetization of the core and hence for the existence of the *geomagnetic field*.

Above the outer core lies the *mantle*, consisting of rock similar to granite (called *basaltic rock*). The consistency of the mantle would appear solid if you could take a piece of it and hold it in your hand. However, on a long time scale and considered in its entirety, it is a fluid mass. As the eons pass, the mantle flows much like hot tar or molasses, rising up from the center of the planet toward the surface in some places and descending in other zones. One theory holds that this is a mechanism for the transfer of heat from the hot core regions to the surface, where the heat energy ultimately is transferred to the atmosphere and radiated into space. The up-and-down currents result in lateral movement near the upper reaches of the mantle.

The outermost layer, called the *crust*, floats on top of the mantle and, as the ages pass, moves around on it. The lateral movements of the mantle carry chunks of crust along. The crust is not a uniform, continuous mass but instead has regions where it is deep (about 30 km, or 20 mi) and other regions where it is shallow (perhaps as thin as 10 km, or 6 mi). The thickest parts of the crust form the continents and larger islands. The thin regions lie beneath the seas and oceans. Figure 8-3 is a simplified cross-

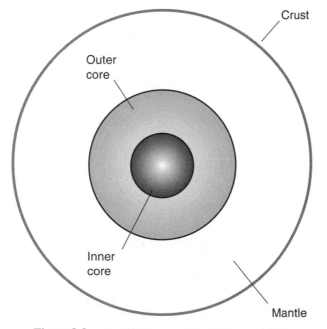

Figure 8-3.   Simplified cross-sectional diagram of Earth.

sectional diagram of the Earth as it would appear if it were sliced in half at the equator.

THE OCEANS

The surface of Earth is largely covered by mineral-rich oceans of water. The largest of these, if seen from a certain vantage point in space, covers almost half the planet. This is the *Pacific Ocean*; it was given this name by some Earth dweller who saw it at one of its more peaceful moments. (*Pacific* means "peaceful" or "tranquil.") However, this ocean is not always calm. Revolving storms, called *hurricanes* in the eastern Pacific and *typhoons* in the western Pacific, churn the waters and transport excess heat from the tropics toward the polar regions.

Hurricanes also occur in the Atlantic Ocean and in the Indian Ocean. The storms form at lower latitudes, and they almost always work their way either onto a land mass, where they dissipate, or toward cold water, where they expire from lack of heat to sustain their winds. Occasionally, one of these storms strikes a human-made anthill. Some of the planet's most elaborate anthills are built directly in known hurricane tracks. I wonder why the human Earth dwellers construct so many of their communities in such places?

The waters flow in slow currents around and around the oceanic basins, generally clockwise in the northern hemisphere and counterclockwise in the southern. This gives rise to warm ocean currents along the eastern shores of the continents and cold currents along the western shores. This has a profound influence on the distribution and movement of weather systems in the planet's atmosphere. When something happens to upset the regularity of these currents, the climate changes over much of the planet. These cycles are natural and have been taking place for millions of human-defined years. However, almost every time such a cycle recurs, especially the sort known as *El Niño* where the eastern equatorial Pacific waters switch from cold to warm, the humans call the resulting weather a *disaster*.

The oceans are critical to the balance of life on Earth. They are like the lungs and blood of a living organism. Tiny life forms called *plankton* live in the oceans; these are eaten by larger life forms such as the *fish*. Fish are a favored food among the two-legged humans. However, humans dump toxic chemicals and hydrocarbon dregs into the oceans, where they work their way into fish and then into their brothers' and sisters' bodies, causing terrible illnesses and suffering. Humans know about this. We have heard them talk about it in their electromagnetic broadcasts. Why do they continue to knowingly harm themselves in this way?

THE LAND MASSES

The oceans tend to heat up and cool down slowly. They hold heat energy. Land masses are just the opposite. They heat up and cool down rapidly. This contrast, along with the oceanic currents and the prevailing differences in temperature between the tropics and the polar regions, creates the climate and weather variations on this third planet from the star we call Sol. Earth is the only planet with weather varied enough to motivate the evolution of life but not so violent or hostile as to exterminate such life before it gets a chance to evolve.

In some places water accumulates in the atmosphere and then precipitates onto land masses in great amounts. This can happen either as liquid water, in which case it is called *rain*, or as frozen water, known as *snow*. There are other, less common forms of precipitation, such as *hail*, but these do not contribute much to the overall ecological system of the Earth.

In the regions where precipitation is abundant and mostly liquid, forests of tall plants grow. Some of these plants are cut down and used as materials by the two-legged humans in the construction of dwellings in their anthills. It is amazing how many different geometries have been invented for these dwellings! The plants, called *trees*, make ideal building material, and they can be replenished by intelligent management. Unfortunately, in some regions of the planet no attempts are made to replenish the supply of trees. Humans obviously know the supply of trees is not infinite and eventually will be depleted if balance is not maintained. Do they not care about their own future?

LAYERS OF THE ATMOSPHERE

The Earth's atmosphere is 78 percent nitrogen at the surface and 21 percent oxygen. The remaining 1 percent consists of argon, carbon dioxide, ozone, and water vapor. The temperature of the atmosphere varies considerably; it can rise to about 55°C (130°F), or plunge to around −80°C (−112°F).

The lowest layer of the atmosphere, rising from the surface to approximately 16 km (10 mi) of altitude, is the *troposphere*. This is where all weather occurs; most of the clouds are found here. In the upper parts of the troposphere, high-speed rivers of air travel around the planet in a generally west-to-east direction. There can be two or three of these rivers in the northern hemisphere and two or three in the southern hemisphere. The strongest of these rivers, called *jet streams*, carry high- and low-pressure systems from west to east at temperate latitudes, primarily between 30°N and 60°N, and between 30°S and 60°S.

Above the troposphere lies the *stratosphere*, extending up to approximately 50 km (30 mi) of altitude. Near the upper reaches of this level, ultraviolet radiation from the Sun causes oxygen atoms to group together into triplets (O_3) rather than in pairs (O_2), as is the case nearer the surface. An oxygen triplet is a molecule known as *ozone*. This gas has the unique property of being opaque to ultraviolet rays. Thus oxygen atoms form a self-regulating mechanism that keeps the Earth's surface from receiving too much ultraviolet from the Sun. Certain gases are produced by industrial processes carried on by the two-legged humans; these gases rise into the stratosphere and cause the ozone molecules to break apart into their individual atoms. This makes the upper stratosphere more transparent to ultraviolet than it would be if nature had its way. Some humans have demonstrated that if this process continues, it could have an adverse effect on all life on the planet. Other humans do not believe this and continue to produce these potentially dangerous industrial by-products.

Above the stratosphere lies the *mesosphere*, extending from 50 km (30 mi) to an altitude of 80 km (50 mi). In this layer, ultraviolet radiation from the Sun causes electrons to be stripped away from atoms of atmospheric gas. The result is that the mesosphere contains a large proportion of ions. This occurs in a layer that communications engineers call the *D layer* of the *ionosphere*.

Above the mesosphere lies the highest layer of the Earth's atmosphere, known as the *thermosphere*. It extends from 80 km (50 mi) up to more than 600 km (370 mi) of altitude. This layer gets its name from the fact that the temperature is extremely high, even hotter than at the surface of Venus or Mercury. However, these high temperatures do not have the devastating effects they would have if they existed at the surface because the atmosphere at this level is so rarefied. Ionization takes place at three levels within the thermosphere, called the ionospheric *E layer*, *F1 layer*, and *F2 layer*. Sometimes, particularly at night, the F1 and F2 layers merge together near the altitude of the F2 layer; the resulting layer is called the *F layer*. Figure 8-4 is a diagram of the Earth's atmosphere showing the various layers and the ionized regions.

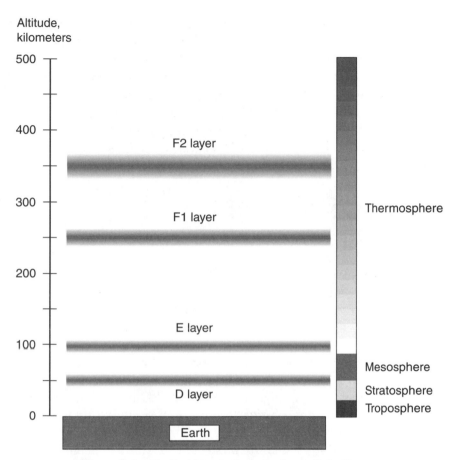

Figure 8-4. The Earth's atmosphere, showing the ionized layers.

Excerpt from a First Officer's Journal

I have just received a stern warning from the authorities back home on Epsilon Eridani 2. Their mandates are as follows:

- Our Earth landing assignments have been reduced from 10 to 3.
- We are not to land within 50 km (30 mi) of any place known to be populated, even sparsely, with the two-legged life forms that call themselves *Homo sapiens*.
- We are to keep our radar and optical cloaking devices activated at all times.

- If we accidentally happen to encounter any *Homo sapiens*, we are to explain to them that we are part of a "Hollywood movie set" and then ask them to leave.

This seems to defeat the most important part of our mission: to find out exactly what makes *Homo sapiens* behave as they do. However, I can't fight the bureaucracy of Epsilon Eridani 2! I will have to be content with looking at these creatures, whom I have decided to call *bipedal ants*, through telescopes while in orbit and analyzing their migration patterns with radar and computer programs.

NORTH ATLANTIC OCEAN

We descend into the middle of the Azores-Bermuda high-pressure system in the North Atlantic Ocean, hoping to find calm conditions. It is the part of the Earth year that the bipedal ants call *April*, when hurricanes are unknown in this part of the planet. Nevertheless, when we reach the ocean, we find that there are huge waves on the surface. The waves come from the north, and we recall that storms can track across the North Atlantic at any time of the year.

In the northern-hemispheric spring, storms follow the jet stream, coming off the North American continent near the mouth of the St. Lawrence Seaway and taking paths at high latitudes toward Europe. Our meteorology expert on the main ship confirms that one of these storms is passing near Iceland, and it is responsible for generating the waves. We are surprised that waves can travel so far and still be so large, but the main-ship radar telescope indicates that their effects reach all the way south to Antarctica.

The atmosphere is perfectly calm; there is no wind as we land and observe a temperature of 23°C (73°F) at high noon. We float like a cork on the swells, which measure 11 m (36 ft) from crest to trough. The surface of the water is smooth except for these sine-wave-shaped swells, a most remarkable phenomenon.

By sunset, the temperature of the atmosphere has hardly changed; it is 22°C (72°F). At midnight, the atmospheric temperature has gone down to 20°C (68°F), and just before sunrise on the day after our landing, it is at its minimum of 19°C (66°F). This small temperature variation between day and night confirms our theory that the oceans keep the lower atmospheric temperature relatively stable. The water temperature is measured at 20°C (68°F).

We remain on the surface of the ocean, examining the abundant life in the water, for exactly one solar day. We see no signs of the bipedal ants, either on the surface of the ocean or in the atmosphere above it, even though we have been told that aircraft will fly overhead and one of them will descend to investigate us. We are relieved when we lift off at noon, exactly 24 Earth hours after we landed. Because of the violent and continuous motion induced by the waves, I have lost 2.5 kilograms (kg) of body weight. This is 5.5 pounds (lb) in the Earth's gravitation. It has taken place because I have been unable to eat or drink anything for the past 24 Earth hours without having it come back up. Many bipedal ants suffer this same malady when they are first introduced to oceanic travel; they call it *seasickness*.

ANTARCTICA

Antarctica is a huge ice cap centered at the Earth's south geographic pole. All the continental land mass, with the exception of a small peninsula that reaches northward toward South America, is covered by water ice throughout the year. This ice extends offshore into the sea for a considerable distance in some places.

It is April, early autumn in the Earth's southern hemisphere. We are concerned about the possibility of high winds upsetting our craft and low temperatures straining our life-support systems if we land on the ice cap itself. We therefore decide to land at the tip of the Antarctic Peninsula that juts northward. This is not only the northernmost point on the continent, but it is largely surrounded by ocean, which, we hope, would keep temperatures from dropping too low.

As we approach the surface, it becomes apparent that this landing is going to make our North Atlantic excursion seem tame by comparison. We see snow (small flakes of water ice) rushing horizontally along the surface, driven by a wind of hurricane force. When we land, a gust almost knocks our craft off its landing gear and onto its side. Despite this wind, small black-and-white bipedal animals, looking like birds but acting more like bipedal ant children, run around, seemingly unaffected by the tempest. They jump in and out of the water, and some of them waddle up to our vessel and then stand there watching us, as if they expect us to come out and play in the water with them. We reject this option. The temperature is −37°C (−35°F).

The gale increases steadily. We decide to return to the main ship before the storm plucks our landing vessel up and rolls it across the bleak, rocky terrain. It never crosses our minds to venture outside into these conditions, which seem, despite the hospitable atmosphere, more severe than the worst storms we have ever seen on *Sol 4* (Mars). Thus we blast off, struggling to maintain stability, and we are relieved when we reach the stratosphere and spot our main ship, the *Dragon*, as a bright dot in the sky.

SAHARA DESERT

The only characteristics that the Sahara Desert shares with Antarctica are wind and dryness. In every other respect the two places are so different that it is hard to believe that they exist on the same planet.

We land at high noon on sandy, rolling terrain that looks like certain parts of Mars but with fewer rocks and boulders. The atmospheric temperature is 49°C (120°F) and rising. There is little wind, but the large dunes give away the fact that strong winds blow regularly in the area. The sky is hazy blue, pinkish near the horizon, again reminiscent of Mars.

By late afternoon, the temperature reaches a peak of 53°C (127°F), which we on Epsilon Eridani 2 consider ideal. The Sun, which has passed the zenith, sets in a ruddy cloud that again reminds us of home. The temperature quickly drops, and a brisk wind comes up. By midnight the temperature is 17°C (63°F), and in the predawn hours it drops to 14°C (57°F). We attribute this large day-night temperature differential to the high altitude of the site we have selected and to the fact that sand does not retain heat very well.

Just before liftoff at sunrise, we see tracks in the sand that appear to have been made by four-legged animals. However, no life is in sight, and we have been strictly warned to avoid the risk of contact with the bipedal ants. According to our Earth sociology books, it is not unknown in the Sahara Desert to see bipedal ants riding four-legged, long-necked animals.

As we blast off, in the distance I see objects moving slowly across the sand. I get my hand telescope and take a magnified look. There is a scene out of a picture book I saw about Earth when I was a child: Four *Homo sapiens*, each riding a four-legged, long-necked, hump-backed animal! I am astounded. Who would have guessed that the bipedal ants of the planet Earth have progressed to such a level of sophistication that they employ nonmechanized, nonpolluting modes of transportation? I expected, if I saw any life at all in the Sahara, to see them riding crazily around in four-wheeled, internal-combustion-propelled vehicles, tearing up what few plants manage to survive in that environment. Maybe the bipedal ants are not as barbaric as we have supposed.

CONCLUSIONS

From what we have seen of Earth, it is a stormy, desolate place. We deliberately chose regions where intelligent life would not likely be found. However, based on these observations, it is hard to imagine how any place on Earth could allow humans to build a civilization without great struggle and sacrifice. The bipedal ants must cooperate closely to build and maintain their anthills. But how can we know what these anthills actually are, what they do, and why they exist until we can visit one of them?

We only looked at three places on Earth, and this is not a sufficient sampling to know the nature of the planet as a whole. Perhaps there are fields of green plants, or undamaged forests, or snowy mountains with small settlements where the bipedal ants can revel in their surroundings and take time to enjoy the art of living. Maybe there are clean lakes and rivers and windy, empty prairies with small individual dwellings separated by enough distance so their occupants do not become mentally and physically deranged. We have heard rumors of such places, and our telescopic observations indicate that they might exist. For now, however, we must content ourselves to visit only desolate regions. We have been told by our security agencies that were these bipedal ants to encounter us, they might think we had come to invade them and react with violence. Even if they did not fear us, they might capture and analyze us, as if we were created by the Cosmos for no other purpose than to arouse and then satisfy their curiosity.

Refer to the text if necessary. A good score is 8 correct. Answers are in the back of the book.

1. The Earth's atmosphere at the surface consists of
    (a) 21 percent oxygen.
    (b) 78 percent oxygen.
    (c) 1 percent carbon dioxide.
    (d) 1 percent ozone.

2. At which of the following latitudes would an observer see the Sun for 24 hours a day during some parts of the year?
    (a) 50°N.
    (b) 23.5°N.

(c) 45°S.

(d) 75°S.

3. The Earth travels most rapidly in its orbit around the Sun during the month of

(a) January.

(b) April.

(c) July.

(d) October.

4. The Earth's core is believed to be

(a) extremely cold.

(b) a rarefied gas.

(c) comprised of basaltic rock.

(d) ferromagnetic.

5. The Earth's axis completes one complete cycle of precession approximately every

(a) 18,000 years.

(b) 12,900 years.

(c) 25,800 years.

(d) 50,000 years.

6. The crust of the Earth is thickest

(a) under continents.

(b) under the oceans.

(c) in the polar regions.

(d) No! The Earth's crust is uniformly thick everywhere.

7. Because of the generally clockwise flow of waters in the northern-hemispheric oceans

(a) the U.S. West Coast gets a warm equatorial current.

(b) the U.S. East Coast gets a cold polar current.

(c) the coast of China gets a warm equatorial current.

(d) the western coast of southern Africa receives a warm equatorial current.

8. Temperatures over the ocean do not change very much between day and night because

(a) the ocean heats and cools slowly so that it tends to keep the air temperature over it fairly constant between day and night.

(b) land masses radiate heat into the atmosphere at night, where it travels over the oceans and keeps the air there from cooling off.

(c) the salt in the oceans regulates the temperature.

(d) No! Temperatures over the ocean change greatly between day and night.

9. The chemical formula for ozone is

(a) O_2.

(b) NO_2.

    (c) O_3.

    (d) CO_2.

10. High- and low-pressure weather systems in the atmosphere are carried from west to east at temperate latitudes by the

    (a) oceanic currents.

    (b) stratosphere.

    (c) ionized layers.

    (d) jet streams.

Test: Part Two

Do not refer to the text when taking this test. A good score is at least 30 correct. Answers are in the back of the book. It is best to have a friend check your score the first time so that you won't memorize the answers if you want to take the test again.

1. The polar ice caps of Mars consist of
   (a) methane and ammonia ice.
   (b) frozen nitrogen.
   (c) frozen water and carbon dioxide.
   (d) white sand exposed by the action of dust storms.
   (e) clouds in the upper atmosphere.

2. A shepherd moon
   (a) tends to grow in size by accumulating stray meteors and comets.
   (b) acts to keep a planetary ring from spreading out.
   (c) has several smaller moons orbiting around it.
   (d) gets its name from the fact that shepherds once used it to keep track of their sheep.
   (e) is another name for a full moon.

3. If an object reflects one-quarter of the light that strikes it, then its albedo is approximately
   (a) 25.
   (b) 2.5.
   (c) 0.25.
   (d) 0.40.
   (e) 4.00.

4. The notion that the Earth is a huge, living cell is known as
   (a) the geobiological theory.
   (b) the Gaia hypothesis.
   (c) the tidal theory.
   (d) the geogenetic theory.
   (e) Nothing! No one has ever had such an idea.

5. The full phase of an inferior planet takes place at and near
    (a) superior conjunction.
    (b) inferior conjunction.
    (c) superior opposition.
    (d) inferior opposition.
    (e) No time; inferior planets never appear in full phase.

6. The Roche limit of a planet is
    (a) the smallest orbital radius a moon can have without being broken up by the parent planet's gravity.
    (b) the minimum temperature at which oxygen in the atmosphere of a planet can exist in a gaseous state.
    (c) the maximum axial tilt a planet can have in order to be a hospitable place for the evolution of life.
    (d) the smallest radius a planet can have and still manage to hold down an atmosphere.
    (e) the maximum amount of ultraviolet radiation that can reach a planet's surface without killing the living things on it.

7. Deimos is
    (a) one of the moons of Mars.
    (b) one of the volcanoes on Mars.
    (c) the highland region on Venus.
    (d) one of the moons of Venus.
    (e) one of the moons of Mercury.

8. The Sun is closest to Earth during the southern-hemispheric
    (a) spring.
    (b) summer.
    (c) fall.
    (d) winter.
    (e) Irrelevant! The Sun is always the same distance from the Earth.

9. The layer of Earth just beneath the crust is called the
    (a) basaltic layer.
    (b) ferromagnetic layer.
    (c) mantle.
    (d) outer core.
    (e) volcanic layer.

10. The so-called greenhouse gases
    (a) help heat to escape from a planet.
    (b) increase ultraviolet radiation reaching a planet's surface.
    (c) block ultraviolet radiation.
    (d) tend to trap heat in a planet's atmosphere.
    (e) keep Earth from becoming like Venus.

11. Ozone gas is known for its
    (a) tendency to block ultraviolet radiation.
    (b) environmentally destructive effects.
    (c) greenhouse properties.
    (d) role in the ice ages.
    (e) presence in low-level clouds.

12. The lack of a substantial magnetic field around Mars
    (a) allows the existence of an ionosphere similar to that of Earth.
    (b) allows an ozone layer to form in the Martian atmosphere.
    (c) lets high-speed subatomic solar particles reach the surface.
    (d) is the result of a magnetically polarized iron and nickel core.
    (e) is the result of intense volcanic activity.

13. Uranus is
    (a) about one-quarter the diameter of Earth.
    (b) slightly smaller than Earth.
    (c) about the same diameter as Earth.
    (d) slightly larger than Earth.
    (e) about four times the diameter of Earth.

14. On an Earth desert at high altitude, the temperature difference between day and
    night is considerable because
    (a) sand retains heat well.
    (b) the wind blows hard.
    (c) sand does not retain heat well.
    (d) there is almost no wind.
    (e) the air is thick.

15. The orbit of Venus is
    (a) retrograde with respect to the orbits of the other planets.
    (b) nearly a perfect circle.
    (c) slanted at 98 degrees relative to Earth's orbit.
    (d) an eccentric ellipse.
    (e) in sync with its rotation, so it always keeps the same side facing the Sun.

16. A manned space vessel would not be advised to enter a low orbit around
    Jupiter because
    (a) the temperature is extremely low.
    (b) there is not enough sunlight to navigate.
    (c) the radiation reaches dangerous or deadly levels.
    (d) there is no ozone layer.
    (e) the planet spins rapidly on its axis.

17. Which of the following pairs of planets are both closer to the Sun than Mars?
    (a) Mercury and Earth
    (b) Earth and Saturn

　　　(c) Saturn and Uranus

　　　(d) Venus and Jupiter

　　　(e) Earth and Neptune

18. The atmospheric pressure on the surface of Venus

　　　(a) is near zero because Venus has almost no atmosphere.

　　　(b) is somewhat less than the pressure at the surface of Earth.

　　　(c) is about the same as the pressure at the surface of Earth.

　　　(d) is slightly greater than the pressure at the surface of Earth.

　　　(e) is much greater than the pressure at the surface of Earth.

19. Which of the following pairs of planets are almost exactly the same size?

　　　(a) Mercury and Jupiter

　　　(b) Venus and Mars

　　　(c) Mars and Jupiter

　　　(d) Uranus and Neptune

　　　(e) Neptune and Pluto

20. When Jupiter is at inferior conjunction,

　　　(a) it appears full.

　　　(b) it appears half full.

　　　(c) it appears as a crescent.

　　　(d) it transits the Sun.

　　　(e) No! Jupiter never attains an inferior conjunction.

21. On a long journey in interplanetary space, artificial gravity might be provided by

　　　(a) special pressure suits.

　　　(b) vigorous daily exercise.

　　　(c) rotation of the living quarters in the vessel.

　　　(d) a massive slab of metal in the back of the ship.

　　　(e) Nothing; there is no such thing as artificial gravity.

22. The term *precession* refers to

　　　(a) the distortion of the geomagnetic field by the solar wind.

　　　(b) the variation of Jupiter's rotation rate with latitude.

　　　(c) the wobbling of the Moon so that we see more than half of it over time.

　　　(d) the wobbling of Earth's axis over long periods of time.

　　　(e) the ionization of Earth's upper atmosphere.

23. A caldera is

　　　(a) a mountain.

　　　(b) a valley.

　　　(c) an escarpment.

　　　(d) a dried-up riverbed.

　　　(e) a volcanic crater.

24. On the planet Venus, Ishtar is

　　　(a) a highland region.

　　　(b) a volcano.

(c) a crater.

(d) an escarpment.

(e) another name for the cloud deck.

25. The layer of Earth's atmosphere in which weather occurs is known as the

    (a) ionosphere.

    (b) isosphere.

    (c) troposphere.

    (d) stratosphere.

    (e) thermosphere.

26. Since the formation of the Solar System, the interior of Mars is believed to have cooled off faster than the interior of Earth because

    (a) Mars is farther from the Sun than is Earth.

    (b) Mars has a thinner atmosphere than does Earth.

    (c) Mars rotates faster than Earth.

    (d) the surface-area-to-volume ratio is larger than that of Earth because Mars itself is smaller.

    (e) No! Mars is believed to have cooled off more slowly than the Earth.

27. The light regions in Jupiter's atmosphere, as seen through a telescope on Earth or from a great distance away in space, are

    (a) whitecaps on the liquid surface.

    (b) snow.

    (c) the tops of high clouds.

    (d) blowing sand.

    (e) volcanic eruptions.

28. Some scientists think that the Pluto-Charon system ought to be classified as

    (a) an asteroid.

    (b) a star.

    (c) a double comet.

    (d) a ring system.

    (e) a shepherd moon.

29. Mercury does not often present itself well for observation because

    (a) it is extremely small.

    (b) it is extremely dense.

    (c) it is close to the Sun.

    (d) it is far from the Sun.

    (e) it always shows us its dark side.

30. The temperature on the dark side of Mars

    (a) is below 0°C.

    (b) is about the same as the temperature on the daylight side.

    (c) is a comfortable room temperature.

    (d) is hot enough to melt lead.

    (e) is near absolute zero.

31. Jupiter's Great Red Spot is
    (a) a volcano.
    (b) a mountaintop.
    (c) an escarpment.
    (d) a crater.
    (e) a weather system.

32. The upper clouds of Venus rush around the planet, completing one revolution in approximately
    (a) one Venus day.
    (b) one Venus year.
    (c) 24 Earth hours.
    (d) four Earth days.
    (e) No! Venus has no clouds.

33. Saturn's brightness, as we see it from Earth, is affected by
    (a) the angle at which Saturn's rings are presented to us.
    (b) the distance of Saturn from the Sun.
    (c) the tilt of Saturn on its axis.
    (d) the temperature on Saturn's surface.
    (e) the distance of Earth from the Sun.

34. The most abundant gas in Earth's atmosphere is
    (a) oxygen.
    (b) hydrogen.
    (c) helium.
    (d) carbon dioxide.
    (e) nitrogen.

35. An escarpment is
    (a) a crater.
    (b) a volcano.
    (c) a tiny planetary moon.
    (d) a moon that has escaped from a planet.
    (e) a cliff.

36. The butterfly effect is another name for the fact that
    (a) minor events always have small consequences.
    (b) major events always have large consequences.
    (c) minor events never have any consequences.
    (d) seemingly insignificant events can have large consequences.
    (e) life cannot exist on Mars.

37. When Mars is at an ideal opposition,
    (a) it is closer to Earth than at any other time.
    (b) it is farther from Earth than at any other time.
    (c) it is brighter than Venus.

(d) it is invisible because the light of the Sun washes it out.

(e) it appears in a crescent phase.

38. The most abundant gas in the atmosphere of Mars is
    (a) oxygen.
    (b) nitrogen.
    (c) methane.
    (d) carbon dioxide.
    (e) carbon monoxide.

39. Which of the following planets has not been observed to have rings?
    (a) Mars
    (b) Jupiter
    (c) Saturn
    (d) Uranus
    (e) Neptune

40. Mercury is believed to have a core consisting of
    (a) granite.
    (b) water.
    (c) solid metallic hydrogen.
    (d) molten silicate rock.
    (e) iron.

PART 3

Solar System Dynamics

CHAPTER 9

Evolution of the Solar System

On a clear night when the Moon is below the horizon, you get a feeling of great depth as you look at the heavens. Some ancient stargazers surely sensed this depth and decided that space must be a huge expanse and that Earth must be insignificant compared with the whole. For a long time, however, this idea was unacceptable. If anyone thought about such things, they kept their notions to themselves. Today's astronomers theorize not only about what the Universe is like now but also about how it and our own Solar System came into being and evolved to its present state.

A Word about Probability

In science, there are certain things we do not know, but we develop beliefs according to what we can see and according to what we can obtain by using mathematics and logical reasoning. We might say "such and such is true" because we've seen it or because we've deduced it based on observations. If we think something is true but aren't certain, it's tempting to say that "such and such is likely." But we're still not sure.

HYPOTHESES AND THEORIES

Most of the material in this book, up until now, has been based on observed facts (except, of course, for our "mind journeys"). The diameter of Earth can be measured, as can the temperature of the solar corona. It is taken as a fact nowadays that Earth revolves around the Sun. However, the way the Solar System was formed is not known with such certainty. No human ever saw nor has any machine ever recorded that sequence of events. The best we can do is propose a *hypothesis*, an idea of what we think took place. Then we can make arguments, based on logic, observation, and computer modeling, so that we can come up with a *theory*.

When people formulate a theory, it is tempting to say that something "probably" happened in the distant past or that there is a "good chance" that such and such exists in the Milky Way galaxy. This is a logical pitfall. It is so easy to make this thought-process error that I have probably committed it somewhere in this book. (Whoops! I did it in that very sentence!)

A THEORY IS A BELIEF

An astronomer can say that he or she *believes* the Milky Way galaxy is moving away from all the other galaxies in the Universe; some will strengthen this into an opinionated statement of fact. If you say, "I think the Universe started with an explosion" or "I believe that the Universe started with an explosion," you are within your rights. Even if you say, "The Universe began with an explosion!" your statement is logically sound, although it must be understood that this is a theory, not a proven fact. However, if you say, "The Universe probably started with an explosion," you have committed the dreaded thought error, which, for lack of a better name, I call the *probability fallacy*. Whatever has been has been! Either the Universe started with an explosion, or else it did not. (There are some people who will argue that either statement can be true depending on how you define the various arrows of time, but I won't get into that right now.)

The same holds true for other unknowns. A good example is the "probability that intelligent life exists elsewhere in the Universe." Assuming that we have defined the meaning of the word *intelligent*, we can confidently say that such beings either exist or they don't. If I say in this book that the probability of intelligent life existing elsewhere is "20 percent," I am in effect saying something like this: "Out of 1,000 observed Universes, 200 of them have been found to have intelligent extraterrestrial life, but I don't know which one of the 1,000 Universes I happen to live in." This is nonsense!

When we think about how the Solar System was formed, we must keep in mind that there is a definite reality, a specific sequence of events, that took place to get us from that place where "all was dark and without form" to where we are now. Our task is to find out the truth and not to try to attach artificial "probabilities" to things that have already happened or to things that never took place at all.

WHEN CAN WE TALK ABOUT PROBABILITY?

Probability can be assigned to an event only on the basis of the results of observations involving a large number of samples. Additionally, probability makes sense only when talking about the future; it is irrelevant when dealing with the past or present.

There are certain theories involving so-called fuzzy truth in which some events can be considered to "sort of happen." These theories involve degrees of reality, and in such a scenario, probability can be used to talk about events in the past, present, and future. The most common example of this kind of theory is quantum mechanics, which involves the behavior of atoms, molecules, and subatomic particles. Quantum mechanics can get so bizarre that some scientists have actually said, "If you claim to understand this theory, then you are lying." Fortunately, we aren't going to be dealing with anything that esoteric in this book.

When statistics is misapplied, seemingly logical reasoning can be used to support all manner of hogwash. It is done in industry all the time, especially when the intent is to get you to do something that will cause someone else to make money. Therefore, keep your "probability-fallacy radar" on. We are about to leap into territory where every good scientist needs it!

If you come across an instance where an author (including myself) slips and says that something "probably happened" or "is likely to take place," think of it as another way of saying that the author, or scientists in general, believe or suspect that something happened or will take place.

Early Theories

Most ancient philosophers believed that all celestial objects were attached to concentric spheres with Earth at the center. It never crossed their minds, apparently, to inquire very much into what might lie outside those spheres. Some thought that the outermost sphere was opaque, with little holes through which an outside light shone. Some thought the stars were fires on the inside of the outermost sphere.

SIMPLE GEOCENTRIC THEORY

Eventually, observers noticed that there were some peculiarities about the con-centric-sphere model of the Universe. The Sun did not always stay in the same position relative to the stars. Some of the stars moved among the stationary majority. These moving stars were called *wanderers*, or *planets*. Each planet was assigned its own sphere. The Moon had a sphere for itself, as did the Sun.

Figure 9-1 shows one of the earliest models of the Solar System out to the planet Saturn, which in ancient times was the most distant known object except for the sphere containing the stars. Earth was believed to be at the center of it all. The theory that Earth was at the center of creation has been called the *geocentric theory*. This theory underwent many variations, refinements, and contortions before the bearers of conventional wisdom saw fit to throw it out.

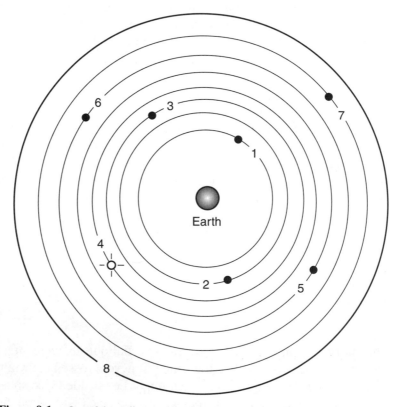

Figure 9-1.   One of the earliest models of the Universe. Earth is at center. Numerical tags: 1 = Moon, 2 = Mercury, 3 = Venus, 4 = Sun, 5 = Mars, 6 = Jupiter, 7 = Saturn, 8 = stars.

Some ancient astrologers (this is what astronomers were called in the olden days) thought that since all these spheres rotated at their various speeds and on their independent axes, there must be friction among them. This friction must, they believed, create cosmic music, perhaps accompanied by the singing of angels—hence the expression "the music of the spheres." Such a noise was considered too faint for ordinary human beings to hear. However, privileged people claimed to hear it, and they assured their contemporaries that it was beautiful. Today's scientists will tell you that this was nonsense, imaginary at best and a bad joke at worst. Sounds made by distant celestial objects cannot reach Earth. Sound does not travel through outer space.

THE PTOLEMAIC MODEL

The model shown in Fig. 9-1 was, after a time, seen to have certain shortcomings. It failed to explain certain things. According to the simple geocentric theory, all the planets would maintain a constant and uniform motion, always in the same direction, with respect to the background of stars. However, observers noticed that the planets do not behave this way.

Once in a while, a planet appears to stop, turn around, go backwards, stop again, and then resume its normal forward motion among the stars. Some planets do this more often than others. Plotting the position of a planet over a period of weeks relative to the stars will reveal a loop (Fig. 9-2). The astronomer Ptolemy, who lived during the second century A.D., developed a model, a variant of the geocentric theory, that explained this phenomenon. It became known as the *Ptolemaic model*, and it endured for centuries.

According to Ptolemy's theory, rather than following a perfectly circular orbit around the Earth, each planet was assigned an orbit consisting of two distinct components, called the *deferent* and the *epicycle*. The deferent was a perfectly circular path around Earth, but the planet in question was believed not to follow it. Instead, the planet was thought to follow the epicycle, or smaller orbit, around a point in the deferent that in turn was theorized to maintain a constant motion around Earth (Fig. 9-3). If this sounds a little strange to you, you are not alone. How can a single object orbit around a point in empty space? Doesn't that violate some principle of physics? Of course it does! However, those principles had not yet been laid down in Ptolemy's time, and the theory of Ptolemy did an excellent job of explaining the observations made by astronomers.

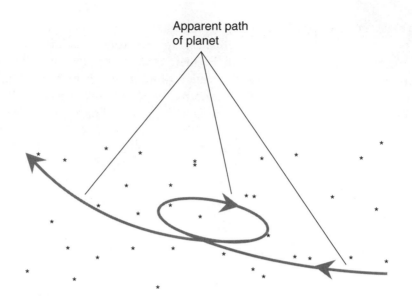

Figure 9-2. The path of a planet, as seen from Earth against the background of stars, occasionally does a "loop-the-loop."

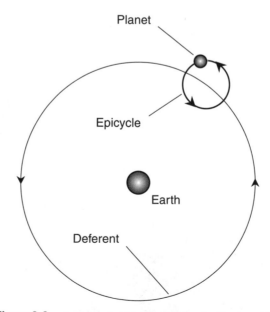

Figure 9-3. An example of a deferent/epicycle planetary orbit.

QUESTIONS AND DOUBTS

Ptolemy's epicycle model was not perfect. On closer and closer scrutiny, it was found that additional epicycles within epicycles were necessary to predict the exact position of a planet at any given future moment. This process could continue *ad infinitum*, with smaller and smaller epicycles superimposed on one another endlessly. The believers in Ptolemy's model began to wonder why God was so perverse when He designed the Universe. (Some scientists still ask this question.) A few people expressed cynicism and veiled sarcasm.

Whenever a scientific theory or model appears to fit observed facts, it is placed under an ongoing attack by skeptics. Their intent is not necessarily malicious. The idea is to test the theory. This is how theories are refined. Once in a while, however, a theory gets so awkward that scientists decide that it should be scrapped and that the whole business ought to start over. It's "back to the drawing board!" This stage was not reached with regard to the Ptolemaic model until the world's attitudes had evolved past the intellectual vacuum known as the Middle Ages.

An unfortunate Italian philosopher named Giordano Bruno was vocal about his doubts in the latter part of the sixteenth century. He was condemned by the powerful church leaders and put to death in the year 1600. A little while later, Galileo Galilei, another open skeptic, was confined to house arrest and told to be silent for expressing similar doubts. The church could not accept the idea that Earth does not sit at the center of all creation.

Heliocentric Theory

The "thought police" of the church held less power in northern Europe than they did in Italy. Proponents of the *heliocentric* (Sun-centered) *theory* were taken seriously in places such as Germany, France, Poland, and England.

THE PIONEERS

Nicolaus Copernicus, a Polish astronomer, published a work in the early sixteenth century suggesting that the Sun, not the Earth, must be at the center of the Universe. (Remember that back in the sixteenth century the Earth, the Moon, the Sun, and the planets basically defined the entire Cosmos. No one knew what the stars were, much less how they were distributed throughout space.) The Earth, thought Copernicus, is a planet just like Mercury, Venus, or Mars insofar as its importance in the overall scheme of things. But Copernicus could not prove his theory to the complete satisfaction of the authorities in his part of the world. If the Earth is moving, asked the skeptics, why don't we feel a constant wind from space? What force could push the Earth? Why should such a force exist?

Another astronomer, Tycho Brahe, was involved with an ongoing meticulous mapping and recording job. He kept careful records of the positions of all the planets over a period of time. Brahe had a German assistant named Johannes Kepler who eventually formulated the three fundamental rules for planetary motion, known as *Kepler's laws*. Isaac Newton put it all together and finally changed mainstream thinking. The Earth had lost its exalted position, replaced by the Sun. The heliocentric theory had survived the test of time and had become the conventional wisdom.

KEPLER'S LAWS

Johannes Kepler published his famous rules of planetary motion early in the seventeenth century. They can be stated briefly as follows:

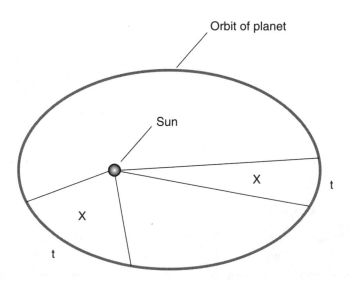

Figure 9-4.   Kepler's first and second laws. The planet's orbit is an ellipse with the Sun at one focus, and equal areas (*X*) are swept out in equal periods of time (*t*).

- Each planet follows an elliptical orbit around the Sun, with the Sun at one focus of the ellipse.

- An imaginary line connecting any planet with the Sun sweeps out equal areas in equal periods of time.

- For each planet, the square of its "year" (sidereal period) is directly proportional to the cube (third power) of its average distance from the Sun.

Theoretically, it is possible for a planet's orbit to be perfectly circular. A circle is an ellipse in which both foci are at the same point. In reality, however, there is always some imperfection, so all planets follow orbits that are slightly oblong.

Kepler did not originally call his rules laws. This label was attached later by others. Kepler came up with his three principles and refined them over a period of several years. The first two rules were finalized in 1609, and the last one came out in 1618. The first two laws are illustrated in Fig. 9-4, and third law is rendered graphically in Fig. 9-5.

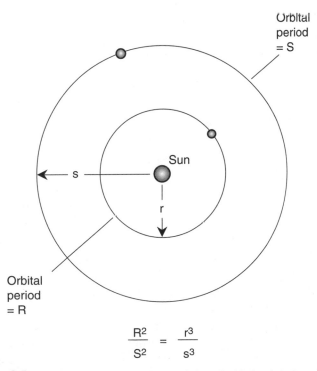

$$\frac{R^2}{S^2} = \frac{r^3}{s^3}$$

Figure 9-5.   Kepler's third law. The squares of planets' orbital periods (R and S) are proportional to the cubes of their mean orbital radii (r and s).

The Tidal Theory

According to the *tidal theory*, the Sun originally had no planets or other satellites. This theory suggests that our Sun formed alone and that the other objects, including the planets and the major asteroids, came later.

RELATIVE MOTIONS OF THE STARS

The Milky Way, the spiral-shaped galaxy in which we dwell, is believed to be 100,000 *light-years* across. A light-year is the distance that light travels in 1 year, approximately 9.5 trillion (9.5×10^{12}) km or 5.9 trillion (5.9×10^{12}) mi. Our galaxy has roughly 200 billion (2×10^{11}) stars, all revolving around the nucleus like an enormous swarm of bees. According to current theories, many of the stars bob up and down, above and below the galactic plane, passing periodically through it. Some stars have highly eccentric orbits around the galactic center.

Although the stars are tiny compared with the space between them, they are in relative motion, and collisions or near misses occur once in a while simply because there are so many stars. On average, however, according to one estimate, an outright collision is an extreme rarity, taking place only about once in every 10 billion (10^{10}) years for a typical spiral galaxy such as ours. This is almost as old as the whole Universe is believed to be! Nevertheless, those people who say that the Sun fell victim to a near catastrophe with another star cannot be discounted completely.

THE SCENARIO

Suppose that another star came close enough to the Sun that it and the other star engaged in a gravitational tug-of-war. What would happen? For one thing, the paths of both stars in the Milky Way would be altered; the two stars would swing around each other. In fact, if they came close enough and the speed was not too great, they would end up in orbit around each other. Suppose, however, that the encounter was extremely close but at high speed so that the two stars did not end up in mutual orbit? They would pull matter from each other and scatter that matter into orbits around either star, where the matter would cool, condense, and form dust, rocky ice chunks, and rocks.

Given time, the particles in orbit around the Sun would coalesce into larger objects because of mutual gravitation. Eventually, several dozen spherical objects, perhaps comparable with the size of our Moon, would be created. These objects would follow all kinds of different orbits because of the chaotic way in which the matter was scattered during the original battle of the stars. The result would be frequent collisions and further coalescing. Computer models can show that the end result would be a few large, massive objects and countless tiny ones. This is, of course, the way we observe the Solar System today.

DIFFICULTIES

There are problems with this so-called *tidal theory*. If this is the way the Solar System formed, the planets would all revolve around the Sun in different planes, and their orbits would be less circular and more elongated than they are (Fig. 9-6). However, the actual state of affairs is far more orderly. The planets all lie in nearly the same plane. With the exceptions of Mercury and the Pluto-Charon system, their orbits are nearly perfect circles. All the planets revolve around the Sun in the same direction. For these reasons, few astronomers today believe that the tidal theory is an accurate representation of what happened. In addition, the fact that such catastrophes in general occur only once every several billion years, in our galaxy at least, is a good reason to doubt that this theory explains how things took place to create our Solar System.

The Nebular Theory

If a star has several times the mass of the Sun, ultimately it will explode in a violent outburst called a *supernova*. These events leave entrails in space—clouds of gas, dust, and rocks of various sizes —in their vicinity. Such mass of debris can appear either light or dark through a telescope depending on how the light of nearby stars shines on it. The cloud is called a *nebula*.

THE STUFF OF STARS

Most nebulae form near the plane of our spiral-shaped Milky Way galaxy. They are clearly visible in other spiral-shaped galaxies when those galaxies

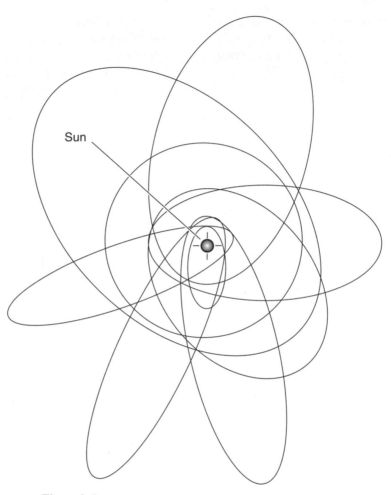

Figure 9-6.   If the tidal theory were correct, the planets would have elongated orbits (gray ellipses) in various planes.

present themselves edgewise to us. Some spiral galaxies are so thick with nebulae that they appear split in two when we see them from within the planes of their disks. Our Solar System is near the plane of the Milky Way, and our galaxy, like all spirals, has plenty of nebulae. This keeps the sky dark at night. If it were not for these obstructing clouds, the sky would be almost as bright when the Sun is "down" as when the Sun is "up."

According to the *nebular theory*, also called the *rotating-cloud theory*, it is from these clouds that second-generation stars, such as our Sun, are born. Evidence suggests that the Solar System formed approximately 4.6

billion (4.6×10^9) years ago from one of these. The Earth, all the other planets, the asteroids, and the comets are all believed to have formed from a cloud produced a long time ago in one or more supernovae.

As you have already learned, the Sun takes about a month to rotate once on its axis. Because of this, it is logical to suppose that the debris cloud from which the Sun formed had some rotational momentum. Imagine a hurricane forming from the clouds in the tropics. Have you ever seen a time-lapse satellite photo of this process? Think about the eddies or whirlpools that form in the water as you pull a canoe paddle through. According to the rotating-cloud theory, the Sun formed at the center of an eddy in interstellar space.

THE ACCRETION DISK

Astronomers have shown that a cloud of debris, collapsing because of the mutual gravitation among all its particles, would develop one or more vortices, or whirlpools. Near each vortex, the matter would become aligned in a plane, creating a rotating, disk-shaped cloud. It can be demonstrated by computer modeling that the matter in such a cloud would condense into an *accretion disk* and thence into numerous discrete objects: a large central mass (to become the Sun) and other, relatively small masses in orbit around it (to evolve into the planets and their moons). One theory, proposed several centuries ago, took notice of this fact (without the help of computer modeling, of course) and came to the conclusion that the matter orbiting the Sun would congeal into rings before finally developing into solid planets.

Figure 9-7 is a hypothetical illustration of how the Solar System's primordial cloud looked from a distance of about 100 astronomical units (AU). In this example, the disk is viewed at an angle, neither face-on nor edgewise, so that the nearly circular rings appear oblong. The Sun is at the center, and it is about to start up its internal nuclear-fusion furnace. The disk-shaped cloud, and in particular its rings, glow from the Sun's increasing radiance and from the light of other nearby stars. According to the rotating-cloud theory, the particles in the rings gradually pulled themselves together over a period of millions of years into small objects called *planetesimals*, and these ultimately accreted into the planets. Most of the nonsolar matter in the cloud found its way into the planet Jupiter; smaller amounts congealed into the other planets. As the planetesimals aggregated into larger objects, the matter in them swirled just as had the original parent cloud. This explains why the planets rotate. It also explains why most

Figure 9-7.   The primordial gas-and-dust cloud according to the rotating-disk theory of
solar system formation.

planetary moons orbit in the same sense as all the planets orbit around the
sun and why most (but not all) planetary moons orbit near the plane of the
planets' orbits.

The original version of this nebular theory is credited to two men who
lived during the eighteenth and early nineteenth centuries: Immanuel Kant,
a German philosopher, and Pierre-Simon Marquis de Laplace, a French
astronomer and mathematician. In particular, Laplace went into detail concerning the motions of the various planets and moons. In recent decades,

the nebular theory has been refined, especially in an attempt to explain why the Sun rotates only once a month and not much faster. In addition, the existence of the rings in the primordial accretion disk has been questioned. Many astronomers believe that the matter simply clumped together into larger and larger "particles," ending up with the system of planets we now have. The asteroids in the belt between Mars and Jupiter were prevented from accreting into a planet because of the powerful gravitational influence of Jupiter.

DIFFICULTIES

Certain questions remain difficult to answer—in particular the extreme tilts of Venus, Uranus, and the Pluto-Charon systems on their axes. You will remember that the axis of Venus is tilted nearly 180 degrees; another way of saying this is that its rotation is retrograde. The nebular theory does not specifically forbid this, although it suggests that most planets will end up rotating in the same sense as they orbit around the Sun. In the case of Uranus, some astronomers think that it was struck by an object so massive that its rotational axis was "knocked flat" by the encounter. In that scenario, both objects were nearly shattered; in the end, however, Uranus survived, and the other object did not. The same thing may have taken place with Venus and Pluto-Charon.

ARE WE TOO HOPE-DRIVEN?

The nebular theory explains why the planets orbit the Sun in a comparatively uniform manner. In addition, assuming that this theory is correct, we have good reason to believe that there are many such systems in our Milky Way galaxy, as well as in other galaxies, especially those of the spiral type with their abundant interstellar gas and dust.

Astronomers have found evidence of other planetary systems. Flat, circular clouds or rings, thought to be accretion disks, have been observed by the Hubble Space Telescope. If we actually are looking at stars with planets forming around them or in orbit around them, then it means that our Solar System is not a freak cosmic accident. If the Universe is teeming with planetary systems like ours, it is tempting to believe that there are many Earthlike planets too and that some of these planets have evolved intelligent life.

Critics of the nebular theory use the foregoing speculations against it. They say that hope drives the thinking of the proponents of the theory and that this emotion interferes with rational reasoning. If our Solar System is the only one of its kind in the whole Universe, they say, then so what? We are here to bear witness to the miracle of life on Earth simply because we are one of its products!

DECLINE AND DEATH

Just as there are theories about how the Solar System was formed, there are notions concerning its long-term future. The ultimate fate of the Solar System depends on its parent star, the Sun. Most astronomers believe that the Sun eventually will swell into a red giant, burning up or vaporizing Mercury, Venus, Earth, and Mars and perhaps blowing the gas away from Jupiter, Saturn, Uranus, and Neptune. Then the Sun will shrink down and die out like an ember in a dying fire around which living beings were once encamped. Where those creatures, our distant descendants, will be by then is a question that no theory can answer.

 **Quiz**

Refer to the text if necessary. A good score is 8 correct. Answers are in the back of the book.

1. Imagine an alien star system in which Planet X has a mean orbital radius of 100 million (10^8) km from Star S and Planet Y has a mean orbital radius of 2×10^8 km from Star S (twice the mean orbital radius of Planet X). Suppose that the "year" for Planet X is equal to exactly one-half Earth year (0.500 yr). How long is the "year" for Planet Y?
   (a) 2.000 years
   (b) 1.414 years
   (c) 1.000 year
   (d) It can't be figured out from this information.

2. The imperfections in Ptolemy's theory were "corrected," without rejecting the whole theory, by
   (a) adding epicycles within epicycles until the theory fit observed facts.
   (b) placing the Moon at the center of the Solar System.

(c) ignoring the distant stars.

(d) considering all the planets except Earth to orbit the Sun.

3. The planet Saturn orbits the Earth according to
    (a) the geocentric theory.
    (b) the heliocentric theory.
    (c) Kepler's theory.
    (d) Newton's theory.

4. Which of the following is *not* a good argument against the tidal theory?
    (a) The planets all orbit the Sun in nearly the same plane.
    (b) The planets all orbit the Sun in the same direction.
    (c) Stars can never pass so close that they pull matter from each other.
    (d) None of the planets have extremely elongated orbits.

5. How likely is it that beings like us exist elsewhere in the Milky Way galaxy?
    (a) Not likely
    (b) Somewhat likely
    (c) Very likely
    (d) This is a meaningless question. Either there are such beings or there aren't.

6. Which of the following statements is implied by Kepler's laws?
    (a) A planet moves fastest in its orbit when it is farthest from the Sun.
    (b) Planets far from the Sun take longer to complete their orbits than planets closer to the Sun.
    (c) All the planets' orbits lie in exactly the same plane.
    (d) All the planets' orbits are perfect circles.

7. According to the Big Bang theory of Solar System formation,
    (a) The planets formed when the primordial Sun exploded, casting some of its matter into space.
    (b) The planets evolved from a rotating cloud of gas and dust.
    (c) The planets were formed from matter ejected from a huge solar volcano.
    (d) Forget it! There is no Big Bang theory of Solar System formation.

8. According to one theory, Uranus has an axis that is tilted to such a great extent because
    (a) the planet was not massive enough for its equator to align itself with the plane of its orbit.
    (b) a large primordial object smashed into Uranus and tipped it over.
    (c) sooner or later such a tilt will be exhibited by all the planets.
    (d) the gravitational effect of Neptune pulled the axis of Uranus out of kilter.

9. For publishing his theories in the sixteenth century, Giordano Bruno was
    (a) knighted by the Queen of England.
    (b) made the official astronomer of the Vatican.

(c) ignored.

(d) executed.

10. When the earliest models of the Solar System were formulated, the most distant known planet was

(a) Mars.

(b) Jupiter.

(c) Saturn.

(d) Uranus.

CHAPTER 10

Major Moons of the Outer Planets

The so-called outer planets are Jupiter, Saturn, Uranus, Neptune, and Pluto (although some people debate whether or not Pluto qualifies as a planet). All these planets have moons, some of which are bigger than Earth's moon. In this chapter, we'll look at planetary moons larger than 1000 km (620 mi) in diameter.

Most of the moons of the outer planets always keep the same side facing their parent planets. This is the result of long-term tidal effects and is the same phenomenon as that which has happened to Earth's moon. Most of the moons of the outer planets orbit near the equators of their parent planets and in the same direction as the planets' rotations. In the cases of Jupiter, Saturn, and Uranus, the planet-moon systems are thought to have evolved like star systems in miniature.

Now that astronomers have had a chance to closely examine (by means of space probes) the major moons of the outer planets along with the planets themselves, how can anyone not be amazed at the variety of worlds our Solar System has produced? The more we learn about these worlds, the more mysterious they become.

Jupiter's Major Moons

Jupiter, the largest of the planets, has a system of moons that resembles a miniature "Solar System" with Jupiter as the "Sun." Four of these moons can be seen through a good pair of binoculars or a small telescope. Galileo Galilei observed them and carefully recorded their behavior in the early 1600s. Their images were starlike points of light that appeared along a straight line passing through Jupiter's disk. Galileo deduced that these points of light were natural satellites, or moons, because their relative positions changed from night to night in a way consistent with bodies traveling in more or less circular paths around Jupiter. These four little worlds, and only these four—*Ganymede*, *Callisto*, *Io*, and *Europa*—are called the *Galilean moons* in memory of Galileo.

GANYMEDE

This moon, Jupiter's largest, is 5,270 km (3,270 mi) across, a little less than half the diameter of Earth but larger than Earth's moon. Compared with Jupiter, Ganymede is tiny (Fig. 10-1). The satellite orbits Jupiter in a nearly perfect circle at a distance of about 1.1 million km (660,000 mi). Like virtually all planetary moons, Ganymede orbits near the equator of its parent planet and revolves in the same direction that the planet rotates. Also like most planetary moons, Ganymede keeps the same face toward Jupiter at all times. At the surface of Ganymede, the gravitational field is only about 15 percent as strong as it is on Earth's surface. Thus, if you weigh 140 lb on Earth, you would weigh only 21 lb on Ganymede.

Ganymede is believed to consist of a metallic core surrounded by rocky material, in turn surrounded by a crust of rock mixed with ice. The moon's density is approximately twice that of water. The surface was seen close up by the *Voyager* probe, and a number of features were observed, including many craters, evidence that Ganymede has been bombarded by meteorites. The shallow nature of the craters suggests that the surface of Ganymede is largely made of water ice. At the extreme cold temperatures at Jupiter's distance from the Sun, ice behaves something like rock on Earth's surface. However, over time the ice flows and settles so that the mountains and rims of the craters flatten out gradually. Nevertheless, this ice maintains crater imprints for millions of years.

Ganymede, like most of the other moons of the outer planets, has little or no atmosphere, although Ganymede has enough of a gravitational pull to

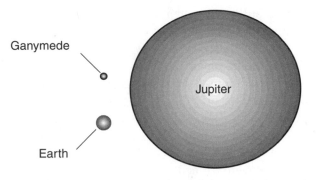

Figure 10-1.   The relative sizes of Jupiter's largest moon Ganymede, the planet Earth, and the planet Jupiter.

hold down trace amounts of oxygen and other gases. The presence of water (in the form of ice) and oxygen in the environment of Ganymede does not imply that this moon bears life; temperatures are too cold and conditions far too tranquil for anything biological to evolve the way it has on our planet Earth. One of the ingredients for the development of life as we know it, interestingly enough, is an environment subject to change.

CALLISTO

The second largest moon of Jupiter, Callisto, is 4,800 km (3,000 mi) in diameter, nearly as large as Ganymede. It is about the same size as the planet Mercury. Callisto orbits Jupiter almost twice as far away as Ganymede: 1.9 million km (1.2 million mi). Callisto is the outermost of Jupiter's four large Galilean satellites. The density of Callisto is similar to that of Ganymede; it has about 1.8 times the specific gravity of water. The interior structure of Callisto appears to differ somewhat from that of Ganymede because Callisto reflects less light. Some regions of Callisto are almost black, whereas others are bright. The oldest geologic features appear in the darker areas.

Callisto is covered by impact craters, as we would expect for a satellite of the planet with the most powerful gravitational field in the Solar System. There are also markings that look like concentric rings when photographs of this moon, taken by the passing *Voyager*, are examined. The most commonly held belief is that these "bull's-eye" features are the remnants of craters produced by gigantic objects that smashed into Callisto in the early ages of the Solar System, just after the formation of Jupiter and its moons, when Callisto was still in a molten or partly molten state. Some of these features are thought to be in excess of 4 billion years old.

Despite the fact that Callisto is heavily cratered, the depths of the craters and the heights of the mountains associated with them are fairly shallow. This suggests a smoothing-out process that has taken place since most of the craters were formed. This is the same sort of thing that seems to have occurred on Ganymede. Callisto also has features that are believed to be eroding; these were first seen in a 2001 space-probe flyby and have been described as "spires" or "knobs" several hundred meters high. It is not clear how these structures were formed.

Some astronomers speculate that Callisto has a salty sea beneath its outer crust of rock and ice. This idea has come from the fact that there are numerous bright spots on the surface consisting of water ice that looks as if it flowed out of some of the craters and froze on the exterior. If this model is correct, then large meteorites striking Callisto occasionally have punctured the solid crust and created holes through which the subsurface ocean spilled out and froze solid when exposed to the cold. After such a crater formed, it would fill up with ice, creating the bright spot. Observations and analyses of the magnetic field, or *magnetosphere*, surrounding Callisto lend further support to the subsurface-ocean theory. The *Galileo* space probe, which began observing Callisto in the 1990s, has detected the presence of a magnetic field that behaves in a way consistent with the presence of a conductive liquid beneath the surface. Salty water is a fairly good electrical conductor and could carry currents sufficient to generate the magnetic field that has been observed.

Io

Many astronomers think that Io, the third-largest moon of Jupiter, is the most interesting object in the Solar System other than our own Earth. When the first photographs of this moon were returned by space probes, scientists could hardly believe their eyes. Rather than a desolate, cratered scape typical of most other moons and asteroids in the Solar System, Io's surface looked like a pizza.

Io is 3,630 km (2,260 mi) in diameter, somewhat smaller than either Ganymede or Callisto. Io is also the innermost major moon of Jupiter. Because of its proximity to Jupiter and its orbital position with respect to the other major moons, Io is constantly bearing the forces of a gravitational tug-of-war. This heats up the whole moon because of tidal flexing in much the same way as a piece of wire heats up if you bend it back and forth. The heat boils away water from the surface. As a result of this, and also because Io appears to have an iron core, the density of this moon is relatively high.

Io's constant internal heating produces volcanic activity that has been photographed by space probes. If you could stand on the surface of Io, you would weigh only about 18 percent of your weight on Earth, but you would, if you didn't know better, think it reasonable to suppose that you had gone to hell. Sulfur and other molten rock compounds are abundant on this little world. There are no impact craters visible from space; the constant flow of volcanic lava on the surface erases them before they can become old. There are plenty of volcanic calderas, however. Mountains several kilometers high also have been seen. Io has a thin atmosphere. This is the sort of thing we would expect from constant emission of gases from volcanoes.

EUROPA

Europa is the smallest of the Galilean moons, with a diameter of 3,140 km (1,950 mi). It orbits only 670,000 km (420,000 mi) from its parent planet.

Europa has a high *albedo* (proportion of light that it reflects), unlike almost all other planetary moons. When this moon was photographed close up by the space probes and the light from its surface was analyzed, scientists concluded that it is water ice. Europa has high density, however, suggesting that this water is only a thin veneer over a predominantly rocky planetoid. The surface is crisscrossed with fracture lines that resemble those in the Earth's Arctic and Antarctic ice shelves. This suggests that the ice is the frozen surface of an ocean that covers the entire moon. There are very few impact craters visible on the surface. Astronomers think that liquid water from underneath the ice periodically floods the surface and freezes over, erasing old surface features and creating new ones. There is a thin atmosphere consisting largely of oxygen.

Europa has a weak magnetosphere. There are two theories concerning the origin of this magnetism: electric currents in a salty sea or electric currents in an iron core. The intensity of this field is affected by Jupiter's vastly more powerful magnetic flux.

Whenever scientists find water on an extraterrestrial body, they are tempted to speculate concerning the evolution of life. There is only one way to find out if there are strange, fascinating marine organisms beneath

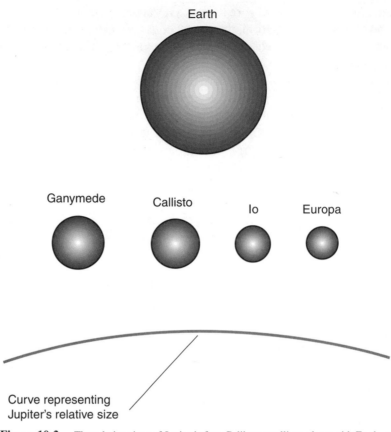

Figure 10-2.   The relative sizes of Jupiter's four Galilean satellites, along with Earth, as they would appear if placed size by side. Jupiter (gray curve) dwarfs them all.

the ice of Europa: Send a probe down to land on the surface and drill holes in the ice! Someday, if astronomers get their way, this will be done.

Figure 10-2 shows the four major satellites of Jupiter, along with the Earth and the curvature of Jupiter itself for size comparison.

Saturn's Major Moons

Saturn has more known natural satellites than any other planet. Most of Saturn's moons are ice-covered orbs; the smaller ones are irregular chunks, some of which are doubtless asteroids that were captured by Saturn long after the planet and its main moon system were formed. Only five of Saturn's moons exceed 1,000 km (620 mi) in diameter. These are *Titan*, *Rhea*, *Iapetus*, *Dione*, and *Tethys*.

TITAN

The largest and most interesting satellite of Saturn is Titan, measuring 5,150 km (3,200 miles) in diameter. It is almost as large as Ganymede, Jupiter's largest moon. Nevertheless, the gigantic, gaseous planet Saturn dwarfs it (Fig. 10-3). Titan is the only planetary moon that has a significant atmosphere. As viewed through the most powerful telescopes, and even from space probes flying by, Titan looks something like an orange little sister of Venus. The cloud layer is so thick that it hides the surface features from visual view.

The atmosphere of Titan is comprised mainly of nitrogen and methane and is cold by Earthly standards, far below 0°C at the surface. The atmospheric pressure at the surface is about half again as great as the normal atmospheric pressure at the surface of the Earth. Thus, although we would not be able to breathe Titan's "air," we would at least not have to worry about being crushed to death by its pressure, as would be the case on Venus.

The main reason scientists find Titan so interesting is that it contains an abundance of organic chemicals. The term *organic* does not mean that these chemicals were produced by or are necessarily indicative of living things in the environment. Methane and ethane, hydrocarbons similar to natural gas, are considered organic because they have the potential to give rise to *amino acids* under the right conditions. The impact of a large mete-

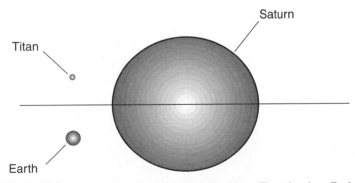

Figure 10-3.   The relative size of Saturn's largest moon Titan, the planet Earth, and the planet Saturn. The rings are shown edgewise, in proportion.

orite or comet or an electrical discharge caused by a thunderstorm creates the high temperatures necessary for the formation of amino acids, which are the building blocks of life.

Titan is a candidate for exploration by humans. The main problem to be overcome in such a visit, besides the enormous distance that separates the Saturnian system from Earth, is the powerful magnetic field surrounding Saturn, which accelerates subatomic particles from the Sun, producing intense belts of ionizing radiation. Although this radiation is less intense than that in the vicinity of Jupiter, it is still much greater than the intensity of the *Van Allen radiation belts* surrounding the Earth. Anyone who lands on Titan also would have to be prepared for the possibility of hitting a turbulent, liquid hydrocarbon surface, perhaps with floating icebergs of frozen methane and heavy methane rain or snow blowing down out of the red sky.

RHEA

Rhea is 1,530 km (950 mi) in diameter. It orbits in an almost perfect circle 530,000 km (330,000 mi) from its parent planet. This moon is only a little more dense than water, and this fact has led astronomers to conclude that it must be comprised mainly of ice and very little rocky material.

Rhea, like most moons, keeps the same face toward Saturn at all times. As a result, one side of the moon "leads the way" through space, whereas the opposite side "trails behind." There is a considerable difference in the appearance of the leading side of Rhea compared with the trailing side. The leading side is as densely packed with craters as any part of Earth's moon, even though the surface of Rhea is mostly water ice. The trailing side has far fewer craters. This would be expected because the leading side would be more exposed to bombardment by meteorites.

Most of the rock in Rhea is believed to be contained in a small core. The moon's small size and its relatively large distance from Saturn prevent heating from tidal effects, keeping the surface far below 0°C and hardening the ice so that it resembles granite and can maintain crater and mountain formations for a long time. Rhea has essentially no atmosphere.

IAPETUS

Iapetus orbits at a great distance from Saturn: 3.6 million km (2.2 million mi). Its diameter is about 1,450 km (900 mi). Like Rhea, Iapetus is only a little more dense than water, and analysis of light reflected from its surface indicates that this moon is made up mostly of water ice.

The leading side of Iapetus is much darker than the trailing side. This is the opposite of the situation with Rhea. The contrast is great; the leading side is nearly as white as snow, whereas the trailing side is nearly as dark as tar. Also in contrast with Rhea, most of the craters on Iapetus are on the trailing side. This has caused some befuddlement among astronomers. Did something stain the leading side of Iapetus and cover up the craters there? Did this "dye" come from inside Iapetus, or did it come from space? Or is it the result of some reaction of material on the surface with ultraviolet light or high-speed particles from the Sun?

Iapetus is the only major moon of Saturn that does not orbit almost exactly in the plane of Saturn's equator. Instead, Iapetus is inclined by 15 degrees. One theory concerning this inclination is that Iapetus did not form along with the Saturnian system but instead was once a huge wandering protocomet or planetoid that was captured by Saturn's gravitation. Another theory holds that Iapetus originally orbited in the plane of Saturn's equator but was knocked out of kilter by a large asteroid.

DIONE

Dione has a diameter of 1,120 km (690 mi) and orbits Saturn in an almost perfect circle at a distance of 377,000 km (234,000 mi). Dione's density is about 1.4 times that of water. This fact and the analysis of the light reflected from its surface indicate that Dione, like most of the other moons of Saturn, is made up largely of water ice. It is thought that the proportion of ice to rock is higher near the surface and lower near the core.

There are variations in the reflectivity of the surface of this moon, but the demarcation is not as great as is the case with Iapetus. The leading side is generally brighter than the darker side. The trailing side has wispy markings that suggest that volatile material, perhaps water vapor, has escaped from the interior and fallen back on the surface to freeze. Some areas of Dione are heavily cratered, whereas other regions contain virtually no craters.

Dione exhibits a property that is sometimes found in the satellite systems of large planets: *orbital resonance* with one of the other moons. In this case the other moon is *Enceladus*, one of the minor satellites of Saturn. While Dione takes 66 Earth hours to orbit once around Saturn, Enceladus takes 33 hours, exactly half that time. Orbital resonance effects are caused by mutual gravitation between celestial objects, such as moons, when they both orbit around a common, larger object, such as a planet. This resonance effect is believed to be responsible for tidal forces on Enceladus that cause it to generate heat from inside.

TETHYS

Tethys has a diameter of 1,060 km (660 mi) and orbits Saturn at a distance of 290,000 km (180,000 mi). Like Dione and Rhea, this satellite is believed to consist mainly of water ice with some rocky material mixed in.

Tethys is noted for its long surface canyon and for a crater that is gigantic compared with the size of the moon itself. The size of this crater and the presence of the fracture suggest that a large asteroid smashed into Tethys and nearly broke the moon in two. Gravity, however, pulled the object back together again. According to one theory, Tethys was liquid at one time, and if this was the case when the violent impact took place, it might have saved the moon from being pulverized.

Like Dione, Tethys is involved in an orbital resonance with one of Saturn's minor moons, *Mimas*. Mimas orbits the planet twice for every orbit of Tethys. There are also two tiny moons that orbit in exactly the same path around Saturn as Tethys but 60 degrees of arc (one-sixth of a circle) ahead and behind it. This is a common phenomenon for major satellites of planets and stars that have nearly circular orbits and is a result of gravitational interaction between the moon and its parent planet or between the planet and its parent star. The points 60 degrees ahead and behind an object in a nearly circular orbit are known as the *Lagrange points*.

Figure 10-4 shows the five major satellites of Saturn, along with the Earth and the curvature of Saturn itself for size comparison.

Uranus' Major Moons

Uranus has numerous moons. Four of them can be considered major satellites, in the sense that they have diameters greater than 1,000 km (620 mi). These four moons are *Titania*, *Oberon*, *Umbriel*, and *Ariel*. As we have already learned, Uranus is tilted on its axis so much that its equator lies almost perpendicular to the plane of its orbit around the Sun. The moons of Uranus orbit in or near the plane of the planet's equator, rather than the plane of the planet's orbit around the Sun. This is to say that the entire planet-moon system is tipped almost perfectly on its side.

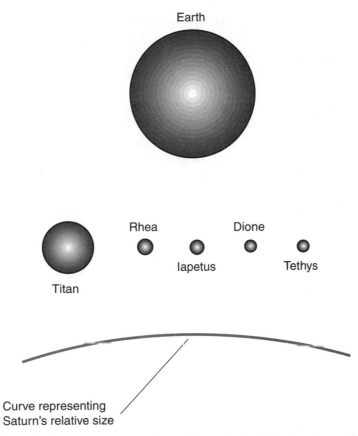

Earth

Rhea

Dione

Iapetus

Tethys

Titan

Curve representing
Saturn's relative size

Figure 10-4.   The five largest satellites of Saturn, along with Earth, as they would appear if placed in close proximity. They are all tiny compared with Saturn (gray curve).

The major moons of Uranus are believed to resemble "dirty snow-balls," a mixture of water ice and rock. The minor moons, with the exception of *Miranda*, are much smaller than the major ones and in some sense can be considered captive comets, containing a higher proportion of ice and less rocky material. None of the moons of Uranus has any appreciable atmosphere. Like most of the moons of major planets, they each keep the same side facing their parent at all times, and their orbits are nearly perfect circles.

TITANIA

Titania, the largest moon of Uranus, is only 1,580 km (980 mi) in diameter. It orbits its parent planet at a distance of 436,000 km (271,000 mi). Titania is much smaller than Uranus and between one-eighth and one-ninth the diameter of the Earth (Fig. 10-5). Observations of this moon and analysis of the light reflected from its surface indicate that it is made up of approximately half water ice and half rocky material.

In addition to the usual craters, the surface of Titania has long cracks or valleys. The reason for the existence of these fracture zones is unknown, but one popular theory holds that Titania was liquid at one time and then it froze from the outside in. As the water beneath the surface froze, the ice above cracked because water expands when it freezes. Another theory suggests that heat from the interior produces occasional eruptions of hot liquid or gas that penetrates the surface.

OBERON

Oberon is just a little bit smaller than Titania, with a diameter of about 1,520 km (950 mi). It orbits Uranus at a distance of 583,000 km (362,000 mi). This moon has a composition similar to that of Titania, but there is some indication that the surface features are older. Fracture zones exist, and their origins suggest that Oberon was geologically active for a while after it formed, but it appears as if Oberon has been a "dead world" for much of its existence.

One of the most interesting features of Oberon is dark material inside many of its craters. The surface consists largely of water ice. At Uranus' distance from the Sun, ice is as hard as granite unless heating occurs as a result of some other action such as tidal forces or internal activity. Neither of these factors seems to play a role on Oberon, and this makes the origin of the dark material somewhat mysterious. Some astronomers think that the dark material is volcanic lava, but there is little evidence to support this

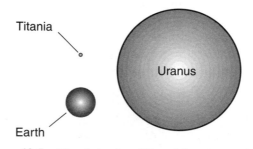

Figure 10-5.   The relative size of Uranus' largest moon Titania, the planet Earth, and the planet Uranus.

kind of activity on Oberon. Another theory holds that the floors of these craters are relatively smoother than the surrounding terrain and that this is why they appear darker. When a large meteorite strikes an icy body such as Oberon, the heat of impact melts the ice in and around the point of impact. The liquid water pools inside the crater and then refreezes, producing a smooth landscape that reflects relatively little light. You have seen this effect if you have ever looked at a smooth, well-kept outdoor skating rink surrounded by snow.

UMBRIEL

This satellite has a diameter barely large enough to satisfy our arbitrary minimum size limit to qualify it as a major moon: 1,170 km (730 mi). It orbits 266,000 km (165,000 mi) above its parent planet and takes a little more than 4 Earth days to revolve once around.

Umbriel is notable for its low albedo. The surface is almost entirely charcoal black. The only reason we can see it at all is that it isn't a perfect light absorber; it reflects approximately one-fifth of the light it receives from the Sun. The orb resembles a gigantic dirty ice ball. Most of the solid material on the surface is water ice mixed with rocky material, but there also appears to be some frozen methane and trace amounts of other frozen elements and compounds that are gases in the familiar environs of our planet Earth.

The entire surface of Umbriel is pitted with craters. One feature, a bright ringlike mark, is thought to be the outline of a crater in which the rim mountains have more exposed ice than either the interior or exterior lowlands. Umbriel shows no signs of geologic activity in the recent past (meaning within the last several million years). Some scientists believe that it has not undergone much change since it was formed as part of the Uranian system.

ARIEL

Ariel is, in terms of size, practically a twin of Umbriel. It measures 1,160 km (720 mi) in diameter. It is much closer to Uranus, orbiting at a mean altitude of 190,000 km (120,000 mi). It takes only $2^1/_2$ Earth days to orbit once around the planet. Like all the other moons of Uranus, Ariel orbits in a nearly perfect circle and keeps the same face toward Uranus all the time.

Ariel reflects about twice as much light as Umbriel, leading astronomers to surmise that its surface consists of relatively more icy material and less rock. The whole surface is cratered, but there are huge rift valleys and canyons too. There is evidence that a mixture of liquid ammonia and methane once flowed across the surface of this moon.

The long cracks in the surface of Ariel suggest that the moon has expanded or contracted since it was formed, resulting in fault lines. Some of the canyons have ridges inside them, as if liquids once flowed out from the interior and then froze solid when exposed to the chill of space in the outer Solar System.

Figure 10-6 is a size comparison of Titania, Oberon, Umbriel, and Ariel, along with the Earth and the curvature of Uranus.

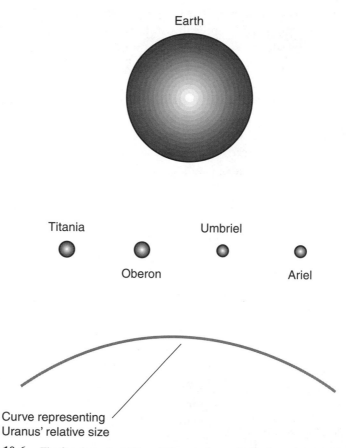

Figure 10-6. The four largest satellites of Uranus, along with Earth, as they would appear if placed in close proximity. The relative size of Uranus is represented by the gray curve.

Moons of Neptune and Pluto

The two outermost known planets in our Solar System, Neptune and Pluto, each have only one moon that is more than 1,000 km across. Neptune's lone major satellite is called *Triton*, and Pluto's is called *Charon*.

TRITON

Neptune's dominant moon has a diameter of 2,700 km (1,680 mi). It orbits at 385,000 km (240,000 mi) from Neptune, just about the same distance as Earth's moon is from Earth. Figure 10-7 compares Triton in terms of size with Earth and Neptune.

This little world has the distinction of possessing the chilliest surface of any known planet or moon, approximately −235°C (−390°F). Triton is also unique in another way, for it is the only known moon in our Solar System that orbits its parent planet in the opposite direction from that of the

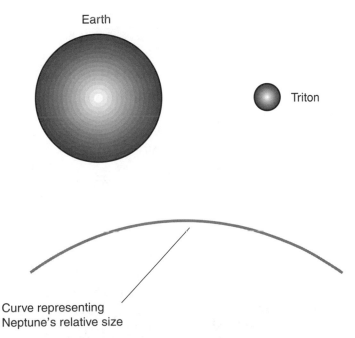

Figure 10-7.   Triton as it would appear if placed next to our own planet Earth.
The relative size of Neptune is shown by the gray curve.

planet's rotation. As if this does not make Triton peculiar enough, its orbit is greatly tilted with respect to the plane of Neptune's equator.

Measurements of Triton's density indicate that it is made up of relatively less ice and more rock than the other major moons of outer planets. This fact, along with the retrograde orbit, has given rise to the theory that Triton was not originally a moon at all but instead was a planet in its own right when the Solar System was formed. It ventured too close to Neptune, and the large planet captured it. Strangely, however, Triton's orbit is essentially a perfect circle, and this fact can be used to argue against the "once it was a planet" hypothesis. Triton keeps the same side facing Neptune all the time, but this generally takes place with planetary moons as the parent planets' gravitational fields create tidal bulging over millions of years, pulling the rotation and revolution rates into synchronicity.

Triton has an atmosphere, but it is so thin that it would make a good laboratory vacuum for most Earthly purposes. The surface is believed to contain frozen methane along with nitrogen ice because of the pinkish cast to much of its surface. Clouds form occasionally, and these apparently consist of tiny particles of frozen nitrogen. There is evidence of wind erosion on the surface.

Volcanic activity apparently occurs on Triton, but the eruptions are entirely different from the volcanoes we know on Earth or from those that dominate the surface of Jupiter's restless satellite, Io. Instead of hot lava from the interior, the ejected material is believed to be liquid nitrogen or methane that freezes as soon as it comes into contact with the bitter-cold surface environment.

CHARON

Charon, with a diameter of 1,190 km (740 mi), is the only known satellite of Pluto. Charon is small in absolute terms, but it is significant compared with Pluto. Figure 7-11 (in Chap. 7) compares Charon for size with Earth and Pluto. Charon orbits Pluto at an average distance of approximately 20,000 km (12,500 mi). Charon is unique not only in that it is the largest moon in size relative to its parent planet, but it is also extremely low in its orbit. In fact, the two bodies tidally affect each other to the extent that they always keep the same sides facing each other.

Charon is believed to be composed of essentially the same stuff as Pluto, a combination of rocks and ices. Charon's surface, however, is mainly frozen water, unlike that of Pluto, which is largely frozen nitrogen with traces of methane ice. While Pluto has a pinkish or reddish tinge when observed in visible light, Charon appears gray.

Some astronomers think that Charon and Pluto are surviving members of an originally much larger group of icy, comet-like bodies in orbits outside that of Neptune. According to one theory, most of these objects congealed to form Pluto and Charon. During this process, there were countless collisions. Some objects got hurled in toward the Sun and became comets in the classic sense as they got close enough to the Sun to develop tails. The collisions also gave the Pluto-Charon system its eccentric orbit around the Sun. While major collisions of these outer denizens of the Solar System are a thing of the past, smaller collisions and gravitational interactions still take place, and every few years a new comet happens across the watchful eye of some comet-seeking astronomer.

The Pluto-Charon system seems distant and insignificant as you read about it in a book, but Pluto and Charon are relatives of objects that have played crucial roles in the evolution of life on our planet. Some scientists think that a comet brought the first primitive life forms to Earth or produced the energy necessary for amino acids to form. A comet is believed to have struck the Earth in the present-day Gulf of Mexico about 65 million years ago, leading to the extinction of dinosaurs and an upsetting of the terrestrial equilibrium that, had it not been perturbed, would still give sanctuary to the giant ruling lizards today.

 Quiz

Refer to the text if necessary. A good score is 8 correct. Answers are in the back of the book.

1. Most planetary moons rotate on their axes
   (a) once for every orbit they complete around their parent planets.
   (b) in a retrograde manner.
   (c) perpendicular to the axes of their parent planets.
   (d) in synchronicity with the parent planet's rotation on its axis.

2. The greatest danger that will face astronauts who plan to visit any of the inner moons of Jupiter or Saturn is
   (a) the parent planet's radiation belts.
   (b) the moon's gravitation.
   (c) the moon's lack of an atmosphere.
   (d) the lack of sufficient visible light.

3. Tethys and Mimas are notable because they
   (a) orbit each other.
   (b) are in orbital resonance with each other.
   (c) have retrograde orbits.
   (d) both have exceptionally high albedo.

4. Continuous, moonwide volcanic activity is observed on
   (a) Ganymede.
   (b) Io.
   (c) Charon.
   (d) no planetary satellite.

5. Umbriel is a moon of
   (a) Jupiter.
   (b) Saturn.
   (c) Uranus.
   (d) Neptune.

6. Which of the following is a Galilean moon of Saturn?
   (a) Ganymede
   (b) Titan
   (c) Triton
   (d) There are no Galilean moons of Saturn.

7. The Lagrange points in a satellite's orbit are the result of
   (a) radiation from the Sun.
   (b) volcanic activity on the parent planet.
   (c) thermal heating from inside the satellite.
   (d) gravitational interaction.

8. Which of the following moons is noted for its difference in albedo between the leading side and the trailing side?
   (a) Charon
   (b) Titan
   (c) Iapetus
   (d) Callisto

9. The planetary moon with the most extensive atmosphere is
   (a) Titan.
   (b) Charon.
   (c) Deimos.
   (d) Rhea.

10. On the surfaces of moons of the outer planets, water ice has a hardness and consistency similar to that of
    (a) molasses at room temperature.
    (b) putty at room temperature.
    (c) rock at room temperature.
    (d) steel at room temperature.

CHAPTER 11

Comets, Asteroids, and Meteors

Eight known substantial bodies orbit the Sun: Mercury, Venus, Earth, Mars, Jupiter, Saturn, Uranus, and Neptune. Since its discovery, a ninth object, the Pluto-Charon system, also has been considered as a planet. If there are other planets in the Solar System besides these, they have thus far evaded the observation of watchful amateur and professional astronomers.

The planets, along with the Sun and our own Moon, capture most of the visual attention of hobby astronomers. Saturn with its rings, Mars with its ruddy glow, and Jupiter with its visible moons and surface features are perennial favorites. However, insofar as the fate of humanity is concerned, these planets have no direct effect. They stay in their orbits, and we stay in ours. Jupiter, even with its powerful magnetosphere and ferocious radiation belts, does not pose a threat to us here on Earth. Mars, named after the ancient war god, has not sent any living beings to maliciously attack civilizations on Earth (or on any other planet), although there is evidence that rocks have landed here that had their origins on Mars.

The Small Stuff

There are celestial objects that present a threat to life on Earth or, perhaps better stated, have had and will continue to have a profound effect on the way life evolves here. They are pipsqueaks in the Solar System, none of which can be recognized without a telescope and many of which remain unknown to this day. They are *comets*, *asteroids*, and *meteoroids*, remnants of the primordial Solar System. They are leftovers that did not congeal into the planets. Originally, according to the most popular theories, the entire Solar System consisted of objects like them.

CLOUDS OF COMETS

Comets are one of the greatest mysteries in astronomy. They have awed and terrified people since the beginning of recorded history. Even today, some people think comets have supernatural characteristics. Yet comets are, as one astronomer has said, more numerous than fishes in the sea. They are, in the long term, more dangerous too.

A large comet smashing into the Earth would be, for the human race, the equivalent of a shark attack on an individual human being. However, the risk shouldn't be overblown. It is not necessary to lose sleep worrying about what is going to happen when the next major comet comes down. We can say with confidence that such an event will take place, but we can't say when, other than to note the fact that the time intervals between massive impacts are measured, on average, in tens of millions of years.

Beyond the orbit of Neptune, the *Kuiper belt* consists of comets that orbit the Sun at distances from 30 to 50 astronomical units (AU) (Fig. 11-1). The existence of this zone, which circles the Sun in more or less the same plane as the planets, was only a hypothesis until 1992 when the first object there, a large comet, was positively identified. More objects were found after that. Objects found between 1992 and 1994 were all gigantic compared with the comets that have made history, such as Halley's Comet, which last appeared in 1985–1986. Since 1995, some "normal sized" comets have been found in the Kuiper belt.

Surrounding the Kuiper belt and extending out to a distance of about 1.5 light-years (one-third of the way to the nearest star other than the Sun) is a vast spherical halo of comets called the *Oort cloud*. The exact dimensions of this cloud are unknown because of its great size and distance. Nevertheless, it is believed to contain millions upon millions of comets.

Normally, the comets in the Kuiper belt and the Oort cloud stay in their parts of the Solar System, the "suburbs" and the "surrounding countryside." All of them would remain there too if it were not for the fact that our Milky Way galaxy is an unstable and chaotic place. Once in a while, a star comes close enough to the Solar System so that its gravitation has an effect on

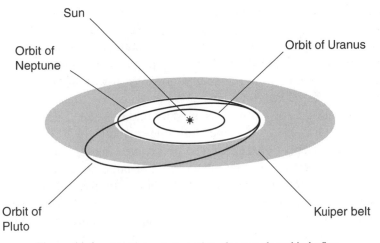

Figure 11-1.   The Kuiper belt consists of comets that orbit the Sun
beyond the orbit of Neptune, out to approximately 50 AU.

objects in the Oort cloud, and some of these comets are deflected into paths
that take them into the inner Solar System. Some astronomers suspect that
there is an as-yet unknown massive planet beyond Pluto that does similar
things to comets in the Kuiper belt.

THE ASTEROID BELT

In 1772, a German mathematician named Johann D. Titius noticed that the
orbits of the planets seemed to have radii, or average distances from the
Sun, that fell into a neat mathematical progression. At that time, Saturn was
the outermost known planet. Titius noticed that the radii of the planetary
orbits, in astronomical units (AU), could be found by applying the progression. Later, the astronomer Johann E. Bode expounded on this number
sequence, which has become known as *Bode's law*.

The progression is defined as follows. Start with the number 3, and double each number, getting the sequence 0, 3, 6, 12, 24, 48, 96, and so on.
Then add 4 to each of these numbers so that you get 4, 7, 10, 16, 28, 52,
100, and so on. Finally, divide each of these numbers by 10. This gives the
following sequence:

$$S = 0.4, 0.7, 1.0, 1.6, 2.8, 5.2, 10.0, ...$$

These numbers, with the conspicuous exception of 2.8, correspond almost
exactly to the orbital radii of the planets Mercury (0.4 AU), Venus (0.7 AU),

Earth (1.0 AU), Mars (1.6 AU), Jupiter (5.2 AU), and Saturn (9.5 AU). The question naturally arose among scientists: Is this significant? Did the planets form at these distances from the Sun for some physical reason? Today, most astronomers doubt that Bode's law is anything more than a coincidence. Back in the eighteenth century, however, it was suspected that there was some *modus operandi* to this correspondence and that an undiscovered planet must lie 2.8 AU from the Sun. Why had it escaped notice?

A search was begun for the "missing planet." In 1801, the Italian astronomer Giuseppe Piazzi found an object orbiting the Sun at the correct distance. However, it was small, certainly not large enough to be a planet. It appeared starlike, a mere point of light, but its motion relative to the distant stars gave it away as a resident of the Solar System. It was called an *asteroid* (meaning "starlike") and was given the name *Ceres*. More asteroids were soon found, also orbiting the Sun at distances of approximately 2.8 AU. Ultimately, thousands were discovered, and they all orbit in or near the zone corresponding to the missing slot in the Titius-Bode sequence. This zone has become known as the *asteroid belt*.

Most astronomers in Piazzi's time concluded that the asteroids, also called *planetoids*, were objects that had failed to congeal into a planet or else were the remnants of a planet that was shattered by a cosmic catastrophe. Today, the prevailing theory holds that these objects are part of the original cloud of boulders, rocks, and dust that surrounded the Sun in the earliest part of the Solar System's lifespan and that the gravitation of Jupiter prevented their getting well enough organized to condense into a planet.

Most of the asteroids follow nearly circular paths around the Sun and orbit between Mars and Jupiter. However, some known asteroids follow orbits that take them inside the orbit of Mars, and others wander outside the orbit of Jupiter. A few maverick asteroids cross the orbit of the Earth and, every few million years, pass even closer to our planet than the Moon. There is plenty of evidence, in the form of visible craters, that asteroids have crashed into the Moon. Such craters are erased in time by erosion when they are formed on Earth, but a few craters have been found on our planet that strongly suggest that we, along with every other object in solar orbit, are on the "asteroid visitation list."

MYRIAD METEOROIDS

The largest known asteroid, Ceres, measures hundreds of kilometers in diameter. The smaller we go, the more asteroids there are, in general. There isn't any definitive limit at which we say, "This thing is an asteroid, but if it gets 1 milligram less massive, then we'll call it a meteoroid." In a general way, we can say that if it's too big to be called a boulder, then it's an asteroid; boulders and rocks in space can be called *meteoroids*.

Even then, it's not that simple. Meteoroids and comets are substantially different. Comets consist of rocky and icy material combined, but meteoroids have essentially no ice. Some meteoroids are stony, something like the granite we know on Earth. Others are mixtures of stony and metallic stuff. Still other meteoroids are mostly metal, largely iron. Some resemble pieces of amber or glass.

How do we know what meteoroids are made of? They're too small to be observed through telescopes; space probes have never been sent specifically to visit them, although some of them have struck our space vehicles. We know about meteoroids because they often fall to Earth's surface. Technically, when a meteoroid enters Earth's atmosphere, it becomes known as a *meteor*. If a meteor survives to reach the surface, either as one object or as fragments of the original meteor, then it becomes a *meteorite*. Meteorites are abundant on the continent of Antarctica, where the ice preserves them, and where they are easy to differentiate from Earth's surface.

There are believed to be more rock-sized meteoroids than boulder-sized ones, more pebble-sized ones than rock-sized ones, and more sand-grain-sized ones than pebble-sized ones. Then, as we keep getting smaller, we have to call them *micrometeoroids*. In the extreme, they can be called *interplanetary dust* or *meteoric dust*.

The asteroid belt between Mars and Jupiter is home to a huge number of "space rocks," but there are plenty of such objects that orbit outside this zone. Some orbit the Earth instead of the Sun. It is believed that there are myriad rock-sized objects in orbit around each of the planets, as well as around the major moons of the planets, including our own Moon. There is also evidence that some rock-sized objects are entirely independent of the Solar System and that they come into our little corner of the Universe as a result of the movement of the Sun within the Milky Way galaxy. Technically, a space rock becomes a meteoroid when it attains a solar orbit such that it has the potential to be pulled into the Earth by our planet's gravitation.

Anatomy of a Comet

Before the first physical visit in 1985–1986 when celestial robots swarmed around Halley's Comet, scientists were not quite certain what comets are made of. People could only observe them from Earth through telescopes and various nonvisual instruments. The observable parts of a comet became known as the *nucleus* or *core*, the *coma*, and the *tail* (Fig. 11-2).

We now know that comets are solid objects composed of ice, rock, and dust. When they get close enough to the Sun, some of the ice evaporates into space, and some of the dust is blown away by high-speed subatomic particles from the Sun (the *solar wind*). The evaporated ice and blown-off dust stream outward from the core of the comet, away from the Sun.

EARLY THEORIES

Before humanity got a close look at a comet, there were two theories concerning their anatomy, called the *Whipple model* and the *Lyttleton model*. According to astronomer Fred Whipple, comets were believed to consist of a "dirty snowball" core, and the coma and tail were thought to arise as a result of solar heating and the solar wind. Today this model is widely accepted because scientists have actually seen and analyzed some comets close up. They have found objects that look like huge irregular boulders with jets of bright gas spewing out.

According to Raymond Lyttleton, the cores of comets were thought to be swarms of small meteoroids held together by their mutual gravitation. When such swarms ventured close enough to the Sun, some of the material was blown off and away by the solar wind, and this accounted for the

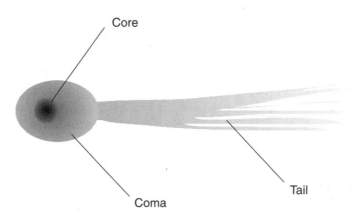

Figure 11-2.   A crude representation of the visible parts of a comet.
(The brightness in this illustration is inverted.)

coma and tail. Few scientists today accept this as the normal state of comet anatomy.

Before modern astronomy, comets aroused superstition, aggression, and fear. Because they were seen rarely, and because of their ghostly appearance, comets were seen as signs from heaven. Some comets have curved tails that look like sabers or scimitars. This led a few emperors and kings to conclude that the appearance (called an *apparition*) of a comet meant war was at hand. In the year 1066, Halley's Comet became visible just before the Norman French invaded England, an event whose effects have shaped the course of history throughout the world. A famous work of art known as the *Bayeaux Tapestry* shows people pointing up in fear at the comet while soldiers approach.

THE NUCLEUS

The actual size of a comet nucleus can vary from a few meters across to hundreds or thousands of kilometers in diameter. It's difficult to determine the shape and diameter of any particular comet's nucleus using a telescope from Earth because by the time the comet gets near enough to us to be visible, the glowing coma obscures the core.

One way to see the comet's nucleus clearly is to view it when, and if, it passes directly in front of the Sun. Obviously, this opportunity does not present itself very often, but it took place in 1890, and the only thing astronomers could conclude about that comet is that it was either solid but too small to present a substantial image or else it was something so tenuous that it didn't cast a visible silhouette at all.

Astronomers have now seen some comets at close range, and we know that the cores of these objects are solid. Whipple has been vindicated. Another point of argument in favor of Whipple's solid-core model is the fact that the major *periodic comets* (those that return at regular intervals) that come very close to the Sun survive their passage around our parent star. A nucleus comprised of tiny objects would be completely vaporized by the Sun's heat and by the solar wind as it neared and then passed its perihelion.

THE COMA

When a comet is far away from the Sun, assuming that it is of the Whipple type, it resembles a gigantic dirty snowball. By the time it is seen by the first astronomers looking for a comet, it has begun to glow because solar energy causes some of the icy material to vaporize and form a sort of "atmosphere" around the nucleus. This material glows because its atoms get excited and give off radiation at well-known wavelengths in the spectrum of visible light.

A typical periodic comet's orbit is a greatly elongated ellipse. Nonperiodic comets, or those with periods so long that they have never been measured, have orbits so eccentric that they nearly resemble parabolas. In addition, some comets orbit the Sun outside the general plane in which the planets orbit (Fig. 11-3). Such an object spends most of its time in the outer reaches of the Solar System. As it comes in from the distant, dark, cold depths of space, it picks up speed. The coma begins to appear as the comet nears the orbit of Jupiter, although the exact point is impossible to define and would vary from comet to comet even if we could define it.

By the time the comet gets closer than 1 AU to the Sun—that is, its distance becomes smaller than the radius of Earth's orbit—the coma has become several times as large, in diameter, as the core. Some of the gas and dust from the core is blown off with such force that it streams away from the core like smoke from a fire in the wind.

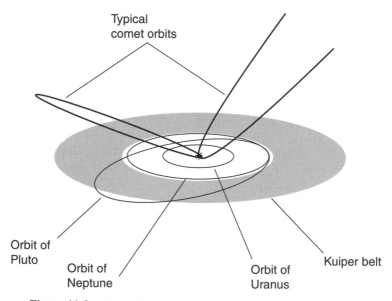

Figure 11-3. The orbits of most comets are elongated ellipses, and in the extreme, can be parabolic in shape.

THE TAILS

The Sun radiates energy in many forms. The visible light we see and the heat we feel arise from *photons*, or packets of electromagnetic energy, that travel at the speed of light (about 299,792 km/s or 186,282 mi/s). Photons have only a tiny bit of momentum, but because of their extreme speed, they have enough physical impact to affect matter in the vicinity of the Sun. Our parent star also emits dense particles, such as *protons* and *alpha particles* (helium nuclei), at slower speeds. These more than make up for their sluggishness by their momentum, many times greater than that of photons. The tail of a comet is produced by the physical motion of these particles through space.

Actually, comets have two tails, known as the *gas tail* and the *dust tail*. Both point generally away from the Sun so that as the comet approaches the Sun, its tails follow behind, and as the comet leaves the vicinity of the Sun, its tails precede the nucleus. The gas tail is thin and almost perfectly straight because the gases are ejected from the comet at high speed. The dust tail is more spread out and often appears curved because its particles travel more slowly, allowing the nucleus to "get ahead of it" (Fig. 11-4).

Many people think that comets travel head-first and that the tails stream behind. This is natural to conclude intuitively, but it is only the case as the comet first appears on its way in toward the Sun. By the time it comes within 1 AU of the Sun, a typical comet's tails are already flowing out at a significant angle relative to the direction of its motion. As the comet passes perihelion, the tails blow away sideways to the comet's forward motion.

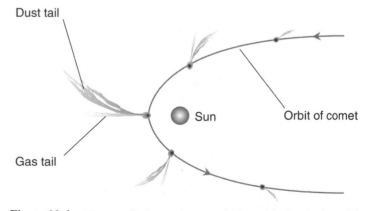

Figure 11-4.   The gas tail of a comet appears fairly straight, but the dust tail is curved, especially as the comet passes perihelion.

Then the angle increases still further until, by the time the last of the tails fade from view, they are pointing almost straight out ahead of the core.

Comet Personalities

No two comets look the same. Any given comet appears different to Earthbound observers every time it returns to the Sun. Halley's Comet, one of the most well-known of all comets in history, was brilliant when it appeared in 1910, but it was a disappointment to observers who saw and tracked it in 1985–1986.

OBVIOUS VARIABLES

Comets vary in size and shape. They are also believed to vary in composition; some are thought to have relatively more ice, whereas others have less. The distribution of the matter in different comets also should be expected to vary. Comets have countless different orbits around the Sun. The relative positions of Earth, the Sun, and the comet at any given time affect the way a comet looks to people watching it from our planet.

The most spectacular comets are those with long tails that stand out against the evening or morning sky. Some comets are bright enough to be seen in broad daylight in a cloudless sky. A classic long-tailed comet, *Ikeya-Seki*, appeared in 1965. This comet passed close to the Sun, and it did so when the viewing angle from Earth was favorable. The dust tail extended millions of miles away from the nucleus when the comet was near perihelion. The high speed of the comet near perihelion produced a dramatic curvature in the dust tail.

Some comets pass so close to the Sun that they enter our parent star's outer corona. This increases the size of both the gas and the dust tails. Such comets are called *Sungrazers*. In the extreme, a few comets pass so close to the Sun that they are broken apart or vaporized by the intense heat and subatomic-particle bombardment. And once in a while, a comet falls into the Sun.

CELESTIAL BRIGHTNESS

Comets vary in luminosity, or brightness, just as much as do the stars. To define how bright a celestial object appears, astronomers use a numerical scale called *visual magnitude*. The lowest numbers represent the brightest things, and the highest numbers represent the dimmest things. Stars such as Spica and Pollux are of the *first magnitude*. Stars noticeably less bright than this are of the *second magnitude*. A change of brightness of 1 unit on the magnitude scale represents an increment of 2.5 times, or 250 percent. Thus a second-magnitude star is 2.5 times as bright as a star of the *third magnitude* and 2.5 × 2.5, or 6.25, times as bright as a star of the *fourth magnitude*.

It turns out that according to the modern definitions of magnitude, certain celestial objects have numbers less than 1 or even less than 0. Fractional numbers are the rule, not the exception, and they are usually expressed in decimal form. The "dog star," Sirius, has magnitude −1.43. Venus, the Moon, and the Sun have even lower magnitudes.

The human eye can detect the presence of objects down to magnitude 5 or 6 without the aid of binoculars or a telescope, provided viewing conditions are optimal. A good hobby telescope can see down to magnitude 12 or 13. Using time-exposure photography, astronomers can look into the heavens and see objects far dimmer even than this. Ultimately, a limit is imposed by *sky glow*, scattered light in the atmosphere caused by human-made lighting and the general illumination from the other stars, the Moon (if it is above the horizon), and the planets. In space, the background of stars or the presence of nebulae or galaxies affects how far down on the magnitude scale astronomers can resolve things.

HOW BRIGHT ARE COMETS?

Comets can range in brightness from far below the limit of visibility to a brilliance rivaling that of Venus or, occasionally, even the Moon. For a comet to be visible to the unaided eye as something out of the ordinary, it must be brighter than the third magnitude. Although comets can be seen when they are as dim as the fifth visual magnitude, only the glowing coma appears, and it looks like a faint point of light until its luminosity becomes great enough that its diffuse nature can be appreciated.

Some of the most famous comets have attained visual magnitudes low enough (that is, they have become bright enough) to strike genuine fear in some observers. Once every few hundred years a comet makes an apparition that takes it close enough to Earth and in the proper position relative to Earth and the Sun that it becomes visible in daylight and can be seen to cast a shadow on moonless nights. When this happens, the tails appear in great detail. An example of such an apparition was the so-called Great Comet of 1744, which seemed to have six separate tails as it passed between the Earth and the Sun.

When you hear or read about the discovery of a new comet or about the apparition of a known one, you should be equipped with a good pair of binoculars because it will have a relatively high visual magnitude (low brightness). The best binoculars are the wide-angle types with large-diameter objective lenses. The sort used by military commanders are ideal. This will let you scan the sky with ease, and the stereoscopic viewing will help to enhance the contrast between the comet and other celestial objects. Once you have found a comet using binoculars, then a telescope, using an eyepiece having a long focal length for low magnification, can be employed to look at it in more detail. *Schmidt-Cassegrain telescopes*, popular among amateur astronomers and casual observers, are available in hobby stores and are good for comet watching.

FLARE-UPS AND BURNOUTS

Comets are always dim, fuzzy objects when they are first detected plunging inward past the outer planets. Once in a while we get a surprise. Rapid changes in brightness are not unknown. Astronomers routinely graph the visual magnitudes of comets versus time as the objects make their way around the Sun. On average, comets get brighter as they draw near the Sun and dimmer as they move away. An example of such a graph, showing the luminosities of three hypothetical comets as functions of their distances from the Sun, is shown in Fig. 11-5. Comets 1 and 2 are more or less typical; comet 3 exhibits a *flare-up* followed by a *burnout*.

Notice that there are gaps in the graphs for comets 1 and 2. These gaps are there because the comets do not actually fall into the Sun. Their minimum distances, in astronomical units, can be approximated by looking at the points on the horizontal scale where the graphs disappear and then reappear again on the opposite side. These points are about the same distance on either side in this example, but this is not always the case. Variations in the relative positions and motions of Earth, the Sun, and the comet can result in the disappearance and reappearance points being at different distances from the Sun.

In this hypothetical example, comet 3 does not appear again after its burnout. Apparently it either disintegrated completely or else it fell into the Sun! The only way to be sure which of these two ill fates represents the truth is to calculate the actual orbital path of the comet around the Sun.

Why might a comet flare up or burn out? If the distribution of matter in the nucleus is not uniform, then as the core of the comet "melts" from

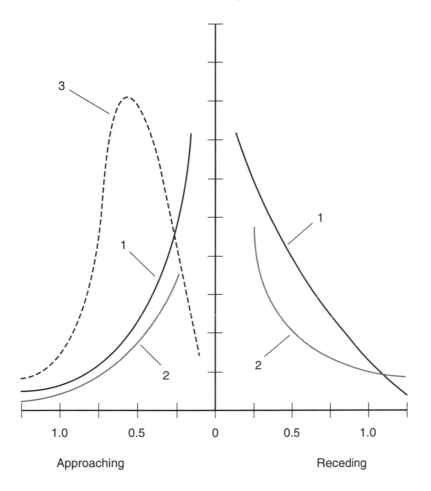

Relative visual magnitude

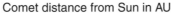

Comet distance from Sun in AU

Figure 11-5.   Brightness-versus-distance graphs for three hypothetical comets. Comets 1 and 2 are typical. Comet 3 shows a sudden flare-up followed by a burnout.

the heat of the Sun, it might break apart when a portion consisting mainly of ice evaporates. This would temporarily expose a large surface of ice to the Sun, accounting for the flare-up; this same exposure would hasten the destruction of the nucleus by exposing more of its surface to bombardment from solar heat and particle radiation. A solar flare also might cause a comet to flare up because such flares are always attended by increases in the emission of subatomic particles from the Sun. Another possible cause for a comet flare-up and burnout is the collision of the nucleus with a large meteoroid or small asteroid, which would break the nucleus apart.

Comets actually have been observed breaking up. One such unfortunate, known as *Beila's Comet*, split apart during one apparition. When it was due to come back and make its next showing, observers saw no comet, but there was a shower of meteors that did not normally take place at that time of year. The conclusion: All the ice from Beila's Comet had been driven away after the breakup, leaving only a swarm of rocks, some of which fell into the Earth's atmosphere and burned up. In a sense, therefore, a Whipple-type comet had become a Lyttleton-type "comet," although the latter was not a true comet because it did not have a defined coma or tail.

COMET LIFESPANS

Comets, like all things in the Universe, cannot last forever. It is reasonable to suppose that as a periodic comet passes close to the Sun again and again, its icy material is depleted, until finally there is nothing left but rocky stuff, and the object loses its ability to develop a coma and tail. Then it is no longer a comet but merely a meteoroid, or if it breaks apart because of the loss of its icy binder, it becomes a swarm of meteoroids.

The longevity of any given comet depends on several factors. Massive comets would be expected to live longer than small ones. Comets that originally contain more ice ought to survive longer than comets that start out with hardly any ice in them. A Sungrazer of a particular size and composition will not last as long as a comet of the same size and composition whose perihelion is farther away from the Sun. Short-period comets that pass close to the Sun often will deplete faster than comets of the same size and composition that have perihelions at the same distance but have long periods and thus pass near the Sun less often.

Halley's Comet, with an orbital period of about 76 years and a perihelion of approximately 88,000,000 km (55,000,000 mi), has been observed periodically for more than 2,000 years. No one knows exactly when prehistoric humans looked up into the sky and saw this comet and, for the first time, wondered what this strange, fuzzy star with a tail actually was. We should think that the comet was, in general, larger and brighter in an absolute sense thousands of years ago than it was on its most recent apparitions in 1910 and 1985–1986 and that gradually, as the eons pass, it will fade into insignificance.

Comet nuclei that orbit far from the Sun, never approaching to less than a few astronomical units, will exist until the Sun swells into its red-giant phase several billion (1 billion $= 1,000,000,000 = 10^9$) years from now. Then thousands or millions of them will all flare up at once. However, humanity won't be around to watch the spectacle. Our species will either have moved to another star system with an Earthlike planet or else passed from the Cosmic scene forever.

Stones from Space

An intense *meteor shower* is an unforgettable spectacle. So is a single large meteor as it lights up the sky, creates a sonic boom, and leaves a lingering trail. The Native Americans called these objects *shooting stars*. Other people called them *falling stars*. This is not surprising; in the absence of information to the contrary, that is what they look like. In the more recent past, some European immigrants to North America in the 1700s theorized that meteorites formed in Earth's atmosphere as a by-product of the lightning in thunderstorms. We know today that meteors and meteorites do indeed come from space.

METEOROID ORIGINS

When the Solar System was much younger and the planets had formed but had not yet reached their present stable states, there was more debris in interplanetary space than is the case now. The most popular theories of the origin of the Solar System involve the accretion of the Sun and planets from a rotating disk of gas and dust. Planets formed at certain distances from the central Sun. Some of the planets developed satellite systems of their own, like Solar Systems in miniature; Jupiter and Saturn are the most notable of these.

Most of the material in the primordial gas/dust disk made its way into the Sun and the planets. Some of the stuff, however, was left over. Some simply remained as gas and dust; some congealed into pebbles, rocks, boulders, and asteroids. Many of these smashed into the young planets and moons, forming craters. However, there are still plenty of space rocks out there. They, like their predecessors, are engaged in an incredibly complicated gravitational dance among themselves and the planets. Jupiter is the "conductor" of a vast "orchestra" of such rocks. Unlike a human orchestra conductor, though, Jupiter does not always maintain rhythm and harmony among its subjects. Often a rock is thrust into an orbit that puts it in a path where it has the potential to strike the Earth. Then it becomes a meteoroid. When we find one of these objects on the surface of our planet, we know the rest of the story!

CRATERS

Craters provide dramatic evidence of past meteorite bombardment on planets having little or no atmosphere. The Moon is the most familiar example. Mars is another. Mercury is still another. Most of the moons of the outer planets have craters.

We don't see many craters on our planet because blowing dust and sand and falling rain have eroded them entirely or at least beyond recognition. No craters have ever been seen on Jupiter, Saturn, Uranus, or Neptune; these gas giants are not believed to have solid surfaces on which craters can form. This is not to say that these planets haven't escaped bombardment by space debris but only that rocks from space leave no signatures on them.

Meteorite craters tend to be much larger than the objects that make them. This is because of the tremendous force that accompanies the crash landing of an object at several kilometers per second. Large meteorites form craters with central hills or small mountains. The rims of large craters form circular mountain ranges that can rise well above the surrounding terrain and several kilometers above the crater floor.

Some meteorite craters have *rays*, which appear as streaks radiating outward from the point of impact, and which can extend from the rim of the crater out to several times the diameter of the crater. Rays are produced by debris hurled up into the sky by the force of the impact and are especially prominent in craters produced by meteorites that struck the surface at a sharp angle.

Figure 11-6*A* is a cross-sectional diagram of a typical large meteorite crater of the type commonly found on the Moon or Mercury. Figure 11-6*B* is a top-view diagram of a similar crater.

OTHER EFFECTS

Some meteorites cause fractures in the surface of the object on which they land. These fractures can take the form of concentric rings around the

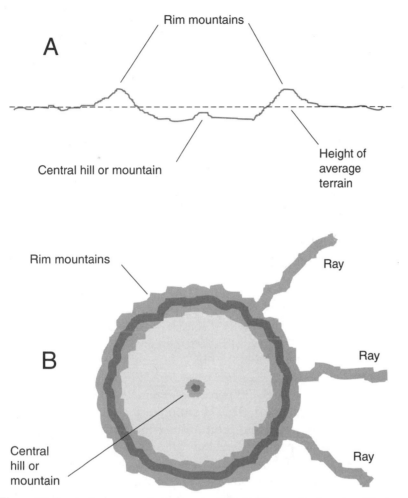

Figure 11-6.   At *A*, a cross-sectional diagram of a typical large meteorite crater of the type commonly found on the Moon or Mercury. At *B*, a top-view diagram of a similar crater.

crater. Jupiter's moon Callisto has a crater with these features. In other cases, there is no particular pattern to the fracture lines; the surface apparently cracks along the weakest points in the crust. Sometimes the fractures cover the whole surface of the object. They can occur as canyons or as *escarpments* (cliffs).

The impact of an especially massive meteorite or a small asteroid can produce volcanic activity all over the object it strikes, provided that there is hot lava inside the object that can rise up through the crust. The *maria*, or "seas," of the Moon are believed to have formed when lava from the interior spread over the surface following one or more violent meteorite landings. The same thing is thought to have taken place on the Earth from time to time, but the features have been modified by weather erosion and by the constant shifting of the *tectonic plates* in the Earth's crust.

In the extreme, a catastrophic impact can shatter a moon or planet. Nothing of this sort is thought to have taken place on any of the known planets or moons in the recent past, but in the early evolution of the Solar System, such events were commonplace. Going all the way back to the formation of the Sun and planets from the rotating disk of gas and dust 4.6 billion years ago, major collisions were the rule and not the exception. The ultimate long-term effect of meteorite landings is a sweeping up of space debris into major objects. Thus, as time passes, catastrophes become less and less frequent.

METEOR SHOWERS

If you look at the sky for a long enough time on any given night, eventually you will see the bright glow and trail of at least one meteor. They usually look like silvery or gold-colored streaks of light lasting from a fraction of a second to perhaps 2 or 3 seconds. An especially brilliant meteor leaves a trail that is visible for several seconds after the object itself has disintegrated or reached the surface.

In theory, approximately 33 percent more meteors should fall near the equator than near the poles. This is simply a matter of the geometry of the spherical Earth versus its orbit around the Sun. However, this rule changes slightly depending on the time of year. At and near the September equinox, the north pole receives slightly more meteors than the south pole. At and near the March equinox, the opposite is true. The poles receive roughly equal numbers of meteors at and near the solstices.

There are certain brief events, known as *meteor showers*, that take place at the same time every year. These showers seem to originate from particular places in the sky. Generally, meteor showers are named according to the constellation (or, in some cases, a star within the constellation) from which the meteors appear to be coming. Part or all of the name of the particular constellation or star is followed by the suffix *-ids*. For example, near the end of October there is a shower called the *Orionids*; these meteors seem to come out of the constellation Orion. Table 11-1 lists some well-known meteor showers that take place each year.

Why do meteors seem to come from a particular spot in the sky during a shower? Why don't they just fall at random? The reason is that we see

Table 11-1. Well-Known Meteor Showers, the Times of Year during Which They Occur, and the Comets with Which They Are Associated

NAME OF SHOWER	TIME OF YEAR	ASSOCIATED COMET
Andromedids	Mid-November	Biela
Delta Aquarids	End of July	Unknown
Draconids	Mid-October	Giacobini-Zinner
Eta Aquarids	Early May	Halley
Geminids	Mid-December	Unknown
Leonids	Mid-November	Temple
Lyrids	Mid-April	1861 I
Orionids	Mid-October	Halley
Perseids	Mid-August	1862 III
Taurids	Early November	Encke
Ursids	Mid-December	Tuttle

them from a certain perspective and the fact that during a shower the meteors tend to fall toward Earth in more or less parallel paths.

Have you ever lain down flat on your back during a rain shower when there was no wind? Bundle up and try it sometime. The raindrops fall down in nearly parallel paths, but they seem to emanate from a point directly overhead as you watch them come toward you. The same thing happens with meteors during a shower. The point from which the raindrops or meteors seem to come is called the *radiant* (Fig. 11-7). The names of meteor showers derive from the positions of the radiants in the sky, which tend to be the same, year after year, for any given meteor shower.

Meteor showers are almost always associated with comets. As its ice evaporates with each apparition, a comet gradually deteriorates, leaving swarms of meteoroids (the rocky stuff of which comets are partly made) that gradually spread out along the comet's orbital path. As the Earth passes through or near the comet's orbit, these meteoroids fall as meteors, and we see a shower.

The most spectacular meteor showers take place when the Earth's orbit precisely intersects a concentrated swarm of meteoroids. This does not always happen for any particular meteor shower. The orbit of the Earth and the orbit of the comet and associated meteoroids might not exactly cross each other, and even if they do, there might not be very many meteoroids in the comet's orbit at that particular point. This is why, for example, the Leonids are spectacular in some years, but in other years they're just so-so.

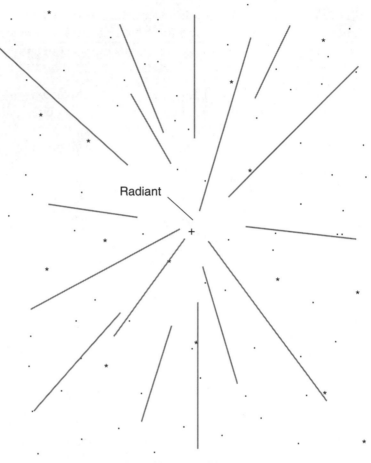

Radiant

Figure 11-7.   A time-exposure negative of a hypothetical meteor shower. In this image, our imaginary camera tracks along with the stars, compensating for Earth's rotation. Meteor paths are straight gray lines; the meteors appear to travel outward.

WHAT ARE METEOROIDS MADE OF?

There's only one way to tell what space rocks are made of, and that is to find meteorites and analyze them. This has been done with plenty of rocks from space. They have been smashed, cut, examined under microscopes, subjected to ultraviolet radiation, heated, cooled, and generally worked over in every conceivable way. Three main types of meteorites have been identified: *aerolites* (stony), *siderites* (metallic), and *tektites* (glassy).

Aerolites bear some resemblance to rocks of Earthly origin. They are made up largely of silicate material and can range in size from pebbles to boulders. The siderites are composed mainly of iron and nickel. Some meteorites are stony with flecks or bands of metal. When a meteoroid enters the atmosphere, the heat of friction causes the outer part of the object to melt. This produces a glassy appearance on the exterior of an aerolite and can blacken the metal on the exterior of a siderite.

Aerolites and siderites are believed to be material left over from the primeval Solar System—stuff that never congealed into planets. If this is true, then they originated in the cores of stars that exploded billions of years ago and scattered their matter throughout the galaxy. This is the only explanation for why these objects exist; otherwise, interstellar space would consist almost exclusively of hydrogen and helium gas. It takes the extreme temperatures inside stars to produce the nuclear fusion reactions that give rise to heavier elements such as silicon, iron, nickel, sulfur, and all the rest.

The tektites tell a different story. These odd, glassy stones resemble rocks of volcanic origin, as if they are parts of a planetary crust or mantle that melted and then solidified again. Tektites have been found in places nowhere near Earthly volcanoes, and they differ dramatically from the composition of the Earth's crust in their vicinity. Because of this, astronomers believe that they came from space. However, they differ from aerolites and siderites in an important way besides their appearance and composition: They are much younger. According to one theory, the tektites were created by one or more catastrophic asteroid impacts on the Moon, events that hurled moon rocks upward with such speed that they escaped the gravitational field of the Moon. Some of these objects, if this took place, would be captured by the Earth's gravitation and would fall to our planet like meteors and meteorites.

The Moon has many craters with prominent rays extending hundreds of kilometers outward. These rays were produced by material ejected from the craters when the impacts occurred. If Moon rocks could be thrown that far, they also could be thrown into orbit or into interplanetary space. The prominent crater *Tycho* has been suggested as a logical candidate for the production of some of the tektites that have been found on Earth.

Recently, objects similar to tektites have been found that are thought to have originated on Mars. It would take a more violent asteroid strike to throw matter into interplanetary space from Mars than it would from the Moon, but calculations show that it is possible. From a statistical standpoint, it is reasonable to suppose that such an event has taken place at least once within the past few million years. After all, scientists believe that only 65 million years ago an asteroid splashed down in the Gulf of Mexico with such force that the resulting environmental disturbances wiped out more than half the Earth's species, including the dinosaurs, and irrevocably shifted the course of evolutionary history.

 Quiz

Refer to the text if necessary. A good score is 8 correct. Answers are in the back of the book.

1. Major asteroid impacts on the planets
   (a) have never occurred in our Solar System.
   (b) take place only on Jupiter and Saturn.
   (c) were once commonplace in the Solar System.
   (d) produce new comets.

2. A comet might be expected to suddenly become brighter if
   (a) all the icy material in the nucleus has evaporated.
   (b) the tail passes through the corona of the Sun.
   (c) a solar flare occurs.
   (d) its perihelion takes it outside the orbit of the Earth.

3. A spherical swarm of millions or billions of distant comets that surrounds the Solar System is known as the
   (a) Van Allen belt.
   (b) Oort cloud.
   (c) primeval Solar System disk.
   (d) tektite belt.

4. A small asteroid or massive meteoroid that crashes into the Moon can produce
   (a) a comet shower.
   (b) a new comet.
   (c) a crater with rays.
   (d) an Oort cloud around the Moon.

5. The "dirty snowball" model for the structures of comets is sometimes credited to
   (a) Fred Whipple.
   (b) Giuseppe Piazzi.
   (c) Johann E. Bode.
   (d) Johann D. Titius.

6. A meteorite
   (a) has the potential to become a meteor.
   (b) is like a meteor, except smaller.
   (c) is a meteor that strikes the surface of the Earth.
   (d) becomes a meteoroid if it is captured by the gravitational field of a planet.

7. The radiant of a meteor shower
   (a) is always straight overhead.
   (b) is fixed with respect to the constellations.
   (c) is opposite the direction of the Earth's motion through space.
   (d) depends on the number of meteors that fall each hour.

8. The asteroids were discovered in part because astronomers were searching for a planet to fit the orbital "slot" at 2.8 AU based on
   (a) the trajectories of fallen meteorites.
   (b) the behavior of the moons of Jupiter.
   (c) the distribution of impact craters on the Moon.
   (d) a mathematical formula developed by Titius and Bode.

9. After a comet has passed perihelion,
   (a) the tail follows behind the nucleus.
   (b) the tail streams out ahead of the nucleus.
   (c) it breaks up into meteoroids.
   (d) its coma grows larger and brighter.

10. The Kuiper belt
    (a) lies outside the orbit of Neptune.
    (b) lies between the orbits of Mars and Jupiter.
    (c) is an intense region of radiation around Jupiter.
    (d) is where comets go to die.

CHAPTER 12

The Search for Extraterrestrial Life

We of the human species have entered the third millennium. We have wondered for a long time whether or not life exists on other worlds, but we do not yet know the answer. There are tantalizing clues, and some astronomers believe that extraterrestrial life exists, but as of this writing, such beliefs are a matter of faith.

What is Life?

In our quest to find life in other parts of the Cosmos, we have assumed, perhaps unconsciously, that such life is similar to life on Earth. We do this partly as a mental crutch to help us get and keep a vision of what we're looking for. We also do it to keep to a scientific course of thought so that we don't fall into pure speculation or into nonscientific thought modes.

WHAT IF . . . ?

Some people suggest that the scientific method forces our minds to take a narrow and conceited view of reality. What if life is "out there" in a form entirely different from life as we know it? Suppose, for example, that some of the science-fiction authors' stories have been true to the mark and that energy-field life forms dwell in the vast tracts between the stars and galaxies? Suppose that the stars and galaxies themselves are life forms that have evolved to levels of sophistication far higher than we humans—so much loftier that we are no more aware of their existence than a bacterium is aware of the elephant in whose ear it dwells? Are there life forms like this? We do not know the answer to this question. We have no idea of how to communicate with such beings. The closest we can come in this respect is to turn things over to the clergy and to approach the problem from a spiritual standpoint.

There exists a psychological split between the church and the scientific community that at times leads one group to criticize the other. Let's not question the beliefs of people who have faith in the existence of life on loftier planes than ours. Nor should we make any claims as to the absolute truth of any scientific theory. Theories are just that. History is full of examples of people or groups of people who turned theory into dogma and later were proven wrong. Our job, as scientists, is to look for good evidence of life on a level similar to life on this planet. By taking this attitude, we have some hope of finding such extraterrestrial life, assuming that it exists, and communicating with it in a meaningful way.

With this in mind, and guarding against the danger of using statistics inappropriately (the "probability fallacy" mentioned at the beginning of Chap. 9), a special group of astronomers is engaged in a pursuit known as SETI (pronounced "SET-eye" or "SEE-tie"), an acronym that stands for *Search for Extraterrestrial Intelligence*. The purpose of SETI is exactly what its name implies: to find another technologically advanced civilization in our galaxy or beyond. We'll encounter estimates of the "probability" or the "chance" that life exists elsewhere in the galaxy or in the Universe. We play this mind game with ourselves to make the nature of our quest comprehensible. In truth, however, we can only say this: Either we will discover life of extraterrestrial origin someday or else we won't. We have a better chance (oops) of finding extraterrestrial life if we search for it than if we don't.

PROPERTIES OF LIVING THINGS

The evolution of intelligent life is a complex process. The molecules of certain chemical compounds, given the right circumstances, develop the ability to make copies of themselves. This process, according to scientific thought, is a necessary basis for life and is part of a common definition of life.

Another necessary property of living beings is the ability to create order from chaos, acting against the *entropy* process at work in the Universe.

Entropy constantly tries to get order to fall into chaos and to distribute all the energy in the Cosmos uniformly so that energy-transfer processes, vital to the existence of life, cannot go on. Living things can reproduce. Living things can gather energy and concentrate it and process it in an orderly fashion. Living things are orderly. Examples of this abound in all human civilizations. Look at the buildings our species has created out of rocks and metal!

A single particle capable of splitting into two other particles identical to itself will rapidly spread over a planet, as long as there is enough raw material and enough energy to sustain the process. In this scenario, the number of such particles increases in a geometric progression: First 1, then 2, then 4, then 8, 16, 32, and so on (Fig. 12-1). This sort of sequence grows rapidly to enormous size.

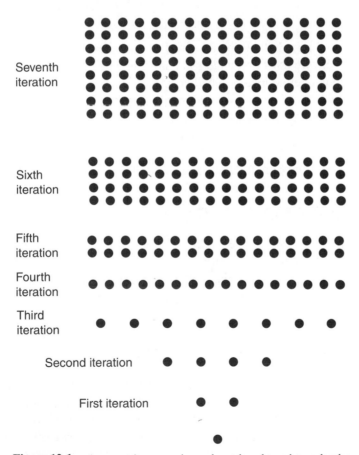

Figure 12-1.   A geometric progression such as takes place when molecules repeatedly copy themselves or when biological cells divide.

Imagine an average reproduction rate of one particle-duplication per day for several weeks. The resulting population, assuming that none of the particles deteriorate or are otherwise destroyed, would dominate the host planet.

In a real-world scenario, the process of replication cannot go on forever; something eventually will limit the number of particles in the population. This "something" is another important property of living things: They die. Death can occur for various reasons: a limited food supply, disease, limited physical space in which to live, changes in the environment, and the effects of cosmic radiation.

Life on Earth

To understand how life can be expected to have developed and evolved on distant planets, we must first understand how the process took place on Earth. Then we might get an idea of whether or not our planet represents a one-of-a-kind miracle, the sole oasis of life in an otherwise sterile Universe.

Let's take an imaginary journey back billions of years in time to the very beginnings of life on our planet. The following is an oversimplification, but it represents a scientific hypothesis for events following the Earth's formation along with the Sun and the other planets in our Solar System.

THE FIRST LIFE FORMS

Several billion years ago, the Earth was much different than it is now. The atmosphere was a noxious mixture of chemicals we would find impossible to breathe: hydrogen, ammonia, methane, and water vapor. There was little or no oxygen. The oceans were less salty than they are today, and they were sterile. Some aspects of Earth would look and sound familiar if we could travel back in time and stand there. Ocean waves broke on rocky shores with the same booming and crashing sounds we know so well, and the land looked like that we see in some places today, such as on newborn volcanic parts of the Big Island of Hawai'i. However, not a single tree graced the horizon. No birds soared over the land or the sea. No grass grew. Somehow, out of this environment, the Earth developed, in about 3 billion years, to a place where life abounds. It is hard to say, after giving the matter serious thought, that this is anything other than the outcome of a miracle. However, if we hope to find life on other worlds, we must hope that such a chain of events is a commonplace thing in the Universe.

According to modern science, life on Earth began with complex groups of particles. Some molecules, called *ribonucleic acid* (RNA) and *deoxyribonucleic acid* (DNA), developed the ability to make copies of themselves. Not long after the first of these molecules appeared, the oceans teemed with them. The presence of a hospitable fluid (water) was an essential ingredient in the proliferation of these particles. No one knows the exact process from which the first of these molecules formed, but it is believed to have involved a burst of energy such as a lightning strike, the impact of a comet or meteorite, or a volcanic eruption. According to some scientists, this process took place in several or many diverse locations on Earth, not necessarily all from the same type of energy burst.

Somehow, the replicating molecules organized themselves into groups and formed animate matter with complex functions. The exact way in which this happened is another mystery. The development of the first living cells took many millions of years. Two major types of cells appeared after about 2 billion years. One type of cell was able to convert sunlight into the stuff necessary to carry on its life processes. This process is known as *photosynthesis*, and the earliest cells capable of it were the first plants on Earth. Among the waste products of such cells was oxygen. Another type of cell developed that was able to use oxygen for its own life processes. These cells were the first animals. The animal cell found, in the oxygen, a more efficient source of energy than sunlight because oxygen is a reactive element. It readily combines with many other types of atoms.

Some of the cells began to stick together in colonies of two, four, five, dozens, or hundreds. The reason why some cells clung to each other and others did not is unknown. Apparently, some cells underwent mutations that caused their outer membranes to be sticky or rough. Mutations occur because the reproductive process is not always perfect (Fig. 12-2). Radioactivity, which is always present to some extent everywhere in the Universe, can cause errors in RNA and DNA duplication. As things turned out, large groups of cells were better able to survive adverse conditions than individual cells. Sometimes an error in the reproductive process actually results in an improvement in the offspring.

NATURAL SELECTION

It's good that reproductive "accidents" happen. Otherwise, according to a popular theory, the Earth would harbor only the rudimentary beginnings of life. Biologist Charles Darwin is credited with developing this idea, known as the *theory of natural selection*.

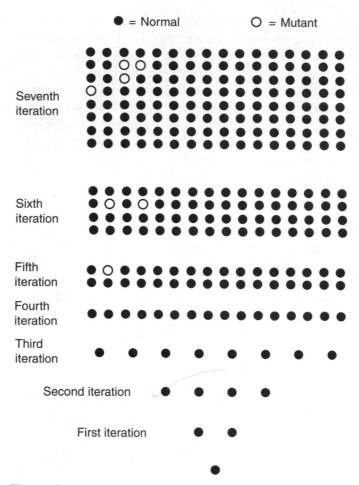

Figure 12-2.   The process of replication is not perfect. Mutations, however rare, are inevitable.

The congregations of cells grew larger. Eventually, groups of cells evolved in which not all the cells were identical. The outer cells became ideally suited to protecting the inner ones from damage or injury. The inner cells were better able to act as food and energy processors. The natural selection process dictated that the outer cells must be physically tough, but this was not required of the inner cells. Congregations with soft outer cells died, whereas those with hardier outer cells survived longer and produced more offspring like themselves.

The theory of natural selection, given sufficient time to operate, results in the evolution of life forms that are better and better suited to the particular environment in which they live. The process is evident only after many generations have passed. However, the available time on Earth is on the order of billions of years; there's no shortage of time! The Sun and the Earth have been around for more than 4 billion years, and our parent star is expected to shine reliably for several billion more years.

Some stars are not as stable as the Sun and do not give their planets time enough for evolution to take place before they use up their fuel and burn out. Other stars are not hot enough to allow for the development of, or the evolution of, life. However, there are plenty of stars that resemble the Sun enough that some of their planets—assuming planetary systems are commonplace in the Cosmos—have temperatures similar to those on Earth. How common are such planets? This is another question that we cannot yet answer definitively because we have no hard evidence. We can only surmise that there are at least a few such planets in our galaxy, and we can see easily enough that there exist millions upon millions of galaxies in the observable Universe. It is difficult to imagine that the known Cosmos does not harbor millions or even billions of planets whose climates resemble that of the Earth.

THE EVOLUTIONARY SPIRAL

Precisely when, in the process of increasing complexity, did the collections of atoms and molecules cease to be a simple matter? At what point can we call something "alive"? Some people say that any RNA or DNA molecule is alive because it can reproduce; other people impose more stringent conditions. There is no well-defined point where we can conclusively say, "Now there is life, but one second ago there was not." If everyone could agree on this, many arguments taking place in such scientific fields as genetic engineering and some of the political dilemmas involving reproduction would be easier to resolve.

As the eons passed, increasingly complex groups of living cells, known as *organisms*, evolved. Thus mutations among individual RNA or DNA molecules became more frequent within any given organism. This is statistically inevitable. Mutations can be expected, in general, to take place twice as often in a congregation of 2 million cells as in a group of 1 million cells. As a result of this, life forms became more varied and more sophisticated. This accelerated the process of evolution, which in turn produced still more diversification. It became an evolutionary positive-feedback system, sometimes called the *evolutionary spiral*.

Some of the cells from the primordial oceans were washed ashore. Most of those cells perished on the dry land for lack of a fluid medium in which to reproduce. Some, however, were able to survive in tidal pools and in puddles left behind by storms. These cells developed into the terrestrial (land-based) plants. As these organisms died, soil was built up on the barren rocky surfaces of the continents. Most of the Earth's living cells remained in the seas, where they developed into marine plants and fishes. The oxygen-burning cell congregations developed the ability to propel themselves from place to place in the never-ending search for food. From this point, the process of evolution reached a climax. Some of the fish developed the ability to live on dry land. These eventually became the dinosaurs, and they reached such a level of perfection that they dominated the continents. A species had emerged that enjoyed a biological monopoly.

Some animals, much smaller than the dinosaurs, also survived, but with difficulty. They had to scurry around at night in their search for food under cover of darkness to avoid being seen and eaten by dinosaurs. Their small bodies gave up heat more quickly than the bodies of the larger dinosaurs, and this problem was exacerbated by the cool nighttime temperatures. The result was the evolution of a new sort of animal that had an increased rate of metabolism, capable of generating enough internal heat to offset the cooling effects of small bodies and chilly surroundings. The earliest of these creatures are believed to have been small rodents, resembling mice, rats, and chipmunks.

CLIMATIC CHANGE

About 65 million years ago, the climate of the Earth cooled. The reason is not known with certainty, but increasingly, scientists suspect that a small asteroid struck the Earth in the Gulf of Mexico. The result was greatly increased volcanic activity for a time, along with the production of ash and dust that was sent into the upper atmosphere, partially blotting out the light and warmth from the Sun. Another theory suggests that the Sun itself cooled off. Still another theory holds that there was a sudden, dramatic change in ocean currents, such as might be produced if the poles of the Earth shifted position or if a land bridge between Siberia and Alaska suddenly appeared or vanished.

Whatever the reason for the climate change, the dinosaurs could not cope with it. The event was too sudden and its magnitude too great for the processes of evolution to keep up, so the dinosaurs perished in a geological "blink of an eye." Within a few million years, nearly all the dinosaurs were gone. However, the small rodents survived because they had the ability to generate internal body heat. In addition, they possessed a characteristic that might best be called *wiliness*—a brain that could figure out how to deal with unusual or complicated situations and crises.

Adversity plays an important role in the evolution of intelligent life. Without problems, there is no need for reasoning power. If the environment becomes ideal and stays that way, evolutionary progress comes to a halt. The dinosaurs are an example of a species that lived in harmony with the environment and adjusted to a status quo that would, were it not for the sudden cooling of Earth 65 million years ago, still exist today. We can curse our cold winters, but we owe our existence to them, according to the most popular theories of evolution and natural selection.

THE FUTURE OF LIFE ON EARTH

Our species, *Homo sapiens*, has roamed Earth for only a moment on the Cosmic scale of time. Imagine watching time pass from a speeded-up perspective so that 1 billion years is represented by a single day. On this scale, the Universe was born 10 to 15 days ago. Suppose that it is high noon on the thirteenth day of the Universe's existence. Our Sun and all the planets that orbit around it are $4^1/_2$ days old. A million years is represented by 86.4 seconds, or a little less than a minute and a half. The sudden cooling of Earth took place at about 10:30 this morning. The earliest known civilizations flourished within the last second.

The process of evolution has not changed the human brain very much (some scientists say not at all) since the dawn of civilization. The ancient Babylonians, Egyptians, Africans, and others were just like us! We did not develop our gadgets and conveniences and weapons of mass destruction because we are smarter than those people were. In fact, as historians delve deeper into the nature of ancient civilizations, evidence mounts to suggest that they were in some ways superior to us. The process of evolution generally requires thousands of centuries to make a significant difference.

What will happen to our species in the future? Are we doomed to destroy ourselves, as the purveyors of gloom keep telling us? Or will we venture out to explore the Universe beyond our Solar System and search for other life forms and civilizations? If the former is our fate, is this also the destiny of other technologically advanced civilizations in the Universe? If so, we cannot expect to communicate with extraterrestrials. However, if *Homo sapiens* can overcome its "suicide seed," or if some species follows ours that does not have this problem, then there is hope. In any case, SETI goes on. A few dedicated scientists are looking for life "out there," if for no other reason than that the alternative—not to search—is unthinkable.

THE MALTHUSIAN SCENARIO

Predictions for the future of any species whose reproduction rate consistently exceeds its death rate, no matter on what planet, can range from extremely pessimistic (certain doom) to extremely optimistic (they will rule their planet and venture into space). According to a scientist named Thomas Malthus who lived in the 1800s, any population that increases at a fast enough rate inevitably will face one or more crises. We, the members of the species *Homo sapiens* who dwell on the planet we call Earth, are already feeling some of the real-life manifestations of Malthus's predictions. There are too many of us, and the population is increasing.

We are intelligent, or so we claim. However, in the collective sense, are we smart enough to control our own numbers and prevent the consequences of unchecked population growth? Until we can overcome this problem, our own Earthly concerns may become so weighty as to overshadow efforts toward reaching for the stars. In fact, an excessive reproduction rate could lead us repeatedly back to stone-age conditions. Malthus's principle operates with mathematical rigor. It can be expected to apply to any matter-based life forms on any planet anywhere. A species that cannot control its own numbers faces catastrophe, perhaps more than once, until it learns to keep its population below the limit that the environment of its host planet can support (Fig. 12-3).

Malthus showed that any population increase must take place in a geometric manner: The rate of growth gets faster and faster. If we plot the number of people in the world as a function of time, we get a graph that looks like the left portion of Fig. 12-3. If there were nothing to stop the process, the planet eventually would become so crowded that people would have to sit on one another's shoulders and would occupy every square centimeter of every continent. However, things happen to keep Earth, or any planet with life whose population grows geometrically, from suffering such a fate. Unfortunately, with the exception of voluntary population control, all these limiting processes are horrible.

Malthus believed that the maximum obtainable food supply can, at best, grow at an arithmetic rate, a straight line on a quantity-versus-time graph.

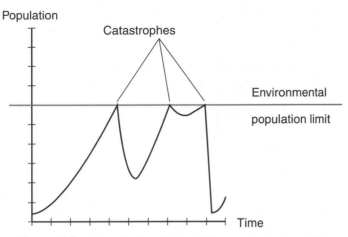

Figure 12-3.   A species faces disaster if the reproduction rate, which grows geometrically, consistently exceeds the death rate.

Today we know that the food supply can't increase indefinitely at a steady rate; it must level off sooner or later (Fig. 12-4) because the planet can support only so much agricultural, fish-breeding, and livestock-raising activity. Thus the population eventually will outstrip the food supply. Mass starvation will occur. This will put a limit on the population by increasing the death rate and by reducing the reproduction rate. The world has not reached this point yet everywhere, but in some countries it is getting painfully close.

Another factor that will limit population growth is disease. When masses of people are crowded, epidemics start and spread much more easily than if people live with plenty of space between each other. New bacteria, viruses, and other pathogens (including the *prions* that cause mad cow disease) develop and evolve rapidly in such conditions, defying attempts at vaccination and literally "learning" how to overcome the effects of antibiotics and other medicines. It is almost as if the planet fights back against further population increase, with disease organisms playing the roles of antibodies and human beings acting as an infectious agent in the "body" of Earth!

Still another limiting factor is the intolerance of humanity for its own kind, manifested in wars and brutal political regimes. New weapons of mass destruction and an increase in the frequency of quarrels leading to wars will tend to limit the population, at least of a species predisposed to violence, as is *Homo sapiens*.

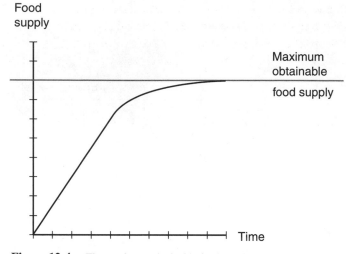

Figure 12-4.　The maximum obtainable food supply grows arithmetically at best, according to the Malthusian model.

This is a grim picture, isn't it? Hunger, disease, and war, with disasters taking place at ever-more-frequent intervals as a planet gets more and more crowded with the species that, were it not for its collective inability to control its reproduction, ought to live lives of peace, comfort, fulfillment, and, to the extent they choose, interstellar adventure. All these things may be ours if only we could learn not to make so many copies of ourselves. If ever an intelligent civilization from another star system sends explorers here, we will know that they got their breeding instinct under control, at least until they found a way to colonize worlds besides their planet of origin.

Communication

The earliest wireless transmissions made by human beings, conducted with spark-gap transmitters at low frequencies, did not escape the ionosphere of Earth and did not propagate into space. However, shortwave radio signals sometimes can penetrate the upper atmosphere, and signals at very high frequencies (VHF), ultrahigh frequencies (UHF), and microwave frequencies inevitably make it into space, where they can travel for unlimited distances. By now, some of our radio and television signals have reached the stars in our immediate neighborhood of the galaxy. Have any extraterrestrial life forms picked these signals up? Have any of these beings sent replies? Have any of these replies actually begun to reach Earth? If so, we ought to be listening!

CQ EXTRATERRESTRIAL

Each and every time a signal escapes into space from our planet, it can be thought of as saying, "CQ alien life"! The expression "CQ" is used by amateur radio operators and means "Calling anyone." CQ signals sometimes are followed by modifiers indicating a preference for the type of station with which communication is desired. An extraterrestrial being, if intercepting one of our standard radio or television broadcast signals, would be smart enough to know that it was not especially intended for his or her (or its) civilization, but the fact that we had allowed the signals to escape into space could be interpreted as an invitation to reply, a call of "CQ extraterrestrial."

Of course, any good search for other stations in a communications medium involves a combination of transmitting and receiving. The equipment for receiving long-distance signals is less expensive than the equipment for sending them, and results can be anticipated sooner. Thus SETI astronomers did their listening first, and reception continues to take priority over transmission. To adopt an age-old principle: We can learn more by listening than we can by talking.

PROJECT OZMA

The first serious attempt to find signals from another civilization was initiated by Dr. Frank Drake using a *radio telescope* at Green Bank, West Virginia, in 1959. Drake and his colleagues called the undertaking *Project Ozma*, named after the fantasy land of Oz. The scientific establishment regarded Project Ozma with interest, amusement, and some skepticism. But Drake believed that if enough stars were scanned with the sensitive radio receivers and large antennas at Green Bank and other radio observatories, it was only a matter of time before signals from an extraterrestrial civilization were picked up.

Some of the stars that Drake investigated were *Tau Ceti* in the constellation Cetus and *Epsilon Eridani* in the constellation Eridanus. Various frequencies were checked, but especially those in the vicinity of the well-known hydrogen emission energy that takes place at a wavelength of 21 centimeters (cm) and corresponds to a frequency of about 1400 megahertz (MHz). This "signal" is prevalent throughout the Universe. Frequencies just above and below it are logical choices for interplanetary and intergalactic radio calling channels. The results were negative, but the amount of time spent on the project was limited. Campaigns similar to Project Ozma continue today as part of SETI.

PROBLEMS AND CHALLENGES

There are difficulties inherent in finding signals from extraterrestrials, as well as in sending signals to them. Here are some of the major challenges that SETI pioneers face.

First, only a tiny part of the sky can be scanned at any given time. A narrow field of reception is necessary because celestial objects generate a lot of radio noise, and this can be minimized only by "focusing in" on very small regions of the sky. Also, in order to get a signal to travel through the vast depths of interstellar space, it must be focused in a narrow beam. We cannot "spray" a signal all over the whole sky; it will become too diluted by the time it reaches the stars.

Second, the apparent direction of a distant star is not always exactly the same as its actual direction. All the stars are moving with respect to each other. The position of a star when its light leaves it, as compared with its position when a signal arrives from an Earthbound transmitter, can change (Fig. 12-5). The transmitted beam must be wide enough to get rid of the possibility that the signal might miss its target. We also must realize that the true target is not the star itself, but a planet in orbit around the star. Radio beams are wide enough so that this timing problem is not significant, but radio waves aren't the only mode of communication that has been suggested. Lasers at visible and infrared wavelengths might be used to focus the beam into as narrow a shaft as possible. If the beam is narrow enough, communicators will have to calculate the actual position of the target planet when the signal is expected to arrive, and this will require precise observations as well as excellent computer programming.

Third, we must decide which wavelength or wavelengths on which to listen and transmit. The resonant hydrogen wavelength at 21 cm is a natural marker in the electromagnetic spectrum and has been recommended as a wavelength near which interstellar communication can be carried out. Intelligent beings, knowing this wavelength and its significance, should be

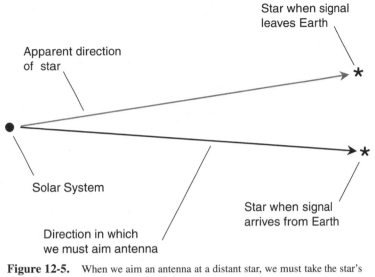

Figure 12-5.   When we aim an antenna at a distant star, we must take the star's motion into account because the signals take years to get there. (The motion in this drawing is exaggerated for clarity.)

expected to send their messages at wavelengths near (but not exactly at) the resonant hydrogen emission signal. In any case, for signals to penetrate great distances in space, they must be concentrated at precise and stable frequencies. Otherwise, the electromagnetic noise generated by some stars, galaxies, and nebulae will overwhelm the communication signals.

All these problems make SETI a task akin to searching for the proverbial pin in a barn full of hay. Nevertheless, the quest goes on, with the hope that the pin is there and that if we roll around in the hay long enough and vigorously enough, it will sooner or later poke us.

UNKNOWN MEDIA

Some physicists, astronomers, and communications engineers think that there are modes of communication we have not yet discovered and that truly advanced extraterrestrial beings are signaling by such esoteric means. Radio signals, infrared waves, and light beams travel at about 299,792 km/s (186,282 mi/s) in space. This seems almost instantaneous in the immediate vicinity of Earth. People who regularly use geostationary satellites for two-way communications and Internet access know about the *latency*, or lag time, which can be upwards of half a second because of the time it takes for signals to travel to and from such satellites. When it comes to interstellar communications, however, latency will be measured in years, decades, centuries, millennia, or eons.

Suppose that we send a message by means of electromagnetic waves to a star system on the other side of our galaxy. If this signal is heard and a reply is sent, we will not receive the reply until 150,000 or 200,000 years have passed. This is, for all practical purposes, just about as good as no reply at all.

Are there particles or effects that travel faster than the speed of light in space? Some recent research suggests that there are. Is it possible to send signals in hyperdimensional levels, somehow short-circuiting the distances among stars and galaxies by cheating on time? This gets into the realm of science fiction, but today's fiction has a way of becoming tomorrow's fact. Some scientists have gone so far as to say that an advanced interstellar civilization would consider electromagnetic communications old-fashioned and quaint, in the same way we think of smoke signals or cannon shots.

A NEW KIND OF PATIENCE

Regardless of the mode, be it radio signals, infrared, visible light, or some thus-far-unknown technology, *Homo sapiens* will have to cultivate great patience to make interstellar communications possible. We will have to be willing to send out signals and realize that they might not be heard until many human generations have come and gone. It is difficult to imagine putting down notes that we do not expect will be read for 50,000 years. But there is no way, as far as anyone knows, that our descendants 500 centuries from now, reading our instructions and adjusting their communications equipment accordingly, can reply to us and ask, "What if your software doesn't run on our computers?"

It is hard enough right now, in most Earthly societies, for parents to communicate with their children. Imagine this generation gap multiplied by several thousand times! If ever a civilization from some distant star system sends its representatives to meet us here on our humble little planet, we will know that they have attained a degree of patience we can only dream about. However, it will inspire us, because the instant we know who they are and where they came from, we will realize that if they can attain such a lofty state of existence, so can we. We will realize that Earth is not a miracle. Or if you prefer, we will come to know that miracles are common in the Cosmos.

Visitations

Communication is an economical way to search for life on other worlds, and it is also the method that we can expect to produce results, if there's anyone out there listening and transmitting with the same intentions. However, communication is not much of an adventure, and if we ever find another civilization "on the radio," we'll want to meet those beings face to face. This can happen in three ways: We can go to them, they can come to us, or we'll run across each other in the vastness of interstellar space.

WE DO THE TRAVELING

Interplanetary travel, within the limits of our own Solar System, has already been done by robotic space probes. We know it's possible for machines to do it, and given reasonably good economic and political conditions in the next several decades, astronauts will someday visit some of the planets and moons, especially the planet Mars and Saturn's largest moon, Titan. It is not necessary to attain fantastic speeds to get to the other planets in our Sun's family. To reach the stars, however, we will have to accelerate our space ships almost to the speed of light.

If the Earth were the size of a marble, the Moon would be a small pea approximately 300 mm (about 1 ft) away. At present space-ship speeds, it takes 2 or 3 days to get to the Moon. On this same scale, the Sun would be about 120 m (or 400 ft) away, roughly the distance from home plate to the center-field fence in a major league baseball stadium. The distance from the Sun to Pluto would be on the order of 5 km (3 mi). We have the ability to span these distances, although it takes years to reach the outermost planets. The nearest star to our Solar System, *Proxima Centauri*, would be more than 32,000 km (20,000 mi) away on this same scale. Even if we can build a ship that will travel at half the speed of light, a round trip to this star will take 18 years. Our Milky Way galaxy is 25,000 times wider than the distance to Proxima Centauri.

If we humans are ever to attempt interstellar travel on a galactic scale, our space vessels will have to reach speeds so high that a peculiar phenomenon, *relativistic time dilation*, takes place. Maybe you've heard about this: the slowing down of time for beings in a vessel traveling at near the speed of light. This would make it possible, in theory, to reach almost anywhere in the known Universe within the span of one human lifetime. Relative to the rest of the Cosmos, however, including the planet of origin

(Earth), such space travelers would be hurled irrevocably into the future by hundreds, thousands, or millions of years.

Long-distance space travel presents all kinds of obstacles and dangers, and we'll look more closely at this subject later in this book. We'll study relativistic time dilation and other effects of extreme speed as well.

THEY DO THE TRAVELING

Despite all sorts of stories and rumors, there is, as of this writing, no conclusive evidence, accepted by scientists in general, that extraterrestrial beings have visited our planet. If it has occurred, the scientific community is not aware of it.

Ever since written records have been kept by humankind, there have been sightings of unidentified flying objects. One such "vessel" is described vividly in the *Book of Ezekiel* in the Christian *Bible*. Public interest in this phenomenon reached a peak in the twentieth century, when the dream of space travel became reality and people's imaginations went wild. Most sightings have been explained in terms of natural phenomena. The planet Venus, when it appears through a certain type of atmospheric haze, appears enlarged and can be mistaken for an approaching aircraft or even a hovering space ship. Lightning discharges, strange lights in the sky accompanying earthquakes, and glowing gases over swampy regions can appear as if they are traveling at high speed through the upper atmosphere. A few sightings have evaded all attempts at logical explanation. Still, scientists and government officials maintain a high degree of skepticism. Until space aliens actually set a vessel down in a public place and say something like, "Take us to your leader," these folks will remain unconvinced that we have been visited by beings from another planet.

Any civilization that has developed the technology to roam among the stars must have patience and other qualities such as the ability to control their social problems at home. Are there extraterrestrial civilizations traveling among the stars as you read this? Have we on Earth been watched, or are we being watched, by such beings?

Some people try to get an idea of what extraterrestrials might do by putting themselves in the place of travelers from other worlds. If you were the captain of a starship and you came across a planet like ours, and if your mission were merely to catalog the behavior of civilizations in various star systems, would you want the beings of planet Earth to know about you? What if you were looking for civilizations with whom you could identify, make friends, and ponder greater questions of Cosmic reality? Would you come down and introduce yourself in Red Square or Central Park? What if your intent were to conquer Earth by violent means? It has been argued that space aliens of that sort won't come here. They'll have annihilated themselves long before they became capable of interstellar travel.

The Green Bank Formula

Suppose that the evolution of intelligent life on our planet Earth is not a miracle. Then extraterrestrial life exists! This part of the problem being solved by faith, let's play a mind game that involves an attempt to calculate the number of other intelligent civilizations we would find if we could travel freely among the stars, galaxies, and clusters of galaxies throughout the known Universe. This mind game involves cheating on the "probability fallacy."

We are going to talk about the *likelihood*, as a proportion, that certain things have happened, are happening, or will happen in our Cosmos. When we say that the chances of some event taking place on a planet are 1 in 100, we really mean that if we could visit a large number of planets, say, 1 million of them, then that event would be discovered on 1/100, or 10,000, of them.

THE CONFERENCE

In 1961, a conference was held at the Green Bank radio telescope observatory, the same place where Project Ozma was conducted. The object of the meeting was to make an estimate of the number of technologically advanced civilizations (defined as capable of communicating by electromagnetic means such as radio) that exist in the Milky Way galaxy. We know this number is at least equal to 1 because we are here and we have radio. Is the number greater than 1? If so, how much greater?

At this conference, the astronomers, led by Frank Drake and Carl Sagan, developed a formula to determine the number of technologically advanced civilizations in our galaxy. It has been called the *Green Bank formula*, the *Drake formula*, or the *Drake-Sagan formula*. Several factors are involved in this mathematical equation; some of them are *probabilities*. The complete formula consists of the product of all these factors. Let's look at the factors one at a time, and then we'll evaluate the entire formula by "plugging in" some educated guesses.

STAR FORMATION: *R*

Let *R* be the average number of new stars that are born in our galaxy each year. There are about 200 billion (2×10^{11}) stars in the Milky Way, and the galaxy is thought to be about 10 billion (10^{10}) years old. This might lead one to suppose that an average of 20 new stars are born every year, that is, that $R = 20$. Let's be a little bit conservative because we can't be certain that stars have always formed at the same rate during the lifetime of the galaxy. For our calculations here, let's use $R = 10$. This value has been suggested as reasonable by many scientists.

PLANETARY SYSTEMS: f_p

Let f_p be the fraction, or proportion, of stars in our galaxy that have planets orbiting around them. Until recently, astronomers had almost no idea of what f_p might be. However, observations with the Hubble Space Telescope and other instruments have shown that planetary formation is not a fluke. It happens with other stars besides our own Sun. Some estimates of f_p range up to 0.5; that is, half of all new star systems include planets. Let's be more conservative and estimate that only 1 in 5 stars have planets. Thus $f_p = 0.2$.

PLANETS SUITABLE FOR LIFE: n_e

Suppose that we look at a large number of star systems with planets. Some of these planets will have environments suitable for the evolution of life as we know it; others (probably most) will not. If we are able to look at a large enough sampling of star systems with planets, we will come up with a number n_e, the average number of life-supporting planets per planetary system. The fact that a planet can support life does not necessarily mean that life exists but only that the environment is such that life *could* exist there. We have seen only one planetary system thus far in enough detail to get any idea of the value of n_e, and statistically, it is nowhere near enough. We might guess that $n_e = 1$ if our Solar System is an average one. However, the more we get to know our planet Earth, the more we realize what a special place it is. Again, let's be conservative and suppose that there exists a planet suitable for life on only 1 out of every 2 star systems that have planets. Then $n_e = 0.5$.

DEVELOPMENT OF LIFE: f_l

If a planet is ideal for the development of life, there is no guarantee that life arises and evolves. A large asteroid or comet impact would cut evolution short if it were violent enough. An unfavorable change in the behavior of the parent star also would snuff out life. A close call with a passing celestial object, such as a neutron star or a black hole, would disrupt the orderly nature of the planetary orbits of the star system. The big question is this: Was life created, and did it get going on its evolutionary way on Earth because of a series of flukes so rare as to have a "combined probability" of almost zero? Scientists have created complex molecules thought to be the precursors of living matter in a laboratory, but this is not the same thing as synthesizing life and demonstrating that its formation is a common thing.

The best we can do with respect to f_l, the proportion of planets suitable for life on which life actually develops, is make a wild guess. Let's call it 0.1, that is, only 1 out of every 10 planets with a good climate can support life.

EVOLUTION OF INTELLIGENCE: f_i

On Earth, life evolved to near perfection in the form of the dinosaurs. However, none of them had brain power approaching that of primates such as monkeys, let alone human beings. If it were not for a supposed small asteroid or comet splashdown around 65 million years ago, the dinosaurs would still be here, and the Earth would be a vastly different place. Many evolutionary scientists think that *Homo sapiens* wouldn't exist. Evolution would never have produced our forebears. Dinosaurs would have eaten them!

Major cosmic collisions, once life has started to evolve on a planet, are not too likely. According to the model of Solar System evolution currently accepted, by the time life was underway on Earth, most of the debris from the primordial solar disk had been swept up into the planets and their moons. But minor collisions are common; we can expect that there will be more of these on Earth yet to come. Interestingly, these minor collisions can serve as a catalyst for evolution, not a fatal blow, as would be the case with a major collision.

What proportion f_i of planets where life has gotten started undergo evolution to the point where intelligence arises? The answer to this question dictates the range of values we can realistically assign to f_i. Evolution and natural selection seem to be relentless processes; we have seen adaptation of species all the way down from ourselves (if we consider *Homo sapiens* a species of animal) to bacteria that develop immunity to antibiotics and viruses that evolve new forms, evading extermination. Given the relentlessness of the evolutionary process and the relative likelihood of changes in the environment that spur the emergence of new evolutionary pathways, we might assign f_i a value of about 0.1. This is a conservative estimate.

TECHNOLOGICAL ADVANCEMENT: f_c

Even if a life form develops intelligence, it might not reach the level where it communicates by radio. On Earth, dolphins (porpoises) and whales can be considered intelligent to a degree. However, they lack hands with which to construct machines. Some porpoises have brains large enough to suggest that their intelligence surpasses that of *Homo sapiens* in some ways, but even if that is true, "dolphin smarts" are qualitatively different from "human smarts."

Suppose that there exist planets covered by oceans teeming with marine animals having intelligence greater than our own. These animals would not have the physical ability to develop radio transmitters and receivers, cars, boats, and airplanes. They would have no need of these devices (as some people argue humanity has no need of radios, cars, boats, and airplanes). What proportion f_c of intelligent species goes on to manufacture the means to communicate among the stars and to conduct their own SETI programs? It is hard to say. Let us suppose that 1 out of 10 planets with intelligent life harbors civilizations capable of communicating by radio; then that proportion f_c is equal to 0.1.

Some scientists lump f_i and f_c together as a product because there is disagreement on exactly what level of brain power constitutes intelligence. Let's get around this problem by estimating that $f_i f_c = 0.01$.

AVERAGE LIFESPAN OF TECHNOLOGICAL CIVILIZATIONS: L

Once a civilization has become intelligent and has developed radio, and once it has turned its electromagnetic "ears" and "voice" to the heavens, how long will such beings remain capable of communicating? It is tempting to suppose that curiosity would drive any intelligent species, anywhere in the universe, to seek out life in other star systems, but we do not know this.

There is a dark point here: As part of our technological "progress," we humans have created weapons of destruction that could annihilate our whole population or at least throw us back to near stone-age conditions. How likely is this? How long can we expect our society to exist as we know it, with radio telescopes and SETI programs? Let's say that a civilization lasts for L years before evolving out of existence. What is the value of L? Again, the best we can do is make a guess. During the 1960s and 1970s, at the height of the Cold War, many people became convinced that human civilization on Earth was doomed to bomb itself out of existence, and soon. Today this view is not as widespread, but unless and until *Homo sapiens* gets rid of its "war gene," the danger remains.

Suppose that a planetary population, at least some of whom can communicate by radio, maintains this level of sophistication for at least 10,000 years before something—war, famine, disease, or asteroid impact—puts an end to it. Then $L = 10,000$. However, even in this case, after the disaster has passed, evolution would continue along its way, and in some cases this would lead to another technologically advanced civilization. This would multiply the value of L. Nevertheless, let's be conservative and set $L = 10,000$.

THE COMPLETE FORMULA

The Drake formula in its entirety consists of the product of all the preceding factors and generates a number N. This is the number of technologically advanced civilizations that we should expect to find in the Milky Way galaxy:

$$N = R f_p\, n_e\, f_l f_i f_c\, L$$

Let's calculate N. Here are the values we have suggested for the variables in the Drake formula:

$$R = 10$$
$$f_p = 0.2$$
$$n_e = 0.5$$
$$f_l = 0.1$$
$$f_i f_c = 0.01$$
$$L = 10,000$$

This yields a final estimate of $N = 10$. That is, based on the guesses made here as to the values of the variables, we can imagine that there are 9 technologically advanced civilizations in the Milky Way besides our own.

If these worlds are more or less evenly spaced in the spiral arms of the galaxy, it is unlikely that we will ever communicate with them by radio because the latency will be too great (unless, by coincidence, one of the other civilizations is within a few light-years of us). However, we have the capacity to hear signals from one of these societies even if we can never send a reply and expect it to be heard within an individual human lifetime, and even if what we hear represents the distant past (an extreme form of time-shifting communication). Thus SETI continues. The "odds" are slim. The potential rewards are enormous, even if it is nothing more or less than the joy of knowing that we are not alone in the Cosmos.

The Final Answer

The estimates made in this chapter for variables in the Drake formula are mine; they don't necessarily reflect those of scientists in general. Many respected academics have come up with values of N much larger than the number obtained in this chapter. However, there are some people who refuse to accept that N can be anything but 1.

A final, definitive answer to the question of whether or not other advanced, intelligent civilizations exist in our galaxy or in the Universe as a whole will elude all of humanity until contact is actually made, if it is ever made. The issue remains open. The jury is out. Some comfort might be gained from reverting to hard-line wariness of the "probability fallacy" and proclaiming that we know but one thing for certain: Either there is advanced extraterrestrial life or else there is not. What do you think?

 Quiz

Refer to the text if necessary. A good score is 8 correct. Answers are in the back of the book.

1. Tau Ceti is a notable star
   (a) because it emits a large amount of energy at 21 cm.
   (b) because it has several known Earthlike planets.
   (c) in the sense that it has been closely examined by SETI scientists.
   (d) for no reason.

2. The hydrogen emission wavelength of 21 cm might be used by extraterrestrial civilizations as
    (a) a reference wavelength for space communications by radio.
    (b) a means of eliminating a planetary energy crisis.
    (c) a method of obtaining high-speed propulsion for space ships.
    (d) nothing; it is a logical wavelength to avoid.

3. The product of all the variables in the Drake equation stands for
    (a) the number of advanced civilizations in our galaxy.
    (b) the number of planets in our Solar System that have life on them.
    (c) the number of stars in the Universe that have planets.
    (d) the probability that extraterrestrial life exists.

4. Suppose that a biological cell is capable of replicating itself and does so exactly once every 24 hours. If we start with one cell, how many cells will we have after 7 days, assuming that none of the cells die?
    (a) 14
    (b) 32
    (c) 64
    (d) 128

5. Suppose that the scenario described in the preceding question is allowed to go on for an indefinite time. Suppose also that the cells can supply their own food by photosynthesis. What will eventually limit the growth of the population of cells?
    (a) The intensity of the light from the Sun
    (b) The amount of available physical space on the planet
    (c) The temperature
    (d) The length of the growing season

6. The value of L in the Drake equation depends on
    (a) how soon, or if, technologically advanced civilizations die off or annihilate themselves.
    (b) how far a planet orbits from its parent star.
    (c) the tilt of a planet on its axis.
    (d) the proportion of stars that have planets with conditions suitable for life.

7. The purpose of SETI is
    (a) to travel to other worlds.
    (b) to encourage space aliens to land on Earth.
    (c) to figure out what unidentified flying objects actually are.
    (d) to find conclusive evidence of an advanced civilization on another planet.

8. The impact of a small asteroid on a planet
    (a) can cause massive species extinctions.
    (b) can spur the emergence of new species.
    (c) can cause a change in the planet's climate.
    (d) More than one of the above

9. The gloomy scenarios painted by Thomas Malthus for the population of a dominant species can be avoided by
   (a) finding new ways to grow food.
   (b) reproducing as much as possible.
   (c) population control.
   (d) food rationing.

10. Which of the following is not one of the necessary conditions for a material thing to be generally considered alive?
    (a) The ability to make copies of itself
    (b) The ability to create order from chaos
    (c) The ability to live forever
    (d) A finite lifespan

Test: Part Three

Do not refer to the text when taking this test. A good score is at least 30 correct. Answers are in the back of the book. It is best to have a friend check your score the first time so that you won't memorize the answers if you want to take the test again.

1. The term *epicycle* refers to
   (a) the motion of a planet with respect to the stars.
   (b) the variation in the Moon's orientation with respect to Earth.
   (c) the speed of a planet's revolution around the Sun.
   (d) the wobbling of Earth on its axis.
   (e) a component of a planetary orbit according to Ptolemy.

2. Which of the following planet has the most known moons?
   (a) Mercury
   (b) Mars
   (c) Saturn
   (d) Uranus
   (e) Pluto

3. An accretion disk is
   (a) a ring around a planet such as Saturn.
   (b) a region around a planet in which orbits are stable.
   (c) the plane of the orbits of the planets around a star.
   (d) a rotating, flat cloud of matter from which planets form around a star.
   (e) a computer disk used for storing position data for celestial objects.

4. Fill in the blank in the following sentence. The names of meteor showers derive from the positions of the _____ in the sky, which tend to be the same, year after year, for any given meteor shower.
   (a) escarpments
   (b) originating comets
   (c) originating asteroids
   (d) radiants
   (e) focal points

5. Fill in the blank in the following sentence. The theory of _____, given sufficient time to operate, results in the evolution of life forms that are better and better suited to the particular environment in which they live.
   (a) natural selection
   (b) biological entropy
   (c) conservation of energy
   (d) Thomas Malthus
   (e) planetary evolution

6. According to the geocentric theory
   (a) the Earth revolves around the Sun.
   (b) the Sun revolves around the Earth.
   (c) the Moon revolves around the Sun.
   (d) the distant stars are reflections of Earth.
   (e) everything is sitting still in the Universe.

7. A laser can be used in long-distance space communication when it is necessary to
   (a) direct the energy in a narrow beam.
   (b) use as wide a frequency range as possible.
   (c) spread the radio waves over a wide angle.
   (d) generate large amounts of intergalactic noise.
   (e) slow down the rate of data transmission.

8. Entropy has a tendency to
   (a) turn chaos into order.
   (b) equalize temperatures.
   (c) create life.
   (d) make stars shine.
   (e) make planets form from cosmic dust.

9. Thomas Malthus is known for his theory that
   (a) the human population increases geometrically, but the food supply increases arithmetically at best.
   (b) the human population increases arithmetically, but the food supply increases geometrically at best.
   (c) both the human population and the food supply increase geometrically.
   (d) war ultimately will bring about the destruction of any technologically advanced civilization.
   (e) a desire to travel among the stars is a characteristic of any intelligent species.

10. The Pluto-Charon system is unique in the sense that
    (a) the planet and the moon always keep the same faces toward each other.
    (b) the planet and the moon are the same size.
    (c) the moon is larger than the planet.

(d) both objects are believed to have once been moons of Neptune.

(e) they are much warmer than the other outer planets or their moons.

11. A comet might burn out because
   (a) it falls into the Sun.
   (b) a solar flare hastens the disintegration of its nucleus.
   (c) it is struck by a small asteroid, which shatters its nucleus.
   (d) either a, b, or c take place.
   (e) No! Comets never burn out.

12. Attempts to determine the number of advanced civilizations in the Milky Way galaxy have been carried out mathematically using
   (a) radio waves.
   (b) star charts.
   (c) optical telescopes.
   (d) the Green Bank formula.
   (e) the Malthusian formula.

13. The planet Venus has a retrograde orbit around the Sun
   (a) because it was knocked out of alignment long ago by a collision with a protoplanet almost as large as itself.
   (b) because it is influenced by the gravitational fields of Earth and Mercury.
   (c) because it is tilted on its axis by almost 180 degrees.
   (d) because it was originally from outside the Solar System and was captured by the Sun's gravity after all the other planets were formed.
   (e) No! The planet Venus does not revolve around the Sun in a retrograde manner, but in the same sense as all the other planets.

14. So-called minor collisions, small asteroids striking planets, have been suggested as a catalyst for
   (a) wiping out all life.
   (b) the evolution of intelligent life.
   (c) creating dinosaurs.
   (d) adding oxygen to the atmosphere.
   (e) creating oceans.

15. The Milky Way galaxy contains about
   (a) 20,000 stars.
   (b) 200,000 stars.
   (c) 2 million stars.
   (d) 20 million stars.
   (e) None of the above

16. Copernicus was one of the earliest astronomers to hypothesize that
   (a) Earth revolves around the Sun.
   (b) the stars revolve around the Sun.
   (c) epicycles exist within other epicycles.

(d) the planets are more distant than the stars.

(e) the Milky Way is not the only galaxy in the Universe.

17. Planetary moons almost always have rotation periods that correspond to their orbital revolution periods
    (a) because of tidal effects between the moon and the parent planet.
    (b) because the moons orbit in the plane of the parent planet's equator.
    (c) because the moons were captured from interplanetary space by the parent planet's gravity.
    (d) because the moons are perfect spheres.
    (e) No! Planetary moons almost never have rotation periods that correspond to their orbital revolution periods.

18. When we say a celestial body has a low albedo, we mean to say that it is
    (a) relatively low in density.
    (b) relatively low in specific gravity.
    (c) nonspherical in shape.
    (d) relatively dark in appearance.
    (e) not a good place for life to evolve.

19. When a planet or moon that has many craters is viewed from a distance, some craters have bright lines that run outward from the center. Such lines are called
    (a) subcraters.
    (b) crateroids.
    (c) escarpments.
    (d) rift valleys.
    (e) rays.

20. The age of the Solar System is believed by most scientists to be approximately
    (a) 4.6 million years.
    (b) 4.6 billion years.
    (c) 10 billion years.
    (d) 200 billion years.
    (e) 4.6 trillion years.

21. The average time interval between major comet impacts on the Earth is estimated to be on the order of several
    (a) decades.
    (b) centuries.
    (c) tens of millennia.
    (d) tens of millions of years.
    (e) tens of billions of years.

22. Orbital resonances can be caused by
    (a) solar flares and sunspots.
    (b) synchronization of a planet's rotation rate with the rotation rate of the Sun.
    (c) mutual gravitation between different moons of the same planet.
    (d) powerful magnetic fields, such as the magnetosphere of Jupiter.
    (e) radio-wave emissions from planets.

23. The "probability fallacy" refers to
    (a) an overestimate of the chance that something will happen.
    (b) an inaccurate estimate of the chance that something took place.
    (c) inappropriate use of statistics.
    (d) a statement of belief rather than an estimate of probability.
    (e) the derivation of a theory based on faulty observation.

24. According to one theory, tektites come from
    (a) the Sun.
    (b) the Moon and Mars.
    (c) the planet Mercury.
    (d) comets.
    (e) beyond the Milky Way galaxy.

25. A mutation is
    (a) an eccentricity in a planet's orbit.
    (b) the formation of a planet from cosmic dust and rocks.
    (c) a star that has been modified by a collision with another star.
    (d) a change in the shape of a crater.
    (e) none of the above.

26. On the surface of an outer-planetary moon, craters can last a long time even
    when the chemical composition is largely frozen water. This is so because
    (a) water periodically boils up from the interior, freezing on the surface.
    (b) tidal interactions with the parent planet heat the moon from within.
    (c) the extremely low temperatures make the ice as hard as granite.
    (d) the moon's atmosphere prevents wind erosion.
    (e) No! Craters on the surfaces of an outer-planetary moon do not last long.

27. A Sungrazer is
    (a) an asteroid that falls into the Sun.
    (b) a star that passes so close to the Sun that the two stars pull matter from
         each other.
    (c) a comet that enters the solar corona at perihelion.
    (d) a comet that passes directly between Earth and the Sun.
    (e) a comet that passes directly behind the Sun as seen from Earth.

28. In order for an astronaut to travel to the other side of the Milky Way galaxy
    within the span of his or her lifetime,
    (a) the space ship would have to be well streamlined.
    (b) the space ship would have to use liquid fuel.
    (c) the space ship would have to travel at almost the speed of light.
    (d) the space ship would have to travel faster than the speed of light.
    (e) No! No astronaut can possibly reach the other side of the Milky Way
         galaxy within the span of his or her human lifetime no matter how fast the
         space ship goes.

29. Project Ozma involved the use of
    (a) the Hubble Space Telescope.

(b) unmanned space probes.

(c) time-exposure photography.

(d) radio telescopes.

(e) manned space flights.

30. Suppose that there are two celestial objects, X and Y. Object X has a visual magnitude of 2, and object Y has a visual magnitude of 3. Which of the following statements is true?

(a) Object X is 2.5 times as bright as object Y.

(b) Object Y is 2.5 times as bright as object X.

(c) Both objects are too dim to be seen without the aid of a telescope.

(d) Both objects are the same brightness, but they are of different colors.

(e) None of the statements is true.

31. According to the tidal theory of the formation of the Solar System,

(a) the planets formed from matter that two passing stars pulled from each other.

(b) the planets condensed from a disk of matter in orbit around the Sun.

(c) the planets formed from tidal interactions among gas and dust clouds.

(d) two stars collided directly, and the debris condensed into planets.

(e) the planets were captured by the Sun's gravity from interstellar space.

32. One of the major moons of the outer planets is believed by some astronomers to have once been a planet itself that came too close to the larger planet and became a natural satellite. This moon is

(a) Io.

(b) Rhea.

(c) Triton.

(d) Ariel.

(e) Miranda.

33. The "dirty snowball" portion of a comet is called the

(a) Lyttleton portion.

(b) Whipple portion.

(c) nucleus.

(d) coma.

(e) tail.

34. Some ancient astronomers talked about "music of the spheres." Today we believe that the phenomenon they described was

(a) a figment of their imaginations.

(b) caused by solar radiation.

(c) caused by Earth's tidal interactions with the Moon.

(d) caused by tidal interactions among Earth, the Moon, and the Sun.

(e) noise produced by falling meteors.

35. The term "shooting star" refers to

(a) a visible meteor.

     (b) a solar flare.

     (c) meteoric dust.

     (d) an asteroid.

     (e) a planetary moon.

36. Ganymede is a moon of

     (a) Mars.

     (b) Jupiter.

     (c) Saturn.

     (d) Uranus.

     (e) Neptune.

37. Titan, one of the moons of Saturn, has been suggested as

     (a) a place where life must exist.

     (b) a place with an Earthlike environment.

     (c) the place with the most violent storms in the Solar System.

     (d) a generator of powerful radio waves.

     (e) a place where future interplanetary travelers ought to go.

38. The major moons of Uranus

     (a) are all believed to be comets that were captured by Uranus's gravitation.

     (b) orbit in the same plane as Uranus orbits the Sun.

     (c) have highly eccentric orbits.

     (d) orbit almost directly over the planet's poles.

     (e) orbit in the same plane as the planet's equator.

39. The Oort cloud is notable because it is a source of

     (a) radiation.

     (b) radio signals.

     (c) infrared energy.

     (d) fuel for the Sun.

     (e) none of the above.

40. Which of the following statements is false?

     (a) Each planet follows an elliptical orbit around the Sun, with the Sun at one focus of the ellipse.

     (b) An imaginary line connecting any planet with the Sun sweeps out equal areas in equal periods of time.

     (c) For each planet, the square of its sidereal period is directly proportional to the cube of its average distance from the Sun.

     (d) In reality, no planet orbits in a perfect circle around the Sun; the orbits are always at least a little bit oblong.

     (e) All the planets and all the planetary moons orbit in the same plane around the Sun.

PART 4

Beyond Our Solar System

CHAPTER 13

Stars and Nebulae

Have you ever heard that stars are "distant suns"? This is true, but it does not tell the whole story. There are plenty of stars that resemble our Sun, but many are larger or smaller, brighter or dimmer, or hotter or cooler. Some stars radiate more energy at shorter wavelengths than does our Sun; other stars favor the longer wavelengths. Stars are as diverse as snowflakes.

Have you ever been told that interstellar space is a "vacuum"? This is true in a relative sense. We could not breathe in space without special equipment, and the pressure is low enough to be considered a vacuum by most laboratory technicians on Earth. However, interstellar space contains matter. We can see some of it directly, whereas other space matter is visible only because it blocks light from stars behind it.

How Bright? How Distant?

You've learned about visual magnitudes of objects in the sky. A difference of one magnitude is the equivalent of a brightness change of 250 percent, that is, $2^1/_2$ times. As the brightness increases, the magnitude number goes down. However, when you look at a dim star in the sky and then you look at a much brighter one, how do you know which of the two really emits more light? After all, if Star X is $2^1/_2$ times brighter than Star Y but is only 1/100 as far away, Star Y is more brilliant in an absolute sense.

ABSOLUTE VISUAL MAGNITUDE

The observed magnitude of a star or other celestial object is called its *apparent visual magnitude*, or simply the *apparent magnitude*. The actual brightness is called the *absolute visual magnitude*, or simply the *absolute magnitude*. Absolute versus apparent: Things are not always as they appear. By definition, the absolute magnitude of a star is the apparent magnitude we would give it if it were 3.09×10^{14} km (1.92×10^{14} mi) away. This is the distance that light travels in 32.6 years. By Earthly standards, or even when measured against the scale of the Solar System, this is a huge distance. Compared with the size of the Milky Way, though, it is not far at all. Compared with the size of the known Universe, it is microscopic.

One of the brightest known stars, in absolute terms, is *Canopus*, which is best observed from the southern hemisphere. People who live north of the thirty-eighth parallel (roughly the latitude of San Francisco) cannot see Canopus. This star has an absolute visual magnitude of -4.4. By comparison, our Sun has an absolute magnitude of $+4.8$; thus Canopus is approximately 5,000 times as brilliant as our parent star. If the Sun were to increase in brightness suddenly to the same absolute magnitude as Canopus, we would all have to wear dark glasses to get around. However, that wouldn't be the only problem. Canopus radiates more energy than the Sun at other wavelengths, too. Its ultraviolet rays would sunburn us in seconds; the heat would incinerate Earth and end all life here in a matter of days, if not hours.

Because of its distance, Canopus is not a remarkable object in the sky. People who live north of the latitude of San Francisco aren't missing much just because they can't see Canopus. Stars like our Sun, when observed from the same distance as we see Canopus, are so faint that they cannot be seen except by people who live far away from the *skyglow* caused by city lights and who have keen eyesight.

THE LIGHT-YEAR

The standard distance that astronomers use for determining absolute magnitude is a staggering number when you ponder it: 3.09×10^{14} km is 309 trillion (309,000,000,000,000 kilometers or 309 quadrillion (309,000,000,000,000,000) meters. These sorts of numbers evade direct human comprehension.

Astronomers have invented the *light-year*, the distance light travels in 1 year, to assist them in defining interstellar distances. You can figure out how far a light-year is by simple calculation. Light travels approximately 300,000 km in 1 second; there are 60 seconds in a minute, 60 minutes in an hour, 24 hours in a day, and about 365.25 days in a year. Thus a light-year is roughly 9.47×10^{12} km. This turns out to be a little less than 6 trillion miles. There we go with incomprehensible numbers again!

Let's think on a cosmic scale. The nearest star to our Solar System is a little more than 4 light-years away. The standard distance for measuring absolute magnitude is 32.6 light-years. The Milky Way galaxy is 100,000 light-years across. The Andromeda galaxy is about 2.2 million light-years away. Using powerful telescopes, astronomers can peer out to distances of several billion light-years (where 1 billion is defined as 10^9 or 1,000 million). Well, the light-year helps us to comprehend the distances to stars within our galaxy, but once we get into intergalactic space, even this unit is not enough to make distances easy to imagine.

THE PARSEC

The actual distances to the stars remained a mystery until the advent of the telescope, with which it became possible to measure extremely small angles. The angular degree represents 1/360 of a full circle. Smaller units are the minute of arc, representing 1/60 of an angular degree, and the second of arc, measuring 1/60 of a minute. These units were introduced in Chapter 1.

To determine the distances to the stars, astronomers had to be clever. Could it be done by triangulation, the way surveyors measure distances on Earth? For this to work, it would be necessary to use the longest possible baseline. What would that be? How about the diameter of the Earth's orbit around the Sun? This presented a problem for people who believed that Earth was the center of creation and that everything else revolved around it. However, by the time astronomers were ready to seriously attempt to determine distances to the stars, the heliocentric theory had gained general acceptance.

Figure 13-1 shows how distances to the stars can be measured. This scheme works only for "nearby" stars. Most stars are too far away to produce any measurable *parallax* against a background of much more distant objects, even when they are observed from opposite sides of Earth's orbit. (In this figure, the size of Earth's orbit is exaggerated for clarity.) The star appears displaced when viewed from opposite sides of the Sun. This displacement is maximum when the line connecting the star and the Sun is perpendicular to the line connecting the Sun and Earth. A star thus oriented, and at just the right distance from us, will be displaced by 1 second of arc when viewed on two occasions as shown in Fig. 13-1. This distance is

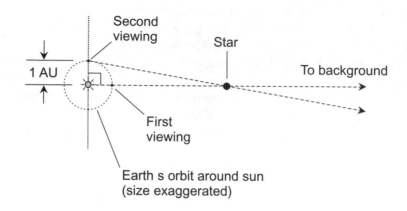

Figure 13-1. The distances to "nearby" stars can be determined by measuring the parallax that occurs as the Earth revolves around the Sun.

called a *parsec* (a contraction of *parallax second*). The word *parsec* is abbreviated *pc* and is equivalent to approximately 3.26 light-years. Sometimes units of *kiloparsecs* (kpc) and *megaparsecs* (Mpc) are used to express great distances in the Universe; in this scheme, 1 kpc = 1,000 pc = 3,260 light-years, and 1 Mpc = 1 million pc = 3.26 million light-years. Now, finally, intergalactic distances become credible.

The nearest visible object outside our Solar System is the Alpha Centauri star system, which is 1.4 pc away. There are numerous stars within 20 to 30 pc of Earth. The standard distance for measuring absolute visual magnitude is 10 pc. The Milky Way is 30 kpc in diameter. The Andromeda galaxy is 670 kpc away. And on it goes, out to the limit of the observable Universe, somewhere around 3 billion pc, or 3,000 Mpc.

Spectral Classifications

On a clear night, especially when the Moon is not above the horizon and there are not many lights to produce skyglow, it's easy to see that stars have different colors, as well as different levels of brilliance. This is so because of differences in the amounts of energy stars emit at various wavelengths.

STELLAR SPECTRA

The wavelengths of visible light are extremely short and are commonly measured in units of *nanometers* (nm), where 1 nm $= 10^{-9}$ m, or in *Ångström units* (Å), where 1 Å $= 10^{-10}$ m = 0.1 nm. These units are microscopic in size. The visible spectrum extends from about 750 nm, representing red light, down to 390 nm, representing violet light. From longest to shortest wavelengths, colors proceed through the spectrum as red, orange, yellow, green, blue, indigo, and violet. The first letters of these colors come out as the odd name *Roy G. Biv.* Some people find this helpful in remembering the order in which the colors of the visible spectrum proceed.

All stars emit energy over a wide range of electromagnetic (EM) wavelengths, from low-frequency radio through microwaves, infrared, visible light, ultraviolet, x-rays, and gamma rays. The EM spectrum is sometimes portrayed in logarithmic form by wavelength (Fig. 13-2). The visible spectrum is a small portion of this.

Astronomers use an instrument called a *spectroscope* to scrutinize the spectra of the Sun, the planets, and distant stars. A spectroscope works according to the same principle by which a prism splits light into a rainbow. The spectroscope, however, is much more sophisticated. It can resolve the rainbow down into tiny slices. It also can examine wavelengths that are not visible to the unaided eye, particularly the *near infrared* or *near IR* (energy at wavelengths slightly longer than 750 nm) and the *near ultraviolet* or *near UV* (energy whose wavelengths are a little shorter than 390 nm).

When the spectra of distant objects are examined with a spectroscope, dark lines appear at certain places. Each chemical element is known to produce a certain pattern of such lines. In this way, astronomers can tell what distant objects are made of. This was first done with the Sun; the dark lines were discovered by accident. They were first studied seriously by a German astronomer named Joseph von Fraunhofer around the year 1800. Today, dark lines in stellar spectra are sometimes called *Fraunhofer lines.* Because they are caused by the absorption of energy at specific wavelengths, they are also known as *absorption lines*.

OBAFGKM

The first serious attempts to study the spectra of stars revealed dark absorption lines, just like the ones observed in the spectrum of the Sun. However, not all stars have the same pattern of lines. In the late 1800s, an astronomer

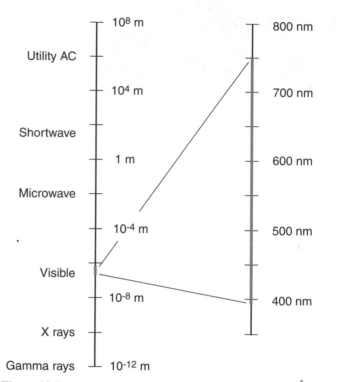

Figure 13-2.   At left, the electromagnetic (EM) spectrum from 10^8 m (300,000 km) to 10^{-12} m (0.001 nm). At right, the visible spectrum from 750 nm to 390 nm. Red is at the top; violet is at the bottom.

at Harvard University named Annie Cannon compiled a record of the spectra of nearly half a million stars. This became known as the *Henry Draper Catalogue*.

There are seven main categories of stars, classified according to the type of spectrum they have. The main categories have been given the names O, B, A, F, G, K, and M. Each of these seven classes is divided into subcategories from 0 through 9. Thus a type A9 star is followed by a type F0 star, which is followed by F1, F2, F3, and so on. In all, there are 70 different spectral types of stars. What do these letters and numbers actually mean?

It turns out that the spectrum of a star tells us the surface temperature. Type O stars are the hottest and appear bluish to the eye. Type M stars have the lowest surface temperatures, and they appear orange or ruddy. Within a particular alphabetic subdivision, the number 0 represents the highest temperature, and the number 9 represents the lowest. The hottest possible star would be symbolized O0 (the letter O followed by the numeral 0); the coolest would be M9. On this scale, our Sun is a type G2 star. This means that it is medium cool. Of course, *hot* and *cool* are relative terms; even an M9 star is scorching hot by Earthly standards.

The surface temperature of a star is related to its absolute visual magnitude. This relationship was found in the early 1900s by a Danish astronomer named Ejnar Hertzsprung and an American named Henry Russell. These two scientists, working independently, graphed the absolute magnitudes of some nearby stars as a function of spectral classification. They found that, in general, as the surface temperature increases, so does the absolute brightness. This is not surprising, but it does not tell the whole story. Temperature is not the only variable in star classification. Size matters too.

Our Sun is a rather small star. The largest stars are called *giants*. The smallest are called *dwarfs*. Some relatively cool stars are bright because they are huge: the *red giants*. Some hot stars are dim because they are tiny: the *white dwarfs*.

THE HERTZSPRUNG-RUSSELL (H-R) DIAGRAM

When the spectral type of a star is graphed along with its absolute magnitude, the star is represented by a single point on a coordinate plane. Figure 13-3 shows what Hertzsprung and Russell found. Such graphs are used by astronomers to this day and are called, appropriately enough, *Hertzsprung-Russell (H-R) diagrams*. On the horizontal axis, the highest temperature is toward the left, and the coolest is toward the right. On the vertical axis, the brightest absolute visual magnitude is toward the top, and the dimmest is toward the bottom. Our Sun is shown by the large dot.

Stars denoted in the upper right part of the H-R diagram are red giants. Blue giants are in the upper left corner, at the top of the *main sequence*. The smallest and coolest stars, the red and orange dwarfs, are at the lower right end of the main sequence. White dwarfs appear at the bottom and are not on the main sequence. There are also supergiants that do not fall onto the main sequence; these are shown at the top middle.

So what exactly is this main sequence? As Hertzsprung and Russell plotted their diagrams, they noticed an interesting correlation. Most stars fall along a curve running diagonally from the upper left to the lower right. This became known as the main sequence because it contains the majority of stars.

When astronomers began investigating the relationship between the location of a star in the galaxy and its position on the H-R diagram, some fascinating discoveries were made. Hot, massive stars seem to be concentrated

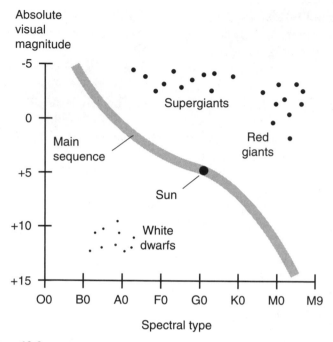

Figure 13-3. Most stars fall along a characteristic curve called the main sequence in the Hertzsprung-Russell diagram. Our Sun (large dot) is a main-sequence star.

mostly in the flat, disk-shaped part of the Milky Way, in the spiral arms but not in the central region. The spiral arms happen to be where most of the interstellar gas and dust is found. It was theorized that stars formed from this material, and we should therefore find more young stars in the spiral arms of the galaxy than near the center. There is not much interstellar material in the central part of the galaxy, and the stars there are much older. This led to the notion that the galaxy has evolved from the center outwards. If this theory is correct, the galaxy looked much different a few billion years ago than it does now.

Stellar Birth and Life

All stars evolve from clouds of gas and dust. If the original material in the Universe were perfectly homogeneous—equally dense at every point in space—stars could, in theory, never form. The slightest irregularity, however, brought about more irregularity, leading to regions where the gas and dust became concentrated. This is known as the *butterfly effect* and was discussed in Chapter 6 when we took a "mind journey" to the planet Mars. This effect breeds new stars.

A STAR IS BORN

Where the clouds of matter were the most dense, the gravitational attraction among the atoms was the greatest. This caused the dense regions to become even more dense and the sparse regions to get more sparse. A vicious circle ensued, which was repeated in countless locations. It is evidently still taking place in the spiral arms of the Milky Way and in other galaxies.

As a cloud of gas and dust contracts, it eventually begins to heat up. The atoms, originally free to move without restriction, get cramped for space and start to collide with one another. This produces outward pressure, but the increased concentration of matter causes a dramatic increase in the gravitational attraction among the atoms. This gravitational force keeps pulling the gas-and-dust cloud tighter and tighter, and it gets hotter and hotter. Finally, the temperature gets so high that hydrogen atoms begin to fuse, forming helium atoms along with great quantities of energy. This causes the star to become extremely hot, and the outward pressure finally rises to meet the inward force of gravitational collapse. Several hundred thousand years, or a few million years, go by between the initial contraction of the gas-and-dust cloud and the start of the hydrogen fusion reaction.

Large stars are born more quickly than small ones. Sometimes two or more stars are formed so close together that they orbit one another; these are *binary stars* and *multiple stars*. Sometimes huge gas-and-dust clouds give rise to clusters of stars.

We can follow the metamorphosis of a young star in an H-R diagram. A *protostar*, as it contracts and heats up, is not very luminous until the fusion reaction starts. Protostars are situated off the scale at the bottom of the graph. As the protostar contracts to the point where fusion begins, the star's position moves upward (Fig. 13-4). Almost every new star comes to rest on the main sequence. The most massive stars end up at the upper left and become blue and white supergiants. The least massive stars reach their stable positions at the lower right and become orange and red dwarfs. Oddly enough, the dim, cool, and least spectacular stars burn longest, and the bright, massive, and hot stars have much shorter lifespans.

Once the fusion process begins and the star shines brightly, the remaining gas in the star's vicinity becomes ionized by ultraviolet (UV) rays and

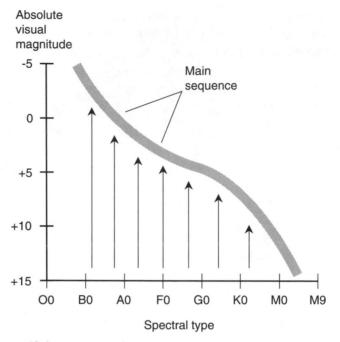

Figure 13-4.   As stars are born and mature, they move onto the main sequence.

x-rays. When we look at the Pleiades, for example, through a large tele-scope, this glowing gas can be seen. Much of the superfluous gas and dust near the star is blown away by the *stellar wind*, high-speed subatomic par-ticles emitted by the star shortly after it begins to shine. This keeps all the gas and dust in the galaxy in a constant state of turmoil, like a room full of smoke in which people are talking and gesturing.

LIFE SUMMARIES

Massive stars lead much different lives than small ones. Our Sun seems to be slightly "on the good side" of the dividing line between stable red and orange dwarfs and unstable blue and white giants.

Eventually, most of the hydrogen in the center of a star gets converted to helium by the fusion process. For nuclear reactions to continue, the core temperature must rise. As the star "runs out of gas," it starts to cool, and gravitation overcomes the outward pressure caused by internal heating. This contraction causes the temperature to rise again, and this time it goes even higher than it did when the star was born. Eventually, it gets so hot that helium atoms begin to fuse, forming carbon atoms plus energy. Our Sun has not yet reached this stage.

Once the helium has been used up, the star contracts again, becoming hotter still; carbon atoms begin to fuse, forming heavier atoms. In this way, it is believed, all the 92 elements up to and including uranium were formed in the interiors of stars long ago. This includes all those atoms in our planet: all the iron ore, all the silicon, all the gold, and everything else. It even includes the atoms in your body, with the exception of hydrogen that forms part of the water that keeps you alive. But how did all these heavier elements get here from the centers of stars? The answer lies in the fate of the most massive stars: They are doomed to blow up.

Large stars have attracted much attention from astronomers. When stars age, they sometimes swell to extreme size, and their surface temperatures cool. They become red giants. This is expected to happen to our Sun in a few billion years. After the red-giant phase, the Sun will contract and gradually fade away. However, more massive stars undergo a sudden and violent death. For a few days they can become as bright as all the rest of the stars in the galaxy combined. Once in a while we see such a *supernova* from Earth. When a big star explodes, it hurls much of its matter into interstellar space, where it cools and becomes the stuff from which future star systems can form. Some of these systems are believed to develop into star-and-planet families similar to our Solar System. This explains how the heavy elements got here. If this theory is correct, we all owe our existence to one or more supernovae that took place billions of years ago.

Variable Stars

Some stars, like our Sun, maintain constant brightness from hour to hour, day to day, and year to year. However, there are many pulsating, or variable, stars in our galaxy and other galaxies.

ECLIPSING BINARIES

Some of the fluctuating "stars" are actually binary systems (pairs of stars in mutual orbit) that eclipse each other. If one of the stars is large and dim and the other is smaller and brighter, and if the orbital plane of the stars appears edge-on to us, we will see a decrease in the brilliance of the system when the dimmer star passes in front of the brighter one. These *eclipsing binaries* are characterized by steady brilliance with periodic sharp dips. The dips always have the same depth; that is, the minimum brilliance is always the same. The dips occur at regular intervals.

Eclipsing binaries can be recognized in another way. The absorption-line spectra of the two stars alternately shift slightly toward the red and blue ends of the spectrum as one star recedes from us and the other approaches. When an object moves away from us, the spectral lines are *red-shifted* (they move toward longer wavelengths); when an object approaches, the spectral lines are *blue-shifted* (they move toward shorter wavelengths). This is a result of the well-known *Doppler effect*, the same phenomenon that makes a car horn sound higher in pitch as the car comes at you and lower in pitch as it passes and then moves away from you. Dual red-blue back-and-forth spectral shifting is a dead giveaway that a variable "star" is an eclipsing binary.

CEPHEID VARIABLES

Certain variable stars, known as *Cepheid variables* or *Cepheids*, are intriguing because of the clockwork regularity of their changes in brilliance. These stars get their name from the fact that one of the first-discovered and most well-known of them is in the constellation *Cepheus*.

Individual Cepheids change brightness at a consistent rate. However, two different Cepheids can have different *periods* (lengths of time from one peak of brilliance to the next). Some Cepheids have periods of less than 1 Earth day; others have periods of several weeks. Some Cepheids vary greatly in brightness, whereas others vary only a little. Polaris, the familiar North Star that has been used by navigators for centuries, is a Cepheid variable. Its brilliance does not fluctuate enough to be noticeable to the casual observer.

After tirelessly searching for coincidences and patterns, astronomers discovered that the periods of Cepheids are correlated with their absolute visual magnitudes. This correlation is so precise that these stars can be used as distance-measuring beacons for determining distances in interstellar and intergalactic space. The longer the period, the brighter the star, averaged over time. Cepheids are massive stars that, when plotted on an H-R diagram, fall in the upper middle region, off the main sequence. These are called *yellow supergiants*. Not all yellow supergiants are Cepheids, but all Cepheids seem to be yellow supergiants.

RR LYRAE VARIABLES

There are other types of variables besides eclipsing binaries and Cepheids. *RR Lyrae variables* are bluish stars and are on average, about 40 times as bright as the Sun. The term comes from the location, in the constellation Lyra, of one of the best-known stars of this kind.

RR Lyrae variables have regular, constant cycles of pulsation, usually with periods of about half an Earth day. They tend to be found in large clusters of stars and in the halo of stars surrounding the Milky Way. Because of their locations, they are always far away and require a telescope to be observed. They, like Cepheids, can be used as distance-measuring beacons.

MIRA VARIABLES

Both the Cepheid and RR Lyrae variable stars physically expand and shrink in size as their brightness changes. Another sort of variable, called a *Mira variable*, oscillates without any apparent change in size. Mira variable stars are named after a certain red giant, *Mira*, in the constellation Cetus. Mira's brightness varies with a period of about a year and fluctuates between the eighth and the third visual magnitudes. Thus, at its peak, it is about 100 times as bright as it is at the minimum in its cycle. Because of their long periods, Mira stars are sometimes called *long-period variables* (LPVs).

Astronomers think that Mira variable stars do not expand and contract in radius, as do the Cepheid or the RR Lyrae variables. The increase in brilliance occurs faster than the decline. While Mira, the first variable of this type to be discovered (in 1596), varies by 5 visual magnitudes, some Mira variables have brightness ranges of only 2 or 3 magnitudes, whereas others can change by up to 10 magnitudes. Mira variables are believed to be old stars that have depleted much of their hydrogen and are fusing helium atoms. These stars contain heavier elements as well; most of them are rich in elemental carbon, and some contain large amounts of elemental oxygen.

Stellar Anatomy and Longevity

It is safe to say that no human being will ever travel inside a star to find out what the interior is like. Conditions inside stars are too hostile for any kind of direct observation or exploration; space ships would be vaporized on reaching the surface, if not before. The science-fiction story "Icarus and Daedalus," written by Russian author Henrik Altov, portrays a hypothetical journey through the center of the Sun. However, such stories will always remain fiction.

MODEL STARS

Astronomers have deduced certain things about the interiors of stars, including the Sun, just as they have been able to gather ideas about the interiors of the planets, including the Earth. Different types of stars have different internal structures. We also know that different types of stars have much different "atmospheres." The "atmosphere" surrounding a star can extend outward for millions of kilometers and is called the *corona*.

Computers have been used to construct *model stars* using programs that simulate events believed to take place in the depths of these fusion furnaces. Known physical laws are applied to denote what happens inside a star with a given mass and material composition. Stars are considered to have concentric layers or "shells," each with a certain temperature, mass, and combination of elements. The whole star is "constructed" by putting the shells together. This can be done outward from the center toward the surface and then to the corona or inward from the corona to the surface and ultimately to the center. Sometimes, when the two modeling schemes (inside-out versus outside-in) are both applied to objects having the same mass and material composition, the results are different.

In the depths of a large main-sequence star such as a blue supergiant, radiation and convection are believed to be mutually responsible for the transfer of energy from the core of the star into space. Near the surface, radiation predominates. Near the center, convection is thought to play a more important role because the matter deep inside stars is so dense that it is opaque to radiation. This model of a large star is shown in Fig. 13-5a. In a much less massive main-sequence star, the reverse is believed to be the case. A red-dwarf star apparently has a convective outer region and a more transparent core. Radiation should predominate deep inside such a star. This model of a red dwarf is shown in Fig. 13-5b.

Both these models are largely speculative. Computers can work only with the data that they get, and these data are based on educated guesswork. Nature is not a computer program. We can't be sure that the models we invent are perfect representations of what really goes on in the Universe. In fact,

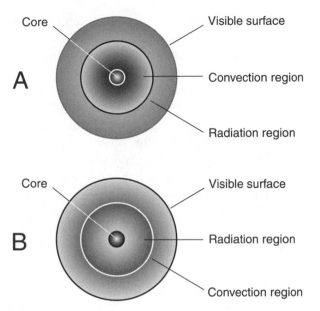

Figure 13-5.   At *A*, simplified model of the interior of a large main-sequence star such as a blue supergiant. At *B*, a model of a small main-sequence star such as a red dwarf.

people ought to know by now, after decades of working with computers, that there are always some differences between the "mind" of nature and the "mind" of a computer program! As observational techniques improve and computers become more enlightened about nature, star modeling should get better and better. We'll never know for sure, however, what the inside of a star is like unless and until we can send some sort of probe down inside one. The Sun is the logical candidate for a first try, but no one has yet figured out how a navigable space vessel could be designed to survive the extreme temperature and pressure.

INTERIOR OF THE SUN

The Sun is an average star in terms of its location on the H-R diagram. It is neither a blue giant nor a red dwarf. What is the interior of the Sun like? Astronomers have one great advantage when they look at the Sun as a star and try to evaluate its internal and external characteristics. This is, of course, the fact that it is much closer than other stars. Telescopes can resolve surface details. Several different theoretical models have been constructed for our parent star.

All the models suggest a solar core temperature of several tens of millions of degrees Celsius. The surface is not nearly as hot, though, only around 5,000 to 6,000°C. Above the visible surface, the temperature rises as the altitude increases, up to a maximum of 500,000 to 600,000°C in the corona. Then the temperature begins to decrease and keeps on dropping as the distance reaches the orbital radius of Mercury, Venus, and so on through the rest of the Solar System. Some solar effects continue on out past the orbit of Pluto, in particular the solar wind, consisting of a continuous outward barrage of subatomic particles. Eventually, the temperature reaches the general level of interstellar space, and the solar wind is overcome by the effects of interstellar winds. This transition zone, where the Sun's dominion ends, is called the *heliopause*.

The interior of the Sun is thought to more closely resemble that of a red dwarf (see Fig. 13-5*b*) than that of a blue supergiant (see Fig. 13-5*a*). This has led to the belief that the Sun's future is bound to be more like that of a small star than a big one. We need not fear that our parent star will ever go supernova on us. But the Sun's lifespan will not be infinite. There will come what has been called a *last perfect day*, after which the Sun will undergo a gradual but radical change. If human beings or other intelligent beings still populate the Earth on the last perfect day, they won't go to bed one night after a benevolent dusk and wake up the next morning to a hostile dawn, but the time of change will be at hand. Someone somewhere will declare: "It is time for us to find a new star system. Everybody start packing."

RED-GIANT PHASE

According to scientists, our Sun has plenty of hydrogen fuel left. Our parent star should keep on shining, just as it does today, for many millions of years to come. On that last perfect day, however, the hydrogen fuel will be depleted to the point that changes begin to take place inside the Sun. The metamorphosis will occur at first in the core and then will migrate outward toward the surface.

At some distant future time, the Sun's core will start to contract because of the depletion of hydrogen fuel. Gravitation will take over for a while, having patiently waited all those billions of years since the hydrogen-fusion process began. As the core contracts, it will heat up again and eventually will become so hot that helium, the material by-product of hydrogen fusion, will begin to fuse. This will generate a new source of energy. The outer layers will continue to burn hydrogen into helium and energy. The core will attain a superhot surface that will cause the outer regions to expand; the result will be an ever-increasing overall solar diameter. The helium in the core will fuse into carbon.

As the Sun expands from continued pressure deep within itself, the surface will cool, but the surface area will increase greatly. The Sun's apparent diameter in the sky, as seen from Earth, will double, then triple, then quadruple. The color will shift from the so-called yellow part of the spectrum into the orange and then into the red. Our parent star will leave the main sequence, migrating upward and to the right as plotted on an H-R diagram. Temperatures on the Earth will increase. First the inland ponds and lakes will boil. Then the rivers and large inland seas will boil. Then the oceans will boil completely away, the atmosphere will be driven off into space, and Earth will become a barren, scorching desert similar to Mercury's current state. Some astronomers think that the Sun will expand so much that its radius will exceed 1 astronomical unit (AU). In this case, our planet will cease to exist.

DEATH OF A STAR

Eventually, all the helium fuel in the core of the Sun will be spent. The core will again begin to collapse under the force of gravitation as the interior pressure drops. The rest of the Sun will follow, and the red giant will grow smaller. The internal and surface temperatures will rise, but not enough to cause further fusion reactions. The only thing stopping the inward drive of gravitation in the end will be the forces inside subatomic particles. It is believed that the Sun will come to rest at a diameter roughly the same as that of today's Earth, and its energy output will fade away like the glow of an ember in a forgotten fire. This fate awaits all stars whose masses are less than, equal to, or somewhat greater than that of the Sun.

Stars that are much more massive than the Sun have more interesting deaths. They produce atoms heavier than carbon, including oxygen, neon, magnesium, silicon, sulfur, and iron. However, there is a limit to this. When iron is formed from fusion, it is the last hurrah. Iron will not fuse. When the iron is spent at the core, it collapses almost instantly and then rebounds, causing an outward shock wave that blows the star up. Then, in the mass that remains, gravitation can become so powerful that it overcomes everything else. The end products of some large stars are thought to be *black holes*, where matter has been crushed by gravitational forces so strong that time and space are altered and nothing can escape, not even the starlight. Black holes were once regarded as the figments of theorists' imaginations, too weird to exist in the real Cosmos. Nowadays, though, there is credible evidence that they indeed exist. We'll delve into the mysteries of these and other "stellar ghosts" in the next chapter.

Star Clusters

Some star systems contain hundreds, thousands, or even millions of individual stars, all held in each other's vicinity by their mutual gravitation. Some huge gas-and-dust clouds condense into *star clusters*. There are three main types of star clusters found within our galaxy and presumably within other galaxies.

OPEN CLUSTERS

The *Pleiades*, also known as the "seven sisters," are a familiar star cluster that can be seen easily with the unaided eye. On a clear night, most people can see six individual stars in this group. People with sharp vision can see seven stars; with a good pair of binoculars, it becomes apparent that there are a lot more than seven stars in this cluster. The Pleiades are an example of an *open star cluster*. This type of star group is also known as a *galactic cluster*. The stars are held together by gravity, but they are not tightly packed. They do not show a striking central concentration nor an orderly pattern or structure.

How do we know that the stars in an open cluster are associated and are not just accidentally close to each other because they all happen to fall along the same line of sight relative to Earth? One way to find out is to measure the radial speeds of the individual stars relative to us and then compare these speeds. If the stars are in a common group, held in each other's vicinity by gravity, then we should not observe much difference in the radial speeds of the stars. If, however, we're just seeing a coincidental lineup of stars, the individuals in the swarm should have much different radial speeds, just like stars chosen at random in the sky. It turns out that the stars in an open cluster all have radial speeds that are nearly identical.

GLOBULAR CLUSTERS

The most spectacular star groups, besides galaxies themselves, are known as *globular clusters*. They get this name because of their symmetrical appearance. Some of these clusters contain more than 1 million stars, 50 percent of which lie inside a spherical shell whose radius is called the *median radius*. The concentration of stars is greatest near the center of the cluster and decreases uniformly in all directions away from the center, as shown in Fig. 13-6.

An excellent example of a globular cluster is M13 in the constellation Hercules. This can be seen in the summer sky from the northern hemisphere. When viewed at low magnification through a large amateur telescope, M13 is spectacular. More than 100 globular clusters have been observed in and around the Milky Way. Globular clusters, unlike open clusters, are arranged in a large spherical halo around the galaxy. There are many globular clusters that lie outside the disk-shaped main part of the Milky Way whose stars are arranged in spirals like the pattern of rainbands in a hurricane.

Globular star clusters are believed to be very old, among the oldest objects in the observable Universe. Such clusters contain large numbers of variable stars, much greater in proportion than the stars in the rest of the

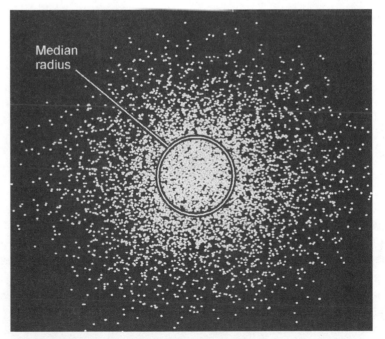

Figure 13-6.　Simplified cross-sectional drawing of a globular star cluster. Half of the stars lie within the median radius.

galaxy. The stars in these clusters are mostly *metal-deficient*. The term *metal* in this context refers to any element other than hydrogen and helium, the two most abundant and oldest elements in the Cosmos. This fact suggests that globular clusters formed from the original stuff of the Universe, the original gases hydrogen and helium, and that few or none of the stars in the clusters are the result of congealed gas and dust from previously blown-up stars.

Nebulae

The space between the stars is far from empty. Gas and dust, some of it left over from the primordial Universe and much of it the remnants of exploded massive stars, is strewn throughout the Milky Way. Most of this material is found in or near the plane of our galaxy.

DARK NEBULAE

When viewed through a good pair of binoculars or a small telescope, the glow of the Milky Way, especially in the direction of the constellation Sagittarius, is resolved in detail. Millions of stars can be seen. In time-exposure photographs, dark rifts, wisps, and tendrils appear. Some of these look like clouds (and they are, but much larger than any clouds on Earth). This motivated astronomers to call such an object a *nebula*, the Latin word for "cloud." The plural is *nebulae*, in true academic fashion.

At first, the *dark nebulae* were mistaken for regions of diluted stars. However, spectral analysis of starlight shining through revealed the truth. When photographed through large telescopes, it became obvious that the dark nebulae were indeed clouds. We cannot observe their motion directly, but these clouds are blowing around in space, having been given momentum by the explosions from which they came and also because they are affected by the stellar winds, gravitation, and magnetic fields produced by stars near them. Sometimes the clouds form vortices like cosmic dust devils or newborn hurricanes. These rotating nebulae collapse into new stars and, in some cases, into planetary systems like our Solar System.

EMISSION NEBULAE

Once astronomers had access to large telescopes, sophisticated cameras, and spectroscopes, a visual world was opened to them that revealed not only dark clouds but also glowing ones. Such clouds are called *emission nebulae* because they appear bright against the dark background of space.

If you have a good amateur telescope measuring 20 cm (8 in) or more in diameter and can set it up for low magnification, try looking at the center star in Orion's sword on a moonless, exceptionally clear night when Orion is high in the sky. You'll see a fine example of an emission nebula, known as the *Great Nebula in Orion*. This cloud of gas and dust glows because some of its atoms are ionized by ultraviolet and x-rays from stars in its vicinity. This nebula is approximately 1,600 light-years from our Solar System and has an actual diameter of some 20 or 30 light-years.

PLANETARY NEBULAE

Emission nebulae can take on almost any shape. One of the more interesting of these is the *Ring Nebula* in the constellation Lyra. There are many other such ring-shaped emission clouds visible in our galaxy through large telescopes. Some appear visually similar to planets and easily can be mistaken for planets by a casual observer; hence they are called *planetary nebulae*. The ring-shaped nebulae are actually shaped like spheres with thick shells. We see them as rings or donuts because we are looking at the most glowing material when our line of sight passes near, but not exactly at, the periphery. The objects appear less bright toward the center and dimmest at the extreme outer edge (Fig. 13-7).

Planetary nebulae always have stars at their centers, and they are always expanding, as can be determined by looking at the emission spectra. The emission lines, the signatures of chemical elements, appear shifted toward the violet end of the visual spectrum near the center of such a nebula and are shifted little or not at all near the edges. This indicates that the whole sphere is growing like a balloon being inflated, although not fast enough for us to observe directly. It is believed that these nebulae are fluorescent gas cast off by stars nearing the ends of their lives. The gas gradually disperses into the interstellar medium, where some of it finds its way into new stars, asteroids, comets, planets, and moons.

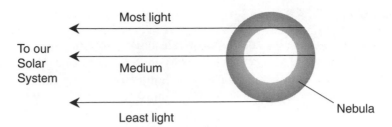

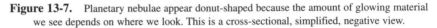

Figure 13-7. Planetary nebulae appear donut-shaped because the amount of glowing material we see depends on where we look. This is a cross-sectional, simplified, negative view.

Quiz

Refer to the text if necessary. A good score is 8 correct. Answers are in the back of the book.

1. Energy from a certain star is observed at a wavelength of 10^{-8} m. This is
   (a) between utility ac and shortwave radio.
   (b) microwave energy.
   (c) visible light.
   (d) between visible light and x-rays.

2. A galaxy is 100 million pc distant. Approximately how far is this in light-years?
   (a) 326 million
   (b) 30 million
   (c) 100 million
   (d) It can't be calculated from this information.

3. If the Sun were viewed from a distance of 32.6 light-years, what would be its apparent visual magnitude?
   (a) +4.8
   (b) −4.4
   (c) 0
   (d) It can't be calculated from this information.

4. Red-dwarf stars
   (a) burn out more quickly than blue giants.
   (b) live for about the same length of time as blue giants.
   (c) live much longer than blue giants.
   (d) never get hot enough for nuclear fusion to occur.

5. A certain stellar object has brightness that is constant, except for dips that occur at uniform intervals. There are two sets of spectral lines; one shifts toward the red, and the other shifts toward the blue. We can say with confidence that this is
(a) a Mira variable.
(b) a Cepheid variable.
(c) an RR Lyrae variable.
(d) an eclipsing binary.

6. Stars in the extreme lower right portion of the H-R diagram are
(a) red giants.
(b) red dwarfs.
(c) white dwarfs.
(d) supernovae.

7. When an object's spectrum is red-shifted, this means that the spectral lines appear
(a) smeared out.
(b) invisible.
(c) longer in wavelength than normal.
(d) shorter in wavelength than normal.

8. A metal-deficient star consists almost entirely of
(a) all elements except the metals.
(b) helium.
(c) gas and dust.
(d) hydrogen and helium.

9. It is believed that someday the Sun's core will
(a) explode.
(b) shrink because its hydrogen fuel has been spent.
(c) collapse down to a geometric point.
(d) condense into liquid water and then freeze into water ice.

10. The absolute visual magnitude of a star is the same as the apparent magnitude at a distance of
(a) 3.26 pc.
(b) 10.0 pc.
(c) 32.6 pc.
(d) 100 pc.

CHAPTER 14

Extreme Objects in Our Galaxy

The Cosmos can do strange things to matter. Strange, that is, according to what is considered "normal" here on Earth. Some scientists think that these extreme manifestations—antiparticles, neutron stars, geometric points with infinite density, and such—are as common in the Universe as any other form of matter.

Matter and Antimatter

The notion of *antimatter* is as old as science fiction. You know how this kind of story can go. A space ship lands on a planet that turns out to be made of antimatter, and the journey comes to an unexpected, abrupt, and catastrophic end. A rogue band of space aliens comes to Earth with an antimatter bomb and sells it to the highest bidder. What is the reality of antimatter? Let's look at the nature of matter first. In simplistic terms, matter consists of three types of particles: the *proton*, the *neutron*, and the *electron*.

THE PROTON

Protons are too small to be observed directly, even with the most powerful microscopes. All protons carry a positive electrical charge. The charge on every proton is the same as the charge on every other. Every proton at rest has the same mass as every other proton at rest. Most scientists accept the proposition that all protons are identical, at least in our part of the Universe, although they, like all other particles, gain mass if accelerated to extreme speeds. This increase in mass takes place because of relativistic effects; you'll learn about this later.

While an individual proton is invisible and not massive enough to make much of an impact all by itself, a high-speed barrage of them can have considerable effects on matter. Protons are incredibly dense. If you could scoop up a level teaspoonful of protons the way you scoop up a teaspoonful of sugar—with the protons packed tightly together like the sugar crystals—the resulting sample would weigh tons in Earth's gravitational field. A little ball made of solid protons would fall into the Earth and cut through the crustal rocks like a lead shot falls through the air.

THE NEUTRON

A neutron has a mass slightly greater than that of a proton. Neutrons have no electrical charge, and they are roughly as dense as protons. However, while protons last for a long time all by themselves in free space, neutrons do not. The *mean life* of a neutron is only about 15 minutes. This means that if you gathered up a batch of, say, 1 million neutrons and let them float around in space, you would have about 500,000 neutrons left after 15 minutes. After 30 minutes, you would have approximately 250,000 neutrons remaining; after 45 minutes, there would be about 125,000 neutrons left.

Neutrons can last a long time when they are in the nuclei of atoms. This is a fortunate thing because if it weren't true, matter as we know it could not exist. Neutrons also can survive for a long time when a huge number of them are tightly squeezed together. This happens when large stars explode and then the remaining matter collapses under its own gravitation.

THE ELECTRON

An electron has exactly the same electrical charge quantity as a proton but with opposite polarity. Electrons are far less massive than protons, however. It would take about 2,000 electrons to have the same mass as a single proton.

One of the earliest theories concerning the structure of the atom pictured the electrons embedded in the nucleus, like raisins in a cake. Later, the electrons were imagined as orbiting the nucleus, making every atom like a miniature star system with the electrons as the planets (Fig. 14-1). Still later, this view was modified further. In today's model of the atom, the electrons

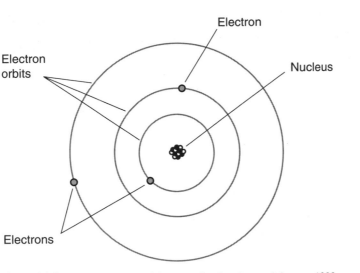

Figure 14-1. An early model of the atom, developed around the year 1900.

arc fast-moving, and they describe patterns so complex that it is impossible to pinpoint any individual particle at any given instant of time. All that can be done is to say that an electron will just as likely be inside a certain sphere as outside. These spheres are known as *electron shells*. Unless there is an external force or force field acting on the atom, all the electron shells are concentric, and the nucleus is at the center of the whole bunch. The greater a shell's radius, the more energy the electron has. Figure 14-2 is a simplistic drawing of what happens when an electron gains enough energy to "jump" from one shell to another shell representing more energy.

Generally, the number of electrons in an atom is the same as the number of protons. The negative charges therefore exactly cancel out the positive ones, and the atom is electrically neutral. Under some conditions, however, there can be an excess or shortage of electrons. High levels of radiant energy, extreme heat, or the presence of an electrical field can "knock" electrons loose from atoms, upsetting the balance.

EQUAL AND OPPOSITE

The proton, the neutron, and the electron each have their own "nemesis" particle that occurs in the form of *antimatter*. These particles are called *antiparticles*. The antiparticle for the proton is the *antiproton*, for the neutron it is the *antineutron*, and for the electron it is the *positron*. The antiproton has the same

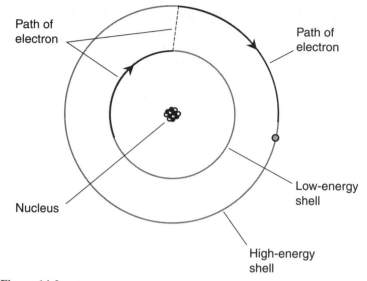

Figure 14-2. Electrons "orbit" the nucleus of an atom at defined levels, each level
corresponding to a specific, fixed energy state.

mass as the proton, but it has a negative electrical charge equal and opposite
to the positive electrical charge of the proton. The antineutron has the same
mass as the neutron. Neither the neutron nor the antineutron have any electrical charge. The positron has same mass as the electron, and it is positively
charged to an extent equal to the negative charge on an electron.

You might have read or seen in science fiction novels and movies that
when a particle of matter collides with its nemesis, they annihilate each
other. This is indeed true. The combined mass of the particle and the
antiparticle is completely converted into energy, according to the same
Einstein formula that applies in nuclear reactions:

$$E = (m_+ + m_-) c^2$$

where E is the energy in joules, m_+ is the mass of the particle in kilograms,
m_- is the mass of the antiparticle in kilograms, and c is the speed of light
squared, which is approximately equal to 8.9875×10^{16} m^2/s^2.

UNIMAGINABLE POWER

If equal masses of matter and antimatter are brought together, in theory, all
the mass will be converted to energy. If there happens to be more matter

than antimatter, there will be some matter left over after the encounter. Conversely, if there is more antimatter than matter, there will be some antimatter remaining.

In a nuclear reaction, only a tiny fraction of the mass of the constituents is liberated as energy; plenty of matter is left over, although its form has changed. You might push together two chunks of uranium-235, the isotope of uranium whose atomic mass is 235 atomic mass units (amu), and if their combined mass is great enough, an atomic explosion will take place. However, there will still be a considerable amount of matter remaining. We might say that the matter-to-energy conversion efficiency of an atomic explosion is low. Of course, this is a relative thing. Compared with dynamite, an atomic bomb is extremely efficient at converting matter to energy. Compared to an antimatter bomb, however, should one ever be devised, an atomic bomb is inefficient.

In a matter-antimatter reaction, if the masses of the samples are equal, the conversion efficiency is 100 percent. As you can imagine, a matter-antimatter bomb would make a conventional nuclear weapon of the same total mass look like a firecracker by comparison. A single matter-antimatter weapon of modest size easily could wipe out all life on Earth. A big one could shatter or even vaporize the whole planet.

WHERE IS ALL THE ANTIMATTER?

Why don't we see antimatter floating around in the Universe? Why, for example, are Earth, Moon, Venus, and Mars all made of matter, not antimatter? (If any celestial object were made of antimatter, then as soon as a spacecraft landed on it, the ship would vanish in a fantastic burst of energy.) This is an interesting question. We are not absolutely certain that all the distant stars and galaxies we see out there consist of matter. We do know, however, that if there were any antimatter in our immediate vicinity, it would have long ago combined with matter and been annihilated. If there were both matter and antimatter in the primordial Solar System, the mass of the matter was greater, for it prevailed after the contest.

Most astronomers are skeptical of the idea that our galaxy contains roughly equal amounts of matter and antimatter. If this were the case, we should expect to see periodic explosions of unimaginable brilliance or else a continuous flow of energy that could not be explained in any way other than matter-antimatter encounters. No one really knows the answers to questions about what comprises the distant galaxies and, in particular, the processes that drive some of the more esoteric objects such as quasars.

Small, Dark, and Massive

After a star such as our Sun has gone through its red-giant phase, the outer materials will drift away into space, and a white dwarf will remain. It will be only about as big as the Earth and will be dim compared with the Sun as we see it today. A similar fate awaits many other stars in the Universe. Some stars have more interesting destinies, though. Some of these end states are such that the term *star* does not seem appropriate.

BLACK DWARFS

Like a glowing cinder in an abandoned fire, a white dwarf becomes dimmer and dimmer until it is a globe having mass comparable with that of a star, temperature near absolute zero, and diameter comparable with that of a terrestrial planet such as Earth, Venus, or Mars. Such a star is called a *black dwarf*.

If space travelers ever come across a black dwarf, it will look something like a planet. However, the gravitation in the vicinity of and on the surface of such an object will give it away. In fact, landing on a black dwarf would be a suicide mission. The force would be so great that the spacecraft would be pulled down into a violent crash, and the bodies of the astronauts, as well as their vessel, would ooze into the surface.

After the Sun becomes a black dwarf, those planets not vaporized during the red-giant phase will continue to faithfully orbit. Much of the Sun's original mass will remain in the dark ball at the center of the Solar System.

NEUTRON STARS

In general, the larger a star is to begin with, the more dense is the final object, the "ghost of a star that once shone brightly." A black dwarf with more than a certain mass finds itself under the influence of gravitation so strong that all the electrons in the atoms are driven into the nuclei. Atoms are mostly empty space; solids don't pass through each other because of the repulsive electrical forces produced by their electron shells. This repulsive force, on an atomic scale, is powerful indeed. However, gravitation can overcome it if sufficient mass is put into a small enough physical volume.

When an electron combines with a proton because of gravitational forces, the result is a neutron. It is believed that some black dwarfs are balls of neutrons stuck together with little or no intervening space between them. This type of object is known as a *neutron star*.

The fate of a massive neutron star challenges even the most vivid imagination. If the original star had spin, the neutron star retains much of that rotational momentum. Because the neutron star is so much smaller than the original star, however, the neutron star spins faster (Fig. 14-3). Some neutron stars spin at dozens or even hundreds of revolutions per second. How can we know this when neutron stars are not visible directly through telescopes? The answer lies in the fact that some of them emit pulsating "signals" at radio, infrared, and optical wavelengths.

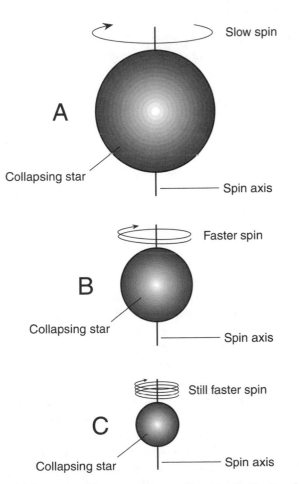

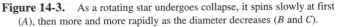

Figure 14-3.   As a rotating star undergoes collapse, it spins slowly at first (*A*), then more and more rapidly as the diameter decreases (*B* and *C*).

Pulsars

At the radio-telescope observatory of the Cambridge University in England, an antenna was specially assembled in the mid-1960s for the purpose of conducting an investigation of the rapid variation, or *scintillation*, of celestial radio-wave sources. The array was made up of more than 2,000 smaller antennas that covered several acres of ground. Scintillations were observed and analyzed by a graduate student, Jocelyn Bell, and her professor, Anthony Hewish. Special electronic circuits were designed to scrutinize the scintillations, making graphs of the radio-wave strength as a function of time. This led to an unexpected discovery.

LITTLE GREEN MEN

As Bell examined the graphs, she noticed strange, inexplicable, extremely regular sequences of pulses. A radio source, having a sharply defined location in the sky, was emitting the bursts of energy at precise intervals. Scintillations caused by interstellar gas-and-dust clouds or by other natural effects were never so well timed. Neither Bell nor Hewish seriously believed that the pulses came from intelligent beings, but when the news of their discovery leaked out, certain people in the media immediately began spreading stories to the effect that the astronomers had received signals from an extraterrestrial civilization. The phenomenon was even given a tongue-in-cheek code name: LGM, an abbreviation for "little green men." The radio source itself was called a *pulsar* and was given the designation CP1919, short for "Cambridge pulsar at right ascension 19 hours, 19 minutes."

Before the results of their discovery could be released, Bell and Hewish had to be sure that the signals were coming from space and were not some form of Earthly radio interference. The position of the pulsar remained at RA 19 h 19 min and was fixed with respect to the stars, moving across the sky with them as the Earth rotated on its axis. Other tests were conducted to rule out the possibility that some internal combustion engine or electronic device was causing the signals, but the conclusion was inescapable: They were coming from somewhere in deep space, beyond the Solar System.

At the same time, tests were conducted in an attempt to find out if the signals were coming from a planet orbiting a star. In such a case, a Doppler shift would occur as the planet first moved away from the Solar System and then toward it in its circumnavigation of its parent star (Fig. 14-4). This

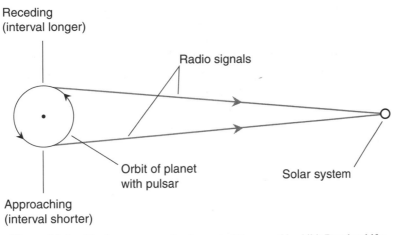

Receding
(interval longer)

Radio signals

Orbit of planet
with pulsar

Solar system

Approaching
(interval shorter)

Figure 14-4.   If pulsars were on planets, most of them would exhibit Doppler shifts
caused by the revolution of that planet around its parent star.

would show up as a regular variation in the pulse interval. No such shift
was found. Only if the planet's ecliptic was at a right angle to our view
would we observe no Doppler shift. (This would be a coincidence, but it
could happen.) If numerous other pulsars could be found and their signals
analyzed for Doppler shift, the question could be answered.

CHARACTERISTICS

More pulsars were found in 1968, and since then, many of the objects have
been catalogued. The periods, or intervals between pulses, vary greatly
from pulsar to pulsar, but they are all regular timekeepers. A few pulsars
show Doppler shifts, but this can be explained by supposing that such an
object is one member of a binary system including one ordinary star and
one pulsar. If pulsar signals really come from intelligent beings inhabiting
planets, then we should expect to see Doppler effects in almost all of
them—and this is not the case.

Although the "ticks" from pulsars always take place at regular intervals,
the waveforms of the pulses are irregular. This is another argument against
the theory that they are signals from extraterrestrial beings. A signal generated by a radio transmitter would be expected to have a smooth pulse shape,
something like that shown in Fig. 14-5a. However, the pulsars have rough
traces; a hypothetical example is shown in Fig. 14-5b.

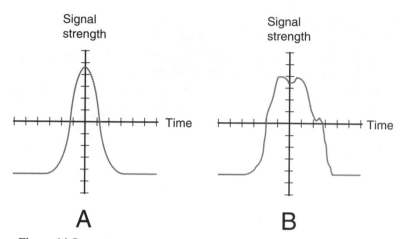

Figure 14-5.　At *A*, a smooth pulse, typical of what would be generated by a radio transmitter. At *B*, an irregular pulse, typical of pulsars.

When the angular diameter of CP1919 was measured, it turned out to be small. If it were visible, it would look like a flickering point of light. This did not come as a surprise to astronomers. The brevity of the pulse period and the shortness of the pulses themselves suggested a generating source that could be no more than a few hundred kilometers across. What sort of star could be that small and yet emit such powerful electromagnetic (EM) bursts? Scientists began to develop theories to account for the observations, but the task proved difficult.

Calculation of the distance to CP1919 indicated that it is about 400 light-years away. This determination was made using an ingenious scheme. At progressively longer wavelengths, the "ticks" from a pulsar arrive on Earth at greater and greater intervals. That is, the period of the pulsar depends on the wavelength to which a radio-telescope receiver is tuned. This takes place because the speed of EM wave propagation depends on the wavelength. In a perfect vacuum, all EM waves, however long or short, propagate at the same speed, approximately 2.99792×10^8 m/s. In interstellar space, however, which is not a perfect vacuum, radio waves travel slower than infrared, which in turn travels slower than visible light. (The effect is not peculiar to pulsars; it occurs for ordinary stars and emission nebulae as well.) This difference in propagation speed is tiny. Sophisticated instruments are required to detect it. This phenomenon is called *dispersion*, and it is the same effect that causes a prism to split white light into the colors of the rainbow.

The greater the distance to a pulsar, the more dispersion takes place, and the greater is the variation in the period as a function of wavelength. Knowing the extent to which dispersion occurs as a function of the wavelength, astronomers can calculate the distance to almost any pulsar because the pulse intervals are as regular as the output of an atomic clock. It is as if these objects were made-to-order distance-measurement devices.

SO WHAT ARE THEY?

Theories concerning the nature of pulsars began to arise after an astronomer named Thomas Gold hypothesized that they are fast-spinning neutron stars generating intense EM fields. The key to the shortness of pulsar periods, which had baffled astronomers and physicists, was the angular-momentum effect described earlier in this chapter and illustrated in Fig. 14-3.

The radio signals we observe coming from pulsars are apparently the result of the rapid spin combined with relativistic effects on the magnetic fields surrounding the object. According to Gold, the intensity of the magnetic field near a collapsing neutron star can grow to several trillion times the intensity of Earth's magnetic field—that is, trillions of gauss. Because the gravitation near the neutron star is so intense, it "warps" space and causes the magnetic lines of flux to "pile up" close to each other. This effect is exaggerated by the rapid spin of the object. These bunches of magnetic flux lines sweep around and around as the neutron star spins. From a distance, to observers located in positions such that the bunches of flux lines sweep across their instruments, the magnetic bursts appear as EM energy. This energy can occur, theoretically, at any wavelength, from low-band radio through microwave, infrared (IR), and even into the visible spectrum.

If Thomas Gold's theory is correct, it is reasonable to suppose that the rotational periods of pulsars should increase gradually. All spinning objects rotate more slowly as time passes because they lose momentum. Careful observations of pulsar periods, made over long stretches of time and with accurate time-measurement devices, have confirmed that this does happen. Pulsar periods invariably get longer as time goes by. This lends support to Gold's theory.

THE MYSTERY GOES ON

The energy from some pulsars arrives in bursts so intense that they can be seen with optical telescopes, and their images can be captured on photographic film. A pulsar in the *Crab Nebula*, which is believed to be the remnant of a supernova explosion approximately 1,000 years ago, has a period of only 0.03 second and emits visible "flashes." The flashing went unnoticed for many years because most optical observation is done using time-exposure photography. In such a photograph, a visible pulsar looks like an ordinary star.

Pulsars remain mysterious, and the mechanisms by which they operate stretch the limits of credibility. Although variations of Thomas Gold's hypothesis have been accepted, no one is completely certain what makes these things tick. It is impossible to manufacture a pulsar in a laboratory. Computer modeling is helpful but not conclusive because the output of a computer can only be as good as the input data. Some pulsars have been observed to suddenly cease their "transmissions" for several seconds or even minutes and then start up again. How can we explain that?

Black Holes

Some neutron stars, as they collapse under their own weight, apparently do not stop even when all the space has been removed from between the subatomic particles. If the mass of such a star is great enough, calculations show that gravitation will become so powerful that no other force in the Cosmos can overcome it, not even the forces within neutrons and other particles.

BASIC ANATOMY

When a neutron star gets going on the runaway frenzy of gravitational collapse that nothing can stop, the object will, in theory, continue to shrink until it becomes a geometric point that contains all the mass of the neutron star from which it formed but zero volume. There is an overwhelming gravitational field at its "surface" and a slowing down of time (because of relativistic effects, to be dealt with in a later chapter) to a complete stop relative to the outside Universe. This object is called a space-time *singularity*.

Surrounding the singularity is a spherical zone within which nothing can escape, not even visible light or other EM radiation. It is a "zone of no return" because the escape velocity is greater than the speed of light in free space. The outer boundary of this zone is called the *event horizon* (Fig. 14-6). To an outside observer, the object would appear as a black sphere having the radius

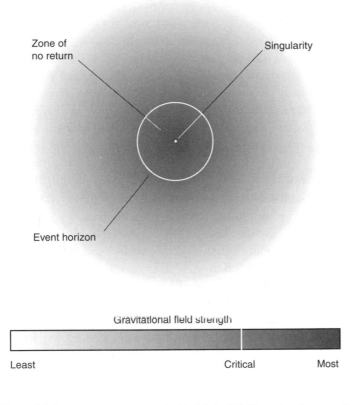

Zone of
no return

Singularity

Event horizon

Gravitational field strength

Least

Critical

Most

Figure 14-6.   Simplified drawing of a black hole. This illustration does not take
relativistic spatial distortion into account.

of the event horizon. The edge of the sphere would glow faintly because of
starlight that has been almost, but not quite, captured and pulled in. The background of stars near such a *black hole* would appear distorted because of
space warping caused by gravitation.

GRAVITY'S ULTIMATE VICTORY

From a simplistic standpoint, gravitation can become so powerful that it
will not let anything escape, not even the energy packets called *photons*
that represent all forms of EM radiation. As the gravitational field at the
surface of a collapsing neutron star becomes increasingly powerful, EM

rays are bent downward significantly as they leave the surface. At a certain point, the rays leaving in an almost horizontal direction fall back. The star continues to collapse, and the gravitational field becomes more powerful still; rays fall back at ever-increasing angles. When the radius of the object gets so small that the gravitational field at the surface reaches critical intensity, only those photons traveling straight up from the surface manage to get away. However, things don't stop there. The collapsar keeps shrinking within the event horizon; then all photons are trapped. No known form of energy can propagate faster than photons, and so the event horizon represents a *one-way membrane*: Things can get in, but nothing can get out.

This idea is not new. As long as there has been a particle theory of light, imaginative scientists have theorized that black holes can exist. When Albert Einstein revolutionized physics with his theory of relativity in the early 1900s, new evidence arose for the existence of black holes. However, nobody had ever seen an object in space that fit the description. The nature of black holes, assuming that they exist, is such that they are invisible at all wavelengths. The fantastic nature of black holes, along with the apparent fact that they can never be observed directly, originally caused some scientists to scoff. How could we say, for example, that angels were not dancing on the surfaces of neutron stars that had collapsed to within their event horizons? After all, no one could take a look and disprove such an idea! Recently, however, most astronomers have come to believe that black holes are not only plausible but real. They're out there.

THE SCHWARZCHILD RADIUS

There is a formula, based on the mass of an object, that allows us to calculate the radius to which any spherical object would have to be squashed in order for the surface to reach the event horizon. This radius is called the *Schwarzchild radius*, named after the astrophysicist Karl Schwarzchild who first came up with the formula in 1916. He based the derivation of his formula on the supposition that if an object got small enough, the energy of photons emitted from the surface of that object would not be sufficient to propel them out of its gravitational field. Sometimes the Schwarzchild radius is called the *gravitational radius*.

Suppose that an object has a mass M (in kilograms). Its Schwarzchild radius r (in meters) would depend directly on the mass. The greater the mass, the greater is the Schwarzchild radius. The formula is remarkably simple:

$$r = 2GM / c^2$$

where c^2 is the speed of light squared (approximately 8.9875×10^{16} m^2/s^2) and G is a constant known as the *gravitational constant*, a characteristic of the Cosmos. The value of G has been measured by painstaking experimentation and has been found to be approximately 6.6739×10^{-11}.

You are invited to perform some calculations with this formula if you like. Do you want to figure out the gravitational radius of your body? Remember, you must use your mass in kilograms to get the answer. You'll come up with an exceedingly small value. You would not survive being crushed to such a submicroscopic size.

The Earth would have to be compressed to the size of a grape in order to fall within its Schwarzchild radius. The Sun would have to collapse to a radius of approximately 2.9 km (1.8 mi). It is believed that a star must originally have at least three solar masses to generate enough gravitational power, on its demise, to collapse into a black hole.

Apparently, our parent star, the Sun, will never achieve the dubious distinction of withdrawing from the Universe in a cocoon of its own gravitation. It's not massive enough. This might bring to mind a new meaning of the old adage, "The bigger they are, the harder they fall." Size does matter!

WARPING OF SPACE

We ordinarily think of space as having three dimensions (height, width, and depth, for example). According to Einstein, however, *three-space*, as it is called, can be curved with respect to some fourth spatial dimension that we cannot see. We expect that rays of light in space should obey the rules of Euclidean geometry. One of the fundamental rules of light behavior, according to Newtonian physics, dictates that rays of light always travel in straight lines. Such a continuum is called *Euclidean space*. However, what if space is *non-Euclidean*?

According to the theory of general relativity, space is distorted, or bent, by gravitation. The extent of this bending is insignificant at the Earth's surface. It takes an enormously powerful gravitational field to cause enough bending of space for its effects to be observed as an apparent bending of light rays. However, the effect has been seen and measured in starlight passing close to the Sun during total solar eclipses, when the Sun's disk is darkened enough so that stars almost behind it can be seen through telescopes. The light from certain distant quasi-stellar sources (*quasars*), when passing near closer objects having intense gravitational fields, has been seen to produce multiple images as the rays are bent.

The bending of space in the vicinity of a strong gravitational field has been likened to the stretching of a rubber membrane when a mass is placed on it. Suppose that a thin rubber sheet is placed horizontally in a room, attached to the walls at a height of, say, 1.5 m above the floor. This sheet is stretched until it is essentially flat. Then a small but dense object such as a ball of solid iron is dropped right in the middle of the sheet. It will be bent in a funnel-like shape. The curvature of the sheet will be greatest near the ball and will diminish with increasing distance from the ball. This is how astronomers believe that space is warped by intense gravitation (Fig. 14-7). We will examine why this happens in Chapter 16.

It's easy to see how objects traveling through warped space can't possibly go in straight lines: There are no straight paths within such a continuum. According to the theory of general relativity, real space is non-Euclidean every-

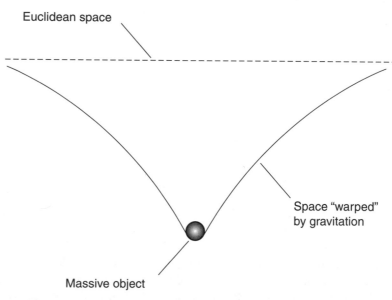

Figure 14-7. Space is warped in the vicinity of an object having an intense gravitational field.

where because there are gravitational fields everywhere. In some regions of space, gravitation is weak, and in other places, it is strong. But there is no place in the Universe that entirely escapes the influence of this all-pervasive force.

DO BLACK HOLES REALLY EXIST?

There are plenty of stars that are more than three times the mass of the Sun. Astronomers agree that many of these stars have long since blown up, gone through their white dwarf phases, and presumably also gone through the collapsing process. This suggests that black holes comprise a significant proportion of the matter in the Universe.

As long as we can't see them, we do not have a good way to find out if black holes exist. Fortunately, however, real black holes are not as black as theory implies. According to the research of the well-known cosmologist Stephen Hawking, black holes gradually lose their mass by emitting energy. This need not necessarily all be in the form of EM radiation; energy also can be lost as *gravitational waves*.

The idea of a gravitational wave was first made plausible when Einstein developed his general theory of relativity. Figure 14-8 is a simplified depiction of a gravitational wave as it leaves a black hole and travels through the space-time continuum. Just as a pebble, when dropped into a still pond, produces concentric, expanding, circular ripples in the two-dimensional surface of the water, so does the black hole produce concentric, expanding, spherical ripples in the three-dimensional continuum of space. Ripples are also produced in time—as hard as this might be to imagine!

Gravitational-wave detectors have been built in an effort to detect ripples in space and time as they pass. Because they involve the very fabric of the Cosmos, such waves can penetrate anything with no difficulty whatsoever. A gravitational disturbance coming from the nadir (straight down as you see it while standing upright) would be every bit as detectable as one coming from overhead.

As of this writing, no conclusive evidence of gravitational waves has been found. However, astronomers are reasonably confident that they exist. The challenge is nothing more (or less) than continuing to refine the detection strategy until ripples in space and time are discovered and can be attributed to some real, however bizarre, celestial object.

When we expand our scope of observation to an intergalactic scale, there is other evidence for the existence of black holes. This will be discussed in the next chapter.

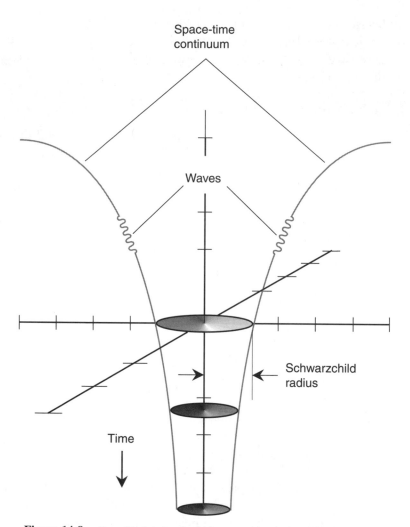

Figure 14-8.   Some black holes should be expected to emit gravitational waves.

Refer to the text if necessary. A good score is 8 correct. Answers are in the back of the book.

1. Photons are
    (a) antiprotons.
    (b) antielectrons.

   (c) antineutrons.
   (d) energy packets.

2. As a black hole swallows up more and more matter,
   (a) its mass decreases.
   (b) the gravitational radius decreases.
   (c) its mass increases.
   (d) None of the above

3. The waveforms of the pulses from a pulsar are
   (a) smooth.
   (b) irregular.
   (c) long.
   (d) short.

4. If an object collapses so that its radius is less than the Schwarzchild radius, then
   (a) things can come out but cannot go in.
   (b) the escape velocity at the surface is greater than the speed of light.
   (c) the escape velocity at the surface is less than the speed of light.
   (d) the object is, by definition, a pulsar.

5. Radio waves travel through interstellar space at a slightly different speed than visible light because of
   (a) magnetic fields.
   (b) differences in photon energy.
   (c) dispersion.
   (d) No! Radio waves always travel through interstellar space at precisely the same speed as visible light.

6. Suppose that a neutron somehow forms, and it is floating all by itself in space. How long can we expect it to last?
   (a) Forever
   (b) Until it collapses into a black hole
   (c) Until it becomes part of a neutron star
   (d) none of the above.

7. The EM energy from pulsars is believed to be a product of
   (a) neutrons decaying into energy.
   (b) matter interacting with antimatter.
   (c) electrons rising into higher orbits.
   (d) none of the above.

8. As a rotating white dwarf dies out and collapses into a neutron star,
   (a) it rotates faster and faster.
   (b) it rotates more and more slowly.
   (c) its rotational speed does not change.
   (d) rotation loses meaning because of the incalculable density of the object.

9. Suppose that a 1,000-kg spacecraft from Earth touches down on a planet and that planet turns out to be antimatter. What will happen?
    (a) There will be an explosion of incalculable violence, and all the matter in both the planet and the spacecraft will be annihilated.
    (b) There will be an explosion, and approximately 90 quintillion (9.0×10^{19}) joules of energy will be liberated.
    (c) There will be an explosion, and approximately 180 quintillion (1.80×10^{20}) joules of energy will be liberated.
    (d) The spacecraft will be quietly swallowed up by the planet and will disappear.

10. When an electron moves into a larger orbit within an atom,
    (a) the electron's energy increases.
    (b) the electron's energy decreases.
    (c) the electron's charge increases.
    (d) the electron's charge decreases.

CHAPTER 15

Galaxies and Quasars

When telescopes became powerful enough to resolve nebulae into definite shapes, one type of nebula presented a conundrum. Many of the *spiral nebulae*, which looked like whirlpools of glowing gas, had spectral lines whose wavelengths were much longer than they ought to be. This phenomenon, called *red shift*, suggests that an object is receding. Red shifts were seen commonly for spiral nebulae, but blue shifts—foreshortening of the waves when an object is approaching—were almost never seen, and when they were observed, they were minimal. Some astronomers thought that the spiral nebulae actually were huge congregations of stars at immense distances and that our own Milky Way was just one such congregation. Until individual stars could be resolved within the spiral nebulae, however, this idea remained an unproved hypothesis.

Types of Galaxies

Today we know that the spiral nebulae do consist of stars, and we call them *spiral galaxies*. Spirals are not the only type of galaxy. Other objects, previously thought to be emission nebulae or globular star clusters within our Milky Way, turned out to be *irregular galaxies* or *elliptical galaxies*. Some of these are billions (units of 1 billion or 10^9) of light-years away from us. Many contain hundreds of billions of individual stars.

ELLIPTICAL GALAXIES

Some of the brightest galaxies, containing the greatest numbers of stars, are ellipsoidal or spherical in shape. These galaxies are classified according to their eccentricity, the extent to which they are elongated from a perfectly spherical shape. (Actually, the sphere or ellipsoid represents the median boundary, the two-dimensional region such that half the stars are inside and half are outside.) Eccentricity zero (E0) represents a perfect sphere; E1 and E2 are egg-shaped. The E3 and E4 elliptical galaxies resemble elongated eggs. When we get to E5, the median boundary is football-shaped. Elliptical galaxies of E6 and E7 classification are even more elongated. Figure 15-1 shows approximate representations for E0 through E7 elliptical galaxies. This scheme was devised by astronomer Edwin Hubble in the 1930s.

Elliptical galaxies contain comparatively little gas and dust. It is believed that this is so because most of the interstellar material has developed into stars. This suggests that elliptical galaxies are old. There are numerous red giants in these galaxies, and this is consistent with the theory that they are old. Some of these must be supergiants, with diameters hundreds of times that of our Sun, and thousands of times brighter. Despite

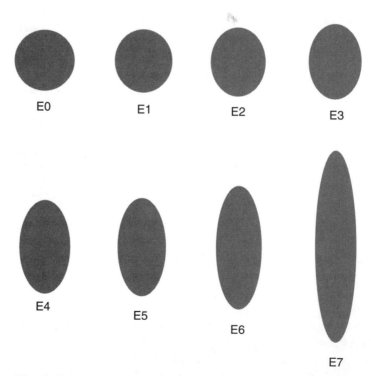

Figure 15-1.   Classifications of elliptical galaxies made by Edwin Hubble in the 1930s. Type E0 has a spherical shape. Type E7 has the greatest eccentricity.

the vast distances separating other galaxies from ours, some of the stars in elliptical galaxies resolve into points of light in large telescopes. Were it not for the red shift, these giant galaxies could be mistaken for globular star clusters within our own galaxy.

IRREGULAR GALAXIES

Irregular galaxies have no defined shape or apparent structure. Our Milky Way has two small irregular galaxies near it. These are the *Magellanic Clouds*, named after the famous explorer who sailed around the world. They can be seen with the unaided eye, but only from the Earth's southern hemisphere. They look like faintly glowing clouds on a moonless night. With a good telescope, it is easy to tell that they are made up of stars.

Some irregular galaxies show signs of coordinated motions among their stars, such as slow rotation around a central axis. In some cases, it is difficult to tell whether such apparent organization is real or an artifact of the *expectation phenomenon*: Sometimes we think we see something only because we expect to see it. In other instances, there is clear evidence of rotation. If the rotation is significant enough, the galaxy can be classified as a spiral.

SPIRAL GALAXIES

The most stunning galaxies, from the standpoint of the visual observer, are the spirals. Their variety is almost infinite. Some spirals appear broadside to us, some appear at a slant, and still others present themselves edgewise. The spiral arms can have many different shapes.

There are two major types of spiral galaxies: the *normal spiral* and the *barred spiral*. There are subclassifications within these two major categories. These are rather subjective and must be judged based on what we see. It is easy to classify spirals when they present themselves nearly broadside to us but difficult when they present themselves edgewise or nearly edgewise. Normal spirals are classified S0, Sa, Sb, and Sc (Fig. 15-2) depending on how tightly their bands of stars are wound around the nucleus. The barred spirals are classified as S0, SBa, SBb, and SBc (Fig. 15-3). The S0 galaxies are shaped like oblate (flattened) spheres, or like donuts with golf balls stuck in their centers. Both the normal spirals and the barred spirals branch off from this common root type. These classifications, like the classifications for the elliptical galaxies, were devised by Hubble.

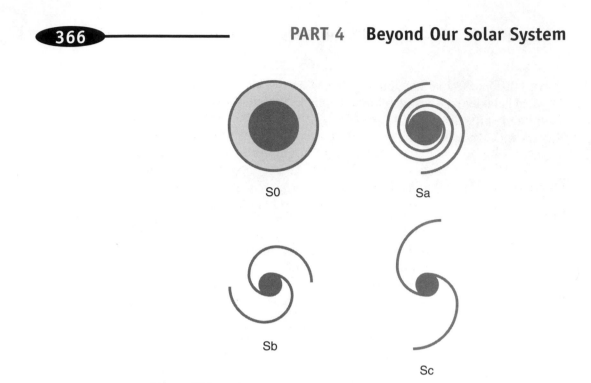

Figure 15-2. Hubble classification of spiral galaxies. Type S0 is shaped like an oblate sphere. Type Sa has tight spiral arms; types Sb and Sc have more open spiral arms.

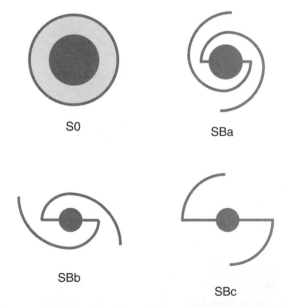

Figure 15-3. Hubble classification of barred spiral galaxies. Type S0 is shaped like an oblate sphere. Type SBa has tight spiral arms and a short bar; types SBb and SBc have more open spiral arms and longer bars.

In a normal spiral galaxy, the arms extend from the bright nucleus and are coiled in a more or less uniform fashion throughout the disk. Some spirals have prominent arms, whereas others have almost invisible arms. In a barred spiral, the central region is rod-shaped. The spiral arms trail off from the ends of the rod, in some cases prominently and in other cases almost invisibly. The barred spirals are especially interesting because the rod-shaped region appears to rotate with constant angular speed at all points along its length, as if it were solid. According to one theory, the nuclei of such galaxies are undergoing catastrophic explosions, and the bars are streamers of gas and dust ejected from the core at high speeds.

The appearance of a spiral galaxy gives the illusion that it is a uniformly rotating system. This is basically true, but the motions of the individual stars within a spiral galaxy are varied, and the overall pattern of motion is quite complicated. The fact that these galaxies rotate is verified by spectral examination of different regions when the disk of the spiral presents itself nearly edgewise to us. On one side of the galaxy, the light is shifted toward the blue end of the spectrum compared with the light from the nucleus. On the other side of the galaxy, the light is shifted toward the red end compared with the light from the nucleus. This indicates that the two sides of the galaxy have different radial speeds, and this can be explained only by rotation around the center. These determinations must be made independently of the overall motion of the galaxy with respect to us; in general, most galaxies are moving away.

Galaxies Galore

Our Milky Way galaxy has several close neighbors in space—close, that is, when we compare their distances to those of the most remote known galaxies. Our "intergalactic township" is called the *Local Group*. This is a bit whimsical when we are talking about millions of light-years, but compared with the whole known Universe, it is local indeed.

CLUSTERS OF GALAXIES

All the galaxies in our Local Group are within a few million light-years of our galaxy. The *Great Galaxy in Andromeda*, also known as *M31*, is almost on the other side of the group from our Milky Way. The spiral galaxy M33, in the constellation Triangulum, is another member of the Local Group. There are several smaller irregular and elliptical galaxies as well.

All the galaxies in our cluster appear tilted at different angles in space. When we look far into space, we find galaxies in every direction. Along the spiral plane of our own Milky Way, it is almost impossible to see exterior objects because of the interstellar gas and dust that obscure the view. The farther out we look, the more galaxies we find. This is to be expected because larger and larger spheres of observation must encompass more and more rapidly increasing volumes of space. Strangely enough, though, the distribution of the galaxies in the Cosmos is not altogether uniform.

Clusters of galaxies are the rule rather than the exception. Our Local Group is a comparatively small cluster. There are clusters with hundreds or even thousands of individual galaxies. Just as stars can vary greatly in their character and the galaxies can exist in many different and unique forms, so the clusters are found in differing shapes, sizes, and constitutions. The galaxies in some clusters are so close together that they collide "often" on a Cosmic time scale measured in millions upon millions of Earth years. After an encounter with another galaxy, a spiral can lose its arms because of the gravitational and electromagnetic (EM) disturbance. Some such spirals become irregular and disorganized.

One of the most dense known clusters of galaxies is found in the direction of the constellation Coma Berenices. Near the center of this rich cluster, any galaxy can be expected to collide with another member several times during its life. What would it be like if our Milky Way were currently in a collision with another galaxy? Such an event occurs over a period of millions of years. However, we certainly would be able to tell if it were happening. There would be two sources of radio noise, not just one, coming from the galactic core, and the noise level would be much greater than we currently observe. If we had the opportunity to view the invading galaxy, its nucleus would be visible with the unaided eye as a diffuse, glowing mass of starlight.

Clusters of galaxies extend as far as we can see with optical telescopes: up to several billion light-years. Such distances must be inferred indirectly; how this is done will be shown a little later in this chapter.

SUPERCLUSTERS, STRINGS, AND DARK MATTER

Between clusters of galaxies, space appears empty. At least there is nothing in these voids that radiates energy we can observe. However, even the clusters such as our Local Group or well-known clusters such as those that lie in the general directions of the constellations Coma Berenices or Virgo (and thus are named after those constellations) appear to exist in larger *superclusters*. You might think of it as the Cosmic urban structure: neighborhoods (galaxies) comprise townships (clusters), which together comprise cities (superclusters).

On a scale larger still, the superclusters are separated by voids of staggering size. When all the known galaxies are mapped using a computer program that simulates three-dimensional space, a foamlike cosmic structure becomes evident. Think of the large bubbles produced in soapy water. The superclusters exist mainly on the filmy surfaces of the bubbles. Inside the bubbles and between them is nothingness—or at least EM darkness. Some astronomers suspect that these dark regions contain some as-yet-unknown "stuff" that has mass and that has a profound effect on the evolution of the Universe. This stuff has even been given a name: *dark matter*. It is a topic of much interest and debate.

RADIO GALAXIES

All galaxies emit energy at radio wavelengths, as well as in the infrared (IR), visible, ultraviolet (UV), x-ray, and gamma-ray portions of the EM spectrum. Usually, the intensity of the radio emissions from galaxies is related to their classification and their observed visual brightness. However, some galaxies emit far more energy at radio wavelengths than we would expect. These objects are called *radio galaxies*. The intense radio source known as *Cygnus A* is one such galaxy. When radio telescopes are used to map the details of Cygnus A, a double structure is found. The radio emission comes from two different regions located on either side of the visible galaxy. Other double galaxies have been observed with radio telescopes.

Several hypotheses have been put forth in an attempt to find out what is taking place in radio galaxies. One theory is that they are pairs of colliding galaxies; the magnetic and electrical fields of the two galaxies interact to produce unusual levels of radio-frequency energy. Iosif S. Shklovskii of Russia theorized that radio galaxies contain more supernovae than do normal galaxies. F. Hoyle and W. A. Fowler have suggested that the tremendous energy of the radio galaxies comes from explosions of the galactic nuclei, following or associated with a catastrophic gravitational collapse. In any case, it is believed that the nuclei of the radio galaxies are undergoing radical changes.

As more becomes known about radio galaxies, astronomers hope to further unravel the puzzles of galactic formation and evolution.

Intergalactic Distances

When attempting to determine the distances to other galaxies and other clusters of galaxies, astronomers use tricks. These schemes rely on two assumptions: (1) the average brightnesses of the stars in all galaxies are similar, and (2) the average brightnesses of the galaxies in all clusters are similar.

USING THE CEPHEIDS

In the early-middle part of the twentieth century, when astronomers began to seriously study galaxies using the newly constructed 100-in (2.54-m) telescope on top of Mount Wilson, they found Cepheid variable stars in some of them. This resolved the question of whether or not the spiral nebulae are, in fact, "island universes." They are—and they are farther away than anything in the Milky Way.

Two astronomers, Edwin Hubble and Milton Humason, made the assumption that the Cepheid variables in other galaxies exhibit the same brightness-versus-period relation as the Cepheids in our own galaxy. On this basis, estimates of distances to some of the spiral galaxies were made. The initial estimate of the distance to the Great Galaxy in Andromeda was 600,000 to 800,000 light-years. (More recently, this figure has been revised upward to about 2.2 million light-years.) Some of the spiral galaxies that Hubble and Humason examined appeared to be 10 million light-years distant. These were the most distant galaxies in which Cepheid variables could be individually observed.

Yet there were galaxies much smaller (in terms of angular size) and fainter than the dimmest ones in which Cepheid variables could be resolved. This, Hubble and Humason reasoned, meant that there are galaxies much farther away than 10 million light-years. Beyond the limit at which the Cepheid variables can be used, astronomers use the brightnesses of galaxies as a whole to infer their distances. This is not an exact science. Observations over great distances are complicated by the fact that what we see is an image of the distant past and not an image of things as they are "right now."

LOOKING BACK IN TIME

Visible light and all EM waves, including radio, IR, UV, x-rays, and gamma rays, travel at a finite speed through space. A galaxy 10 million light-years away appears to us as it was 10 million years ago, not as it is today. If relative brightness is used as a distance-measuring tool, there appear to be galaxies billions of light-years away. We see these as they were billions of years ago—in some cases, as they were before our Solar System existed. Do galaxies maintain the same average brightnesses over time spans this great? We do not know. If they do, then our estimates of their distances are fairly accurate. If not, then our distance estimates are not accurate.

Based on their assumptions, Hubble and Humason found an interesting correlation between the apparent distance to a galaxy and the amount by which the lines in its spectrum are shifted. In general, the farther out in intergalactic space we look, the more the spectra of individual galaxies are red-shifted. This suggests that all the galactic clusters in the Universe are moving away from all the others. The most commonly accepted explanation for this red shift is Doppler effect caused by radial motion away from an observer.

Hubble and Humason made the assumption that the red shifts are caused by a general expansion of the Universe, and based on this, they found that the speed-versus-distance function is linear. On average, objects 200 million light-years away appear to be receding from us at twice the speed of objects 100 million light-years away, objects 400 million light-years away are receding twice as fast as those 200 million light-years away, and so on. This gives astronomers yet another tool for estimating vast distances in the Universe, but it must be based on yet another assumption: The slope of the speed-versus-distance function is constant all the way out to the limit of visibility. When this assumption is made, the conclusion follows that we will never see anything farther away than about 15 billion (1.5×10^{10}) light-years because such objects would be receding from us at a speed greater than the speed of light.

Does the average brilliance of galaxies remain the same over periods of billions of years? Starting in the 1960s, there were reasons to think not.

Quasars

In 1960, the position of a strong radio source was defined with great accuracy, and its angular diameter was found to be less than a second of arc.

Comparing the position of this radio source with various visible objects in its vicinity, this "radio star" was found to be a faint blue star in photographs. There was something especially odd about this star: The astronomers J. L. Greenstein and A. Sandage could not identify the absorption lines in its spectrum. It did not take them long to find the problem. The red shift in the spectrum of this cosmic energy source is so great that the lines are greatly altered, suggesting that the object is receding from us at a sizable fraction of the speed of light.

Soon after the discovery of this "radio star," several other similar objects were found, and they also had very large red shifts in their spectral lines. The objects, because of their visual resemblance to stars and because of their strong radio emissions, were called *quasi-stellar radio sources*. This name has since been shortened to the more palatable term *quasar*.

OBSERVING QUASARS

After the first few quasars were found, many others were discovered and observed. Some quasars had been photographed previously, but in the photographs they had been dismissed as ordinary stars. In one case, when several photographs having been taken over a period of decades were examined, it was found that large changes in brightness had occurred within periods of a few months. This implied that the quasar is a fraction of a light-year in diameter. However, if its red shift is a correct indicator of its distance, its energy output is many times that of a normal galaxy! Quasars are concentrated, as well as intense, sources of energy.

When observed with radio telescopes having high resolution—less than 1 second of arc in some cases—some quasars still appear as point sources. This is also true of the nuclei of certain radio galaxies. Optically, many of the quasars look like point sources of light, and therefore, they resemble stars, until the extreme magnification and resolving power of the Hubble Space Telescope (HST) is put to work on them. Then some quasars show evidence of glowing matter around a central, intense core.

Some quasars can be resolved into components by radio telescopes, but this requires the use of multiple antennas and a baseline of hundreds or even thousands of kilometers. Antennas in diverse locations on Earth are linked by satellite communications systems, and their outputs are combined by computer programs in order to accomplish this. This provides the equivalent resolving power of an antenna much larger than any single structure that could be constructed. The angular resolution goes down to less than 0.001 second of arc. With such sophisticated apparatus, radio galaxies and quasars have been probed in detail.

SCINTILLATIONS

There is another, quite different way to estimate the angular diameter of an object that emits energy at radio wavelengths: observing and measuring changes in intensity, called *scintillations*, that occur as the radio waves pass through turbulent ionized clouds of particles in space.

Everyone has noticed the twinkling of the stars, while the planets appear to shine almost without blinking. The reason for this difference is that the planets have a much greater angular diameter than any star. Small telescopes show the planets as disks, but even the nearest stars resolve only as points of light, even at high magnification. Turbulence in the air, such as that produced on summer evenings as the warm land heats the atmosphere and causes convection currents, make a point of light seem to twinkle because the light rays are refracted more and then less, and then more again by parcels of air having variable density. The charged subatomic particles of the *solar wind*, as they stream outward from the Sun, have a similar effect on radio waves coming from far away in space. Other stars produce "winds" too, making interstellar space a turbulent sea of charged subatomic particles. A source of radio waves with a small angular diameter therefore scintillates.

By observing quasars with a single radio telescope antenna to avoid *diversity effect* (averaging out of the strengths of radio signals as received at different locations), and by carefully recording the intensity of the waves reaching the antenna, it is possible to get an accurate idea of the angular size of a radio object. Quasars always appear as small sources of radio energy, at most a few light-years in diameter. Galaxies, in contrast, are many thousands of light-years across. Other observed properties of the quasars, such as curvature in the spectral lines, have led astronomers to believe that they are small compared with galaxies, even though they emit fantastic amounts of energy.

HOW DISTANT?

The sizes of the quasars, as well as estimates of their energy output, have been determined according to the Hubble relation between red shifts and distances. All quasars show significant red shifts in the absorption lines of their spectra. This has led most astronomers to surmise that they are billions of light-years away from us.

Suppose, however, that the red shifts are being misinterpreted? Are quasars actually local objects of modest size that are thrust outward from the nucleus of our galaxy at tremendous speeds? This is an interesting theory, but it is not widely accepted. If quasars are being ejected from our galaxy, then it is reasonable to suppose that they are ejected from other galaxies too. In such a case, some of the quasars ejected from other galaxies should be observed as approaching us. This would give such objects a pronounced blue Doppler shift. However, no quasar has ever been found that exhibits a blue shift in its spectral lines.

Another attempt has been made to prove that quasars are "local." Albert Einstein showed, in the formulation of his general theory of relativity, that a powerful gravitational field can produce a red shift in the spectrum of the light coming from the source of the gravitation. This effect has been observed and measured, so scientists know that Einstein's theory is correct. Can the red shifts in the spectral lines of quasars be explained in terms of the relativistic effect of gravitation? A super dense object with extreme gravitation near its surface could produce a large red shift. This remains an open question. Still, an affirmative answer would not constitute conclusive proof that quasars are "local."

Recent observations of quasars using the HST have begun to resolve the riddle. Evidence is accumulating to support the theory that quasars are among the most distant objects we can see in the Cosmos and that they therefore present a picture of the Universe as it was when it was much younger than it is now. This gives astronomers a way to look back in time to whatever extent they want simply by observing galaxies and quasar objects at various distances as indicated by their red shifts.

A severe blow was dealt to the local quasar theory when scientists calculated that the first quasar that was discovered, called *3C4S*, would have to have a mass the same as the Sun, be only 10 km (6 mi) in diameter, and be in the Earth's atmosphere in order to account for the radiation intensity it possesses. Even if 3C4S has thousands of times the mass of the Sun, calculations show that it still must reside within our Solar System, and this obviously is not the case. The derivations in these terms for other quasars give similar results.

The determination of the distances to quasars represents a good example of the devil's-advocate method of lending support to a theory by discrediting all its plausible refutations. The quasars, even after attack by the devil's-advocate scientific method of inquiry, appear to be distant and ener-getic cosmic phenomena.

ANATOMY OF QUASARS

The internal anatomy of quasars is still largely a mystery. It seems that the quasars are very distant and also very powerful sources of energy. Suppose that we accept this hypothesis without further question. If we are willing to do this, then certain things can be deduced about quasars.

The quasars are much farther away than all the nearby galaxies. (In this sense, *nearby* means within several hundred million light-years.) Quasars are not only distant, but they are extremely distant. Without exception, they show large red shifts in their spectral lines. In the Cosmos, distance is time; when we look at something 2 billion light-years away, we are looking 2 billion years into the past. Whenever we look at a quasar, we gaze into a past so remote that the Earth itself was much different than we know it today. Suppose that the quasars are—or were—a common phenomenon of the Universe in its younger age? This is a tempting proposition.

The present estimate of the age of our Universe is on the order of 12 billion to 15 billion years. Some of the quasars, at distances approaching 10 billion light-years, are thus images of the Universe at less than half its present age. Many stars have lifespans of much less than 10 billion years. Suppose that the quasars are young galaxies?

Observations of quasars and radio galaxies often reveal striking similarities, so some astronomers believe that quasars and radio galaxies are in fact the same sort of object. The nuclei of radio galaxies have diameters much less than that of a typical galaxy, but they put out vast amounts of energy. They share this characteristic with quasars. When we look at the most distant known radio galaxies with visual apparatus, we see only the brilliant nuclei; the peripheral glow is washed out by the light from the core.

When large amounts of matter are concentrated into a small volume of space, the gravitation can have profound effects. Do dense congregations of stars, such as exist in the centers of spiral galaxies, gravitationally seal themselves off from the rest of the Universe? That is, do they become black holes? The stars near the periphery of the congregation will, in this case, orbit the central region at great speeds before being pulled forever into the mass. Their high velocity and the accompanying magnetic fields would produce large amounts of EM energy at visible and radio wavelengths. Are quasars active black holes, like cosmic tornadoes?

Yet another theory concerning the origin and anatomy of the quasars suggests that they are points in space through which new matter is entering from some other space-time continuum. Such objects, in the vernacular of the cosmologist, are called *white holes*. As matter bursts into our space-time continuum, having been pulled from another Universe by overwhelming gravitational forces, the flash of radiant energy would outshine any typical galaxy. Direct evidence to support this theory is lacking, but it is one of the most fascinating in all cosmology. It implies that there exist other universes with space-time singularities connecting them with ours.

Huge Black Holes

In recent years, the idea that black holes exist at the centers of many, if not most, spiral galaxies has been gaining acceptance. This is so in part because of an interesting twist in the formula for the Schwarzchild radius as a function of mass.

BLACK-HOLE DENSITY VERSUS MASS

The radius that an object must attain in order to become a black hole is directly proportional to the mass. From elementary solid geometry, you will recall that the volume of an object is proportional to the cube of the radius. If an object's mass is doubled, the size of its Schwarzchild radius doubles too. However, its volume becomes eight times as great. This means that the more massive a black hole happens to be, the less dense it is. In fact, the density of a black hole is inversely proportional to the square of its mass (Fig. 15-4).

Suppose that before there were any stars in the Universe, but only hydrogen and helium gas, vast clouds congealed because of gravitation? Suppose that this took place on a much bigger scale than the process of star formation? If such a cloud were large enough, it could become a black hole before it attained a density anywhere near enough to start nuclear fusion. The black hole would continue to pull matter in, becoming larger still and yet less dense. As the atoms approached the event horizon, they would be accelerated to nearly the speed of light. This would give them tremendous kinetic energy, and the result would be an object of brilliance greater than that of

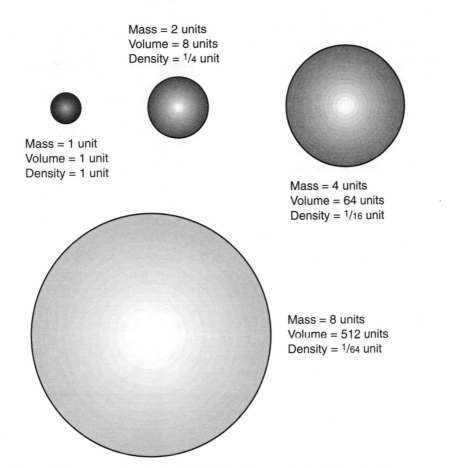

Mass = 2 units
Volume = 8 units
Density = 1/4 unit

Mass = 1 unit
Volume = 1 unit
Density = 1 unit

Mass = 4 units
Volume = 64 units
Density = 1/16 unit

Mass = 8 units
Volume = 512 units
Density = 1/64 unit

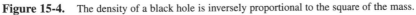

Figure 15-4.   The density of a black hole is inversely proportional to the square of the mass.

any nuclear fusion engine of comparable size. If the cloud had any spin to begin with, that spin would be exaggerated as the gas atoms fell into the black hole, in much the same way as the air circulation around a hurricane gets faster and faster as the molecules are drawn into the eye of the storm.

GALACTIC CENTERS

In the nuclei of spiral galaxies such as ours, the concentration of stars is highest. This is to say, there are the most stars per cubic light-year in and near the center of a galaxy. In the spiral arms, the concentration of stars is lower. The concentration is lower still in regions above and below the plane of the spiral disk and in between the spiral arms. Our Sun is near the plane of the Milky Way's disk, in one of the spiral arms, and approximately halfway from the center to the edge.

The appearance of spiral galaxies, some of which bear remarkable resemblance to satellite photographs of hurricanes and typhoons, makes it tempting to think that they spin around and around. They do, and all the stars move in the direction intuitively suggested by the sense of the pinwheel. However, some stars stay near the plane of the disk, whereas others dip below it and rise above it during each orbit around the center (Fig. 15-5). Stars in the central bulge, which resembles a gigantic globular cluster or a small elliptical galaxy having low eccentricity, orbit in planes that are tilted every which way. Near the center of the bulge, the density of stars increases. If our Sun were one of the stars in this region, our nighttime sky would be filled with many more stars than we see now. Moonless, clear nights would be as bright as a gloomy day.

However, what if the Sun were located at or very near the exact center of the galactic core? Many astronomers think that if that were the case, there would be no life on Earth. The Sun and all the stars in its vicinity would have passed through the event horizon of a black hole and would be "on the inside looking out." This black hole is thought to contain millions, if not billions, of solar masses. If the black hole is big enough, stars can fall through the event horizon and still remain intact. According to this theory, the centers of some, if not most, spiral and elliptical galaxies are "island universes" of a special sort, for they are closed off from the rest of the Cosmos by a one-way gate in time-space. Every time another

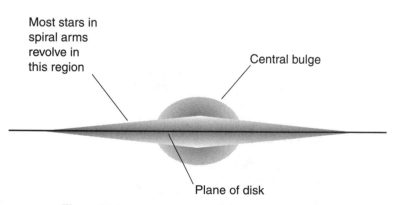

Figure 15-5. Structure of a spiral galaxy as seen edgewise.

star falls in, the mass of the black hole increases, and its density goes down a little more. Given sufficient time, measured in trillions of years, is it possible that these black holes might swallow whole galaxies and then clusters of galaxies?

At this point, we enter the realm of pure speculation. This is a good place to shift our attention to the theory that gave rise to notions of spatial curvature, time warps, and other esoteric aspects of latter-day cosmology.

Refer to the text if necessary. A good score is 8 correct. Answers are in the back of the book.

1. As a black hole pulls more and more matter in,
    (a) the Schwarzchild radius increases.
    (b) the Schwarzchild radius decreases.
    (c) the density increases.
    (d) it gets darker and darker.

2. Quasars with large blue shifts in their spectra
    (a) are receding from us at nearly the speed of light.
    (b) are approaching us at nearly the speed of light.
    (c) are never seen.
    (d) emit large amounts of x-rays compared with other quasars.

3. When a celestial object scintillates, we can surmise that it
    (a) is extremely luminous.
    (b) has a small angular diameter.
    (c) is a great distance from us.
    (d) emits energy mainly at short wavelengths.

4. Radio galaxies
    (a) emit far more energy at radio wavelengths than typical galaxies.
    (b) emit energy only at radio wavelengths.
    (c) have been observed only with radio telescopes.
    (d) are those galaxies toward which we have sent radio signals in an attempt to communicate with alien civilizations.

5. If the spectrum of an object is red-shifted, this means that the emission or absorption lines
    (a) look red in color when examined visually.
    (b) appear at longer wavelengths than normal.

    (c) appear at higher frequencies than normal.

    (d) are most prominent in the red part of the spectrum.

6. A certain galaxy is observed, and its distance is estimated at 10 Mpc. We see this object as it appeared approximately
    (a) 10,000 years ago.
    (b) 10 million years ago.
    (c) 32,600 years ago.
    (d) 32.6 million years ago.

7. Within a cluster of galaxies,
    (a) all the galaxies are of the same type.
    (b) all the galaxies spin in the same direction.
    (c) all the galaxies are approximately the same size.
    (d) None of the above

8. Quasars are believed to be much smaller than typical galaxies based on the observation that
    (a) they emit large amounts of energy at radio wavelengths.
    (b) their spectra are red-shifted.
    (c) they are nearby, in the Milky Way galaxy.
    (d) they can grow dimmer or brighter in short periods of time.

9. The spiral nebulae are large, distant congregations of stars rather than smaller objects within the Milky Way. In order to determine this, astronomers observed
    (a) the colors of the gases comprising them.
    (b) the radio emissions produced by them.
    (c) the rate at which they spin.
    (d) the brightnesses and periods of Cepheid variables within them.

10. A football-shaped galaxy might be classified as
    (a) S0.
    (b) S2.
    (c) E5.
    (d) SBc.

CHAPTER 16

Special and General Relativity

Relativity theory has a reputation as something only geniuses can understand. However, the basics of relativity are no more difficult to grasp than the fundamentals of any other theory. Some of the ideas put forth to explain astronomical observations before Einstein came up with his theory were more esoteric than relativity itself. This chapter contains a little bit of mathematics, but it doesn't go beyond the middle-school level.

There are two aspects to relativity theory: the *special theory* and the *general theory*. The special theory involves relative motion, and the general theory involves acceleration and gravitation.

Before we get into relativity, let's find out what follows from the hypotheses that the speed of light is absolute, constant, and finite and that it is the highest speed anything can attain.

Simultaneity

When he became interested in light, space, and time, Einstein pondered the results of experiments intended to find out how the Earth moves relative to the supposed medium that carries electromagnetic (EM) waves such as visible light. Einstein decided that such a medium doesn't exist—EM waves can travel through a perfect vacuum.

THE ETHER

Before Einstein's time, physicists determined that light has wavelike properties and in some ways resembles sound. However, light travels much
faster than sound. Also, light can propagate through a vacuum, whereas
sound cannot. Sound waves require a material medium such as air, water,
or metal to get from one place to another. Many scientists suspected that
light also must need some sort of medium to travel through space. What
could exist everywhere, even between the stars and galaxies, and even in a
jar from which all the air was pumped out? This mysterious medium was
called *luminiferous ether*, or simply *ether*.

   If the ether exists, some scientists wondered, how could it pass right
through everything, even the entire Earth, and get inside an evacuated
chamber? How could the ether be detected? One idea was to see if the ether
"blows" against the Earth as our planet orbits around the Sun, as the Solar
System orbits around the center of the Milky Way galaxy, and as our galaxy
drifts through the Cosmos. If there is an "ether wind," then the speed of
light ought to be different in different directions. This, it was reasoned,
should occur for the same reason that a passenger on a fast-moving truck
measures the speed of sound waves coming from the front as faster than the
speed of sound waves coming from behind.

   In 1887, an experiment was done by two physicists named Albert
Michelson and Edward Morley in an attempt to find out how fast the "ether
wind" is blowing and from what direction. The *Michelson-Morley experiment*,
as it became known, showed that the speed of light is the same in all directions. This cast doubt on the ether theory. If the ether exists, then according to
the results obtained by Michelson and Morley, it must be moving right along
with the Earth. This is quite a coincidence; it implies that the Earth is stationary relative to absolute space, and no one believed that in 1887. Attempts were
made to explain away this result by suggesting that the Earth drags the ether
along with itself. Einstein could not believe this. He decided that the results of
the Michelson-Morley experiment had to be taken at face value: The speed of
light is constant in every direction. Einstein believed that the Michelson-
Morley experiment would have the same outcome for observers on the Moon,
on any other planet, on a space ship, or anywhere in the Universe.

THE SPEED OF LIGHT IS CONSTANT

Einstein rejected the notion of luminiferous ether. Instead, he proposed an
axiom, a fundamental rule of the Universe: The speed at which light (and

any other EM field) travels in a vacuum is the same no matter what the direction and regardless of the motion of the observer with respect to the EM energy source. Then he set out to deduce what logically follows from this assumption.

Einstein did all his work by using a combination of mathematics and daydreaming that he called "mind journeys." He wasn't an experimentalist but a theorist. There is a saying in physics: "One experimentalist can keep a dozen theorists busy." Einstein turned this inside out. His theories have kept thousands of experimentalists occupied.

THERE IS NO ABSOLUTE TIME!

One of the first results of Einstein's speed-of-light axiom is the fact that there can be no such thing as an absolute time standard. It is impossible to synchronize the clocks of two obervers so that they will see both clocks as being in exact agreement unless both observers occupy the exact same point in space.

In recent decades we have built atomic clocks, and we claim that they are accurate to within billionths of a second (where a billionth is 0.000000001 or 10^{-9}). However, this has meaning only when we are right next to such a clock. If we move a little distance away, then the light (or any other signal that we know of) takes some time to get to us, and this throws the clock's reading off.

The speed of EM-field propagation, the fastest speed known, is approximately 300 million meters per second (3.00×10^8 m/s), or 186,000 miles per second (1.86×10^5 mi/s). A beam of light therefore travels about 300 m (984 ft) in 0.000001 s (one microsecond or 1 μs). If you move a little more than the length of a football field away from a superaccurate billionth-of-a-second atomic clock, the clock will appear to be in error by a microsecond, or 1,000 billionths of a second. If you go to the other side of the world, where the radio signal from that clock must travel 20,000 km (12,500 mi) to reach you, the time reading will be off by 0.067 s, or 67 thousandths of a second. If you go to the Moon, which is about 400,000 km (250,000 mi) distant, the clock will be off by approximately 1.33 s.

If scientists ever discover an energy field that can travel through space instantaneously regardless of the distance, then the conundrum of absolute time will be resolved. In practical scenarios, however, the speed of light is the fastest possible speed. (Some recent experiments suggest that certain effects can propagate faster than the speed of light over short distances, but no one has demonstrated this on a large scale yet, much less used such effects to transmit any information such as data from an atomic clock.) We can say that the speed of light is the speed of time. Distance and time are inextricably related.

POINT OF VIEW

Imagine that there are eight clocks in space arranged at the vertices of a gigantic cube. Each edge of the cube measures 1 light-minute, or approximately 18 million km (11 million mi) long, as shown in Fig. 16-1. We are given a challenge: Synchronize the clocks so that they agree within the limit of visibility, say, to within 1 second of each other. Do you suppose that this will be easy?

Because the clocks are so far apart, the only way we can ascertain what they say is to equip them with radio transmitters that send time signals.

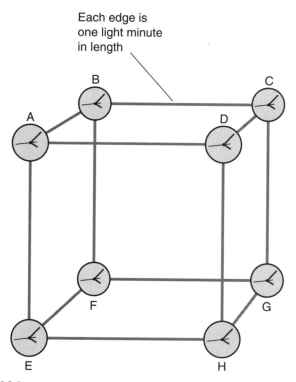

Figure 16-1.   A hypothetical set of eight clocks, arranged at the vertices of a cube that measures one light-minute on each edge. How will we synchronize these clocks?

Alternatively, if we have a powerful enough telescope, we can observe them and read them directly by sight. In either case, the information that tells us what the clocks say travels to us at the speed of light. We get in our space ship and maneuver ourselves so that we are in the exact center of the cube, equidistant from all eight clocks. Then we proceed to synchronize them using our remote-control, wireless two-way data communications equipment. Thank heaven for computers! The task is accomplished in a just a few minutes. It can't be done instantaneously, of course, because our command signals take the better part of a minute to reach the clocks from our central location, and then the signals coming back from the clocks take just as long to get to us so that we can see what they say. Soon, however, everything is in agreement. Clocks A through H all tell the same time to within a fraction of a second.

Satisfied with our work, we cruise out of the cube because there's nothing there of any real interest except a few small meteoroids. We take a look back at the clocks, mainly to admire our work but also because, as technicians, we are instinctively programmed to suspect that something can always go wrong. What do we see? To our dismay, the clocks have already managed to get out of sync. Muttering a curse or two, we take our ship back to the center of the cube to correct the problem. However, when we get there, there is no problem to correct! The clocks are all in agreement again.

You can guess what is happening here. The clock readings depend on how far their signals must travel to reach us. For an observer at the center of the cube, the signals from all eight clocks, A through H, arrive from exactly the same distance. However, this is not true for any other point in space. We have synchronized the clocks for one favored vantage point; if we go somewhere else, we will have to synchronize them all over again. This can be done, but then the clocks will be synchronized only when observed from the new favored vantage point. There is a unique *sync point*—the spot in space from which all eight clocks read the same—for each coordination of the clocks.

No sync point is more valid than any other from a scientific standpoint. If the cube happens to be stationary relative to some favored reference point such as Earth, we can synchronize the clocks, for convenience, from that reference point. However, if the cube is moving relative to our frame of reference, we will never be able to keep the clocks synchronized. Time depends on where we are and on whether or not we are moving relative to whatever device we use to indicate the time. Time is not absolute, but relative, and there is no getting around it.

Time Dilation

The relative location of an observer in space affects the relative readings of clocks located at different points. Similarly, relative motion in space affects the apparent rate at which time "flows." Isaac Newton hypothesized that time flows in an absolute way and that it constitutes a fundamental constant in the Universe. Einstein showed that this is not the case; it is the speed of light, not time, that is constant. In order to understand why *relativistic time dilation* occurs based on Einstein's hypothesis, let's conduct a "mind experiment."

A LASER CLOCK

Suppose that we have a space ship equipped with a laser/sensor on one wall and a mirror on the opposite wall (Fig. 16-2). Imagine that the laser/sensor and the mirror are positioned so that the light ray from the laser must travel perpendicular to the axis of the ship, perpendicular to its walls, and (once we get it moving) perpendicular to its direction of motion. The laser and mirror are adjusted so that they are separated by 3 m. Because the speed of light in air is approximately 300 million (3×10^8) m/s, it takes 10^{-8} s, or 10 nanoseconds (10 ns), for the light ray to get across the ship from the laser to the mirror and another 10 ns for the ray to return to the sensor. The ray therefore requires 20 ns to make one round trip from the laser/sensor to the mirror and back again.

Our laser emits pulses of extremely brief duration, much shorter than 20 ns. We measure the time increment using an extremely sophisticated oscilloscope

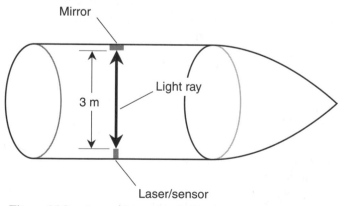

Figure 16-2.   A space ship equipped with a laser clock. This is what an observer in the ship always sees.

so that we can observe the pulses going out and coming back and measure the time lag between them. This is a special clock; its timekeeping ability is based on the speed of light, which Einstein proposed is constant no matter from what point of view it is observed. There is no better way to keep time.

CLOCK STATIONARY

Suppose that we start up the ship's engines and get moving. We accelerate faster and faster, with the goal of eventually reaching speeds that are nearly the speed of light. Suppose that we manage to accelerate to a sizable fraction of the speed of light, and then we shut off the engines so that we are coasting through space. You ask, "Relative to what are we moving?" This, as we shall see, is an important question! For now, suppose that we measure speed with respect to Earth.

We measure the time it takes for the laser to go across the ship and back again. We are riding along with the laser, the mirror, and all the luxuries of a spacecraft that is only 3 m (about 10 ft) wide. We find that the time lag is still exactly the same as it was when the ship was not moving relative to Earth; the oscilloscope still shows a delay of 20 ns. This follows directly from Einstein's axiom. The speed of light has not changed because it cannot. The distance between the laser and the mirror has not changed either. Therefore, the round trip takes the same length of time as it did before we got the ship moving.

If we accelerate so that the ship is going 60 percent, then 70 percent, and ultimately 99 percent of the speed of light, the time lag will always be 20 ns as measured from a *reference frame*, or point of view, inside the ship.

Let's add another axiom to Einstein's: In free space, light beams always follow the shortest possible distance between two points. Normally, this is a straight line. You ask, "How can the shortest path between two points in space be anything other than a straight line?" This is another good question. We'll deal with it later in this chapter. For now, note that light beams appear to follow straight lines through free space as long as the observer is not accelerating relative to the light source. Relative motion does not affect the "straightness" of light rays. (As we will see, acceleration does.)

CLOCK IN MOTION

Imagine now that we are outside the ship and are back on Earth. We are equipped with a special telescope that allows us to see inside the ship as it whizzes by at a significant fraction of the speed of light. We can see the laser, the mirror, and even the laser beam itself because the occupants of the space vessel have temporarily filled it with smoke to make the viewing easy for us. (They have their pressure suits on so that they can breathe.)

What we see is depicted in Fig. 16-3. The laser beam still travels in straight lines, and it still travels at 3×10^8 m/s relative to us. This is true because of Einstein's axiom concerning the speed of light and our own hypothesis to the effect that light rays always appear to travel in straight lines as long as we are not accelerating. However, the rays have to travel farther than 3 m to get across the ship. The ship is going so fast that by the time the ray of light has reached the mirror from the laser, the ship has moved a significant distance forward. The same thing happens as the ray returns to the sensor from the mirror. As a result of this, it will seem to us, as we watch the ship from Earth, to take more than 20 ns for the laser beam to go across the ship and back.

As the ship goes by, time appears to slow down inside it, as seen from a "stationary" point of view. Inside the ship, though, the "speed of time" seems entirely normal. The faster the ship goes, the greater is this discrepancy. As the speed of the ship approaches the speed of light, the *time dilation factor* can become large indeed; in theory, there is no limit to how great it can become. You can visualize this by imagining Fig. 16-3 stretched out horizontally so that the light rays have to travel almost parallel to the direction of motion, as seen from the "stationary" reference frame.

FORMULA FOR TIME DILATION

There exists a mathematical relationship between the speed of the space ship in the foregoing "mind experiment" and the extent to which time is

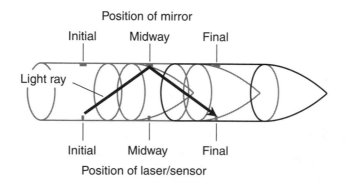

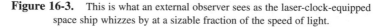

Figure 16-3. This is what an external observer sees as the laser-clock-equipped space ship whizzes by at a sizable fraction of the speed of light.

dilated. Let t_{ship} be the number of seconds that appear to elapse on the moving ship as 1 s elapses as measured by a clock next to us as we sit in our Earth-based observatory. Let u be the speed of the ship as a fraction of the speed of light. Then

$$t_{ship} = (1 - u^2)^{1/2}$$

The 1/2 power is an alternative and popular way to denote the square root. The time dilation factor (let's call it k) is the reciprocal of this:

$$k = 1 / [(1 - u^2)^{1/2}] = (1 - u^2)^{-1/2}$$

Let's see how great the time dilation factor is if the space ship is going half the speed of light. In this case, $u = 0.5$. If 1 s passes on Earth, then according to an Earthbound observer:

$$t_{ship} = (1 - 0.5^2)^{1/2}$$
$$= (1 - 0.25)^{1/2}$$
$$= 0.75^{1/2}$$
$$= 0.87 \text{ s}$$

That is, 0.87 s will seem to pass on the ship as 1 s passes as we measure it while watching the ship from Earth. This means that the time dilation factor is 1/0.87, or approximately 1.15. Of course, on the ship, time will seem to flow normally.

Just for fun, let's see what happens if the ship is going 99 percent of the speed of light. In this case, $u = 0.99$. If 1 s passes on Earth, then we, as Earthbound observers, will see this:

$$t_{ship} = (1 - 0.99^2)^{1/2}$$
$$= (1 - 0.98)^{1/2}$$
$$= 0.02^{1/2}$$
$$= 0.14 \text{ s}$$

That is, 0.14 s will seem to pass on the ship as 1 s passes on Earth. The time dilation factor k in this case is 1/0.14, or approximately 7.1. Time flows more than seven times more slowly on a ship moving at 99 percent of the speed of light than it flows on Earth—from the reference frame of someone on Earth.

As you can imagine, this has implications for time travel. According to the special theory of relativity, if you could get into a space ship and travel fast enough and far enough, you could propel yourself into the future. You might travel to a distant star, return to Earth in what seemed to you to be only a few months, and find yourself in the year A.D. 5000. When science-fiction writers realized this in the early 1900s just after Einstein published his work, they took advantage of it.

Spatial Distortion

Relativistic speeds—that is, speeds high enough to cause significant time dilation—cause objects to appear foreshortened in the direction of their motion. As with time dilation, *relativistic spatial distortion* occurs only from the point of view of an observer watching an object speed by at a sizable fraction of the speed of light.

POINT OF VIEW: LENGTH

If we travel inside a space ship, regardless of its speed, everything appears normal as long as our ship is not accelerating. We can cruise along at 99.9 percent of the speed of light relative to the Earth, but if we are inside a space ship, it is always stationary relative to us. Time, space, and mass appear normal from the point of view of passengers on a relativistic space journey. However, as we watch the space ship sail by from the vantage point of Earth, its length decreases as its speed increases. Its diameter is not affected. The extent to which this happens is the same as the extent to which time slows down.

Let L be the apparent length of the moving ship as a fraction of its length when it is standing still relative to an observer. Let u be the speed of the ship as a fraction of the speed of light. Then

$$L = (1 - u^2)^{1/2}$$

This effect is shown in Fig. 16-4 for various relative forward speeds. The foreshortening takes place entirely in the direction of motion. This produces apparent physical distortion of the ship and everything inside, including the passengers. It's sort of like those mirrors in fun houses that are concave in only one dimension and reflect your image all scrunched up. As the speed of the ship approaches the speed of light, its observed length approaches zero.

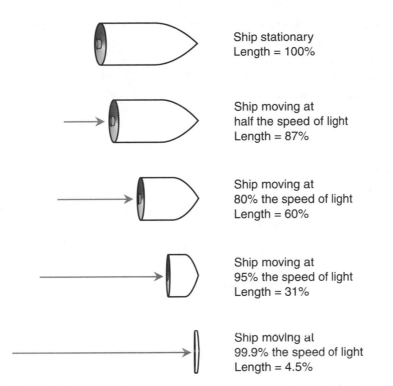

Figure 16-4.   As an object moves faster and faster, it grows shorter and shorter
along the axis of its motion.

SUPPOSITIONS AND CAUTIONS

Spatial distortion is a curious phenomenon. You might wonder, based on
this result, about the shapes of photons. They are the particles of which visible light and all other EM radiation are comprised. Photons travel at the
speed of light. Does that mean they are infinitely thin, flat disks or squares
or triangles hurtling sidelong through space? No one has ever seen a photon, so no one knows how they are shaped. It is interesting to suppose that
they are two-dimensional things and, as such, have zero volume. However,
if they have zero volume, how can we say that they exist?

   Scientists know a lot about what happens to objects as they approach the
speed of light, but it's intellectually dangerous to extrapolate and claim to
know what would happen if the speed of light could be attained by a material thing. We will see shortly that no physical object (such as a space ship)
can reach the speed of light, so the notion of a real object being squeezed

down to zero thickness is nothing more than an academic fantasy. As for photons, comparing them with material particles such as bullets or baseballs is an unjustified intuitive leap. We cannot bring a photon to rest, nor can we shoot a bullet or throw a baseball at the speed of light. As they might say in certain places, "Baseballs and photons ain't the same animals."

Mass Distortion

Another interesting effect of relativistic speeds is an increase in the masses of objects as they move faster and faster. This increase occurs to the same extent as the decrease in length and the slowing down of time.

POINT OF VIEW: MASS

If we travel inside a space ship, regardless of its speed, the masses of all the objects in the ship with us appear normal as long as our ship is not accelerating. However, from the vantage point of Earth, the mass of the ship and the masses of all the atoms inside it increase as its speed increases.

Let m be the mass of the moving ship as a multiple of its mass when it is stationary relative to an observer. Let u be the speed of the ship as a fraction of the speed of light. Then

$$m = 1/\left(1 - u^2\right)^{1/2} = \left(1 - u^2\right)^{-1/2}$$

This is the same as the factor k that we defined a little while ago. It is always greater than or equal to 1.

Look again at Fig. 16-4. As the space ship moves faster, it scrunches up. Imagine now that it also becomes more massive. The combination of smaller size and greater mass produces a dramatic increase in density at relativistic speeds.

Suppose that the *rest mass* (the mass when stationary) of our ship is 10 metric tons. When it speeds by at half the speed of light, its mass increases to a little more than 11 metric tons. At 80 percent of the speed of light, its mass is roughly 17 metric tons. At 95 percent of the speed of light, the ship's mass is about 32 metric tons. At 99.9 percent of the speed of light, the ship's mass is more than 220 metric tons. And so it can go indefinitely. As the speed of the ship approaches the speed of light, its mass grows larger and larger without limit.

SPEED IS SELF-LIMITING

It's tempting to suppose that the mass of an object, if it could be accelerated all the way up to the speed of light, would become infinite. After all, as u approaches 1 (or 100 percent), the value of m in the preceding formula increases without limit. However, it's one thing to talk about what happens as a measured phenomenon or property approaches some limit; it is another thing entirely to talk about what happens when that limit is actually reached, assuming that it can be reached.

No one has ever seen a photon at rest. No one has ever seen a space ship moving at the speed of light, nor will they ever. No finite amount of energy can accelerate any real object to the speed of light. This is so because of the way in which the mass increases as the speed of an object approaches the speed of light. Even if it were possible to move a real object at the speed of light relative to some point of observation, the mass-increase factor, as determined by the preceding formula, would be meaningless. To calculate it, we would have to divide by zero, and division by zero is not defined. (If you tell a theoretical mathematician that "one over zero equals infinity," you will get, at the very least, a raised eyebrow.)

The more massive a speeding space ship becomes, the more powerful is the rocket thrust necessary to get it moving faster. As a space ship approaches the speed of light, its mass becomes arbitrarily great. This makes it harder and harder to give it any more speed. Using a mathematical technique called *integral calculus*, astronomers and physicists have proven that no finite amount of energy can propel a space ship to the speed of light. The mass increases too fast. The function "blows up."

HIGH-SPEED PARTICLES

You've heard expressions such as *electron rest mass*, which refers to the theoretical mass of an electron when it is not moving relative to an observer. If an electron is observed whizzing by at relativistic speed, it has a mass greater than its rest mass and thus will have momentum and kinetic energy greater than is implied by the formulas used in classical physics. This, unlike spatial distortion, is more than a mere "mind experiment." There is little practical concern about spatial distortion in most situations, at least nowadays. (A thousand years from now, when we are roaming among the stars, we should expect that things will be different!) When electrons move at high enough speed, they attain properties of much more massive particles and acquire some of the properties of x-rays or gamma rays such as are emitted by radioactive substances. There is a name for high-speed electrons that act this way: *beta particles*.

Physicists take advantage of the relativistic effects on the masses of protons, helium nuclei, neutrons, and other subatomic particles. When these particles are subjected to powerful electrical and magnetic fields in a device called a *particle accelerator*, they get moving so fast that their mass increases because of relativistic effects. When the particles strike atoms of matter, the nuclei of those target atoms are fractured. It takes quite a wallop to break up the nucleus of an atom! When this happens, energy can be released in the form of infrared, visible light, ultraviolet, x-rays, and gamma rays, as well as a potpourri of exotic particles.

If astronauts ever travel long distances through space in ships moving at speeds near the speed of light, relativistic mass increase will be a dire concern. While the astronaut's own bodies won't seem unnaturally massive, and the objects inside the ship will appear to have normal mass too, the particles whizzing by outside will gain real mass. It is scary enough to think about what will happen when a 1-kg meteoroid strikes a space ship traveling at 99.9 percent of the speed of light. But that 1-kg stone will mass more than 22 kg when $u = 0.999$, that is, at 99.9 percent of the speed of light. As if this is not bad enough, every atomic nucleus outside the ship will strike the vessel's shell at relativistic speed, producing deadly radiation.

EXPERIMENTAL CONFIRMATION

Relativistic time dilation and mass increase have been measured under controlled conditions, and the results concur with Einstein's formulas stated earlier. Thus these effects are more than mere tricks of the imagination.

To measure time dilation, a superaccurate atomic clock was placed on board an aircraft, and the aircraft was sent up in flight to cruise around for a while at several hundred kilometers per hour. Another atomic clock was kept at the place where the aircraft took off and landed. Although the aircraft's speed was only a tiny fraction of the speed of light and the resulting time dilation was exceedingly small, the accumulated discrepancy was large enough to measure. When the aircraft arrived back at the terminal, the clocks, which had been synchronized (when placed right next to each other, of course!) before the trip began, were compared. The clock that had been on the aircraft was a little behind the clock that had been resting comfortably on Earth.

To measure mass increase, particle accelerators are used. It is possible to determine the mass of a moving particle based on its known rest mass and the kinetic energy it possesses as it moves. When the mathematics is done, Einstein's formula is always shown to be correct.

General Relativity

There is no absolute standard for position in the Universe, nor is there an absolute standard for velocity. Another way of saying this is that any reference frame is just as valid as any other as long as acceleration does not take place. The notions of "the center of the Universe" and "at rest" are relative. If we measure position or velocity, we must do so with respect to something, such as Earth or the Sun or a space ship coasting through the void.

ACCELERATION IS DIFFERENT!

Einstein noticed something special about accelerating reference frames compared with those that are not accelerating. This difference is apparent if we consider the situation of an observer who is enclosed in a chamber that is completely sealed and opaque.

Imagine that you are in a space ship in which the windows are covered up and the radar and navigational equipment have been placed on standby. There is no way for you to examine the surrounding environment and determine where you are, how fast you are moving, or what direction you are moving. However, you can tell whether or not the ship is accelerating. This is so because acceleration always produces a force on objects inside the ship.

When the ship's engines are fired and the vessel gains speed in the forward direction, all the objects in the ship (including your body) are subjected to a force directed backward. If the ship's retro rockets are fired so that the ship slows down (decelerates), everything in the ship is subjected to a force directed forward. If rockets on the side of the ship are fired so that the ship changes direction without changing its speed, this too is a form of acceleration and will cause sideways forces on everything inside the ship. Some examples are illustrated in Fig. 16-5.

The greater the acceleration, or change in velocity, to which the space ship is subjected, the greater is the force on every object inside it. If m is

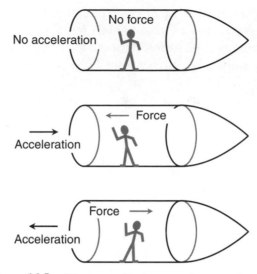

Figure 16-5.   When a vessel in deep space is not accelerating,
there is no force on the objects inside. When the ship accelerates,
there is always a force on the objects inside.

the mass of an object in the ship (in kilograms) and a is the acceleration of
the ship (in meters per second per second), then the force F (in newtons) is
their product:

$$F = ma$$

This is one of the most well-known formulas in physics.

This "acceleration force" occurs even when the ship's windows are covered up, the radar is switched off, and the navigational equipment is placed
on standby. There is no way the force can be blocked out. In this way,
Einstein reasoned, it is possible for interstellar travelers to determine
whether or not their ship is accelerating. Not only this, but they can calculate the magnitude of the acceleration as well as its direction. When it
comes to acceleration, there are absolute reference frames.

THE EQUIVALENCE PRINCIPLE

Imagine that our space ship, instead of accelerating in deep space, is set
down on the surface of a planet. It might be tail-downward, in which case
the force of gravity pulls on the objects inside as if the ship is accelerating
in a forward direction. It might be nose-downward so that gravity pulls on

the objects inside as if the ship is decelerating. It could be oriented some other way so that the force of gravity pulls on the objects inside as if the ship is changing course in a lateral direction. Acceleration can consist of a change in speed, a change in direction, or both.

If the windows are kept covered, the radar is shut off, and the navigational aids are placed on standby, how can passengers in such a vessel tell whether the force is caused by gravitation or by acceleration? Einstein's answer: They can't tell. In every respect, acceleration force manifests itself in precisely the same way as gravitational force.

From this notion came the *equivalence principle*, also known as *Einstein's principle of equivalence*. The so-called acceleration force is exactly the same as gravitation. Einstein reasoned that the two forces act in an identical way on everything, from human bodies to subatomic particles and from light rays to the very fabric of space-time. This is the cornerstone of the theory of general relativity.

SPATIAL CURVATURE

Imagine that you are in a space ship traveling through deep space. The ship's rockets are fired, and the vessel accelerates at an extreme rate. Suppose that the laser apparatus described earlier in this chapter is in the ship, but instead of a mirror on the wall opposite the laser, there is a screen. Before the acceleration begins, you align the laser so that it shines at the center of the screen (Fig. 16-6). What will happen when the rockets are fired and the ship accelerates?

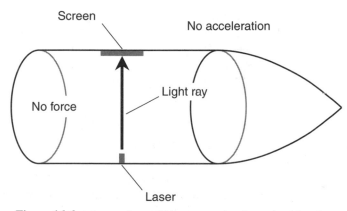

Figure 16-6.   As seen from within a nonaccelerating space ship, a laser beam travels in a straight line across the vessel.

In a real-life scenario, the spot from the laser will not move on the screen enough for you to notice. This is so because any reasonable (that is, non-life-threatening) rate of acceleration will not cause sufficient force to influence the path of the beam. However, let's suspend our disbelief and imagine that we can accelerate the vessel at any rate, no matter how great, without being squashed against the ship's rear wall. If we accelerate fast enough, the ship pulls away from the laser beam as the beam travels across the ship. We, looking at the situation from inside the ship, see the light beam follow a curved path (Fig. 16-7). A stationary observer on the outside sees the light beam follow a straight path, but the vessel pulls out ahead of the beam (Fig. 16-8).

Regardless of the reference frame, the ray of light always follows the shortest possible path between the laser and the screen. When viewed from any nonaccelerating reference frame, light rays appear straight. However, when observed from accelerating reference frames, light rays can appear curved. The shortest distance between the two points at opposite ends of the laser beam in Fig. 16-7 is, in fact, curved. The apparently straight path is in reality longer than the curved one, as seen from inside the accelerating vessel! It is this phenomenon that has led some people to say that "space is curved" in a powerful acceleration field. According to the principle of equivalence, powerful gravitation causes the same sort of *spatial curvature* as acceleration.

For spatial curvature to be as noticeable as it appears in Figs. 16-7 and 16-8, the vessel must accelerate at an extremely large pace. The standard

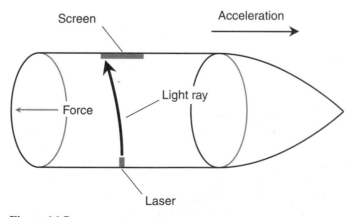

Figure 16-7.   As seen from within a space ship accelerating at an extreme rate, a laser beam travels in a curved path across the vessel.

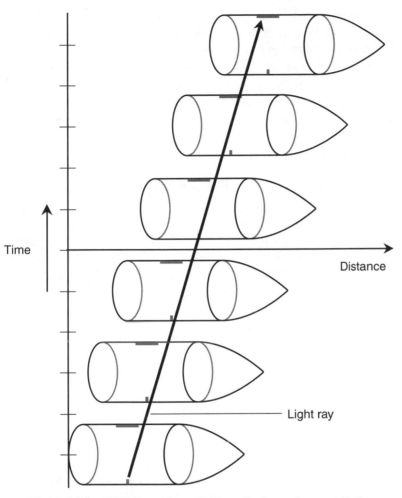

Figure 16-8.   When viewed from a "stationary" reference frame outside the
ship, the accelerating vessel pulls away from the straight-line path of the
laser beam, so the beam strikes the screen off-center.

unit of acceleration is the meter per second per second, or *meter per sec-*
ond squared (m/s^2). Astronauts and aerospace engineers also express acceleration in units called *gravities* (symbolized *g*), where one gravity (1 *g*) is
the acceleration that produces the same force as the gravitational field of
Earth at the surface, approximately 9.8 m/s^2. (Don't confuse the abbreviation for gravity or gravities with the abbreviation for grams. Pay attention
to the context if you see a unit symbolized *g*.) Figures 16-7 and 16-8 show
the situation for an acceleration of many thousands of gravities. If you

weigh 150 pounds on Earth, you would weigh many tons in a ship accelerating at a rate, or in a gravitational field of such intensity, so as to cause that much spatial curvature. In real life, no one could survive such force. No living human being will ever directly witness the sort of light-beam curvature shown in these illustrations.

Is all this a mere academic exercise? Are there actually gravitational fields powerful enough to bend light rays significantly? Yes. They exist near the event horizons of black holes.

TIME DILATION CAUSED BY ACCELERATION OR GRAVITATION

The spatial curvature caused by intense acceleration or gravitation produces an effective slowing down of time. Remember the fundamental axiom of special relativity: The speed of light is constant no matter what the point of view. The laser beam traveling across the space ship, as shown in many of the illustrations in this chapter, always moves at the same speed. This is one thing about which all observers, in all reference frames, must agree.

The path of the light ray, as it travels from the laser to the screen, is longer in the situation shown by Fig. 16-7 than in the situation shown by Fig. 16-6. This is so in part because the ray takes a diagonal path rather than traveling straight across. In addition, however, the path is curved. This increases the time interval even more. From the vantage point of a passenger in the space ship, the curved path shown in Fig. 16-7 represents the shortest possible path the light ray can take across the vessel between the point at which it leaves the laser and the point at which it strikes the screen. The laser device itself can be turned slightly, pointing a little bit toward the front of the ship; this will cause the beam to arrive at the center of the screen (Fig. 16-9) instead of off-center. However, the path of the beam is still curved and is still longer than its path when the ship is not accelerating (see Fig. 16-6). The laser represents the most accurate possible timepiece, because it is based on the speed of light, which is an absolute constant. Thus time dilation is produced by acceleration not only as seen by observers looking at the ship from the outside but also for passengers within the vessel itself. In this respect, acceleration and gravitation are more powerful "time dilators" than relative motion.

Suspending our disbelief again, and assuming that we could experience such intense acceleration force (or gravitation) without being physically crushed, we will actually perceive time as slowing down inside the vessel

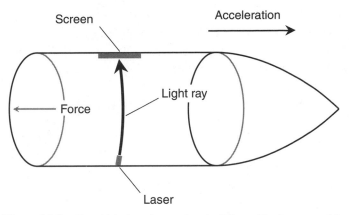

Figure 16-9. Even if the laser is turned so the light ray hits the center of the screen, the path of the ray is curved when the ship accelerates at a high rate.

under conditions such as those that produce spatial curvature, as shown in Fig. 16-7 or Fig. 16-9. Clocks will run more slowly even from reference frames inside the ship. In addition, everything inside the ship will appear warped out of shape.

If the acceleration or gravitation becomes far more powerful still (Fig. 16-10), the spatial curvature and the time dilation will be considerable. You will look across the ship at your fellow travelers and see grotesquely elongated or foreshortened faces (depending on which way you are oriented inside the vessel). Your voices will deepen. It will be like a science-fiction movie. You and all the other passengers in the ship will know that something extraordinary is happening. This same effect will be observed by people foolish enough to jump into a black hole (yet again ignoring the fact that they would be stretched and crushed at the same time by the force).

OBSERVATIONAL CONFIRMATION

When Einstein developed his general theory of relativity, some of the paradoxes inherent in special relativity were resolved. (These paradoxes are avoided here because discussing them would only confuse you.) In particular, light rays from distant stars were observed as they passed close to the Sun to see whether or not the Sun's gravitational field, which is quite strong near its surface, would bend the light rays. This bending would be observed as a change in the apparent position of a distant star in the sky as the Sun passes close to it.

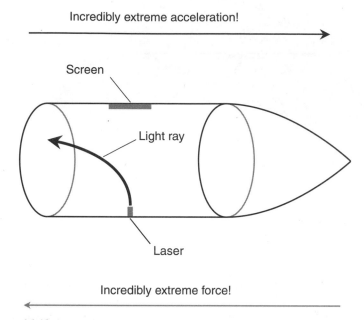

Incredibly extreme acceleration!

Screen

Light ray

Laser

Incredibly extreme force!

Figure 16-10.   If the acceleration is great enough, the spatial curvature becomes extreme.

The problem with this type of observation was, as you might guess, the fact that the Sun is far brighter than any other star in the sky, and the Sun's glare normally washes out the faint illumination from distant stars. However, during a total solar eclipse, the Sun's disk is occulted by the Moon. The angular diameter of the Moon in the sky is almost exactly the same as that of the Sun, so light from distant stars passing close to the Sun can be seen by Earthbound observers during a total eclipse. When this experiment was carried out, the apparent position of a distant star was offset by the presence of the Sun, and this effect took place to the same extent as Einstein's general relativity formulas said it should.

More recently, the light from a certain quasar has been observed as it passes close to a suspected black hole. On its way to us, the light from the quasar follows multiple curved paths around the dark, massive object. This produces several images of the quasar, arranged in the form of a cross with the dark object at the center.

The curvature of space in the presence of a strong gravitational field has been likened to a funnel shape (Fig. 16-11), except that the surface of the funnel is three-dimensional rather than two-dimensional. The shortest distance in three-dimensional space between any two points near the gravita-

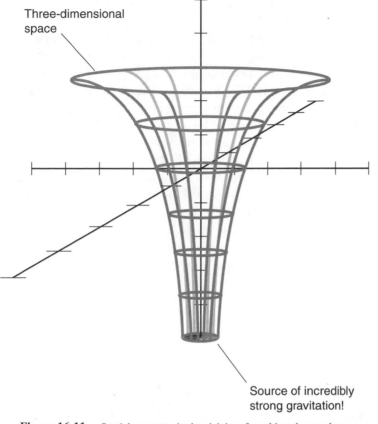

Three-dimensional space

Source of incredibly strong gravitation!

Figure 16-11.   Spatial curvature in the vicinity of an object that produces an intense gravitational field.

tional source is always a curve with respect to four-dimensional space. This is impossible for most (if not all) people to envision directly without "cheating" by taking away one dimension. But the mathematics is straight-forward enough, and observations have shown that it correctly explains the phenomenon.

WHY DOES THIS MATTER?

We have conducted "mind experiments" in this chapter, many of which require us to suspend reality. In real life, scenarios such as these would kill anyone attempting to make the observations. So why is relativity theory important? If space is bent and time is slowed by incredibly powerful gravitational fields, so what?

The theory of general relativity plays an important role in astronomers' quests to unravel the mysteries of the structure and evolution of the Universe. On a cosmic scale, gravitation acquires a different aspect than on a local scale. A small black hole, such as that surrounding a collapsed star, is dense and produces gravitation strong enough to destroy any material thing crossing the event horizon. However, if a black hole contains enough mass, the density within the event horizon is not so large. Black holes with quadrillions of solar masses can exist, at least in theory, without life-threatening forces at any point near their event horizons. If such a black hole is ever found, and if we develop space ships capable of intergalactic flight, we will be able to cross its event horizon unscathed and enter another Universe. We can be sure someone will try it if they think they can do it, even if they are never able to communicate back to us what they find.

According to some theorists, we need not travel far to find the ultimate black hole. It has been suggested that our entire Universe is such an object and that we are inside it.

Quiz

Refer to the text if necessary. A good score is 8 correct. Answers are in the back of the book.

1. A common unit of acceleration is the
   (a) meter per second.
   (b) kilometer per second.
   (c) kilometer per hour.
   (d) gravity.

2. Suppose that you have a spherical ball with mass of a hundred grams (100 g) at rest. If you throw the ball at three-quarters of the speed of light, what will its mass become, as measured from a stationary point of view?
   (a) 100 g
   (b) 133 g
   (c) 151 g
   (d) It cannot be calculated from this information.

3.   Suppose that the ball in Problem 2 has an apparent diameter, as measured laterally (sideways to the direction of its motion), of a hundred millimeters (100 mm) when it is speeding along at three-quarters of the speed of light. What will its diameter be when it comes to rest?
     (a)   100 mm
     (b)   133 mm
     (c)   151 mm
     (d)   It cannot be calculated from this information.

4.   If a space ship is slowing down, that is, losing speed in the forward direction, the acceleration force inside the ship is directed
     (a)   toward the rear.
     (b)   toward the front.
     (c)   toward the side.
     (d)   nowhere; there is no acceleration force.

5.   The Michelson-Morley experiment
     (a)   showed that the speed of light depends on the direction in which it is measured.
     (b)   showed that the speed of light depends on the velocity of the observer.
     (c)   showed that the speed of light does not depend on the direction in which it is measured.
     (d)   proved that the ether passes right through the Earth.

6.   If you are in a spacecraft accelerating at 9.8 m/s^2 through interplanetary space, you will feel the same force as you feel if you are sitting still on the surface of the Earth. This is an expression of
     (a)   a complete falsehood! Traveling through space is nothing at all like being on Earth.
     (b)   the fact that the speed of light is absolute, finite, and constant and is the fastest known speed.
     (c)   Einstein's equivalence principle.
     (d)   the results of the Michelson-Morley experiment.

7.   Suppose that you see a space ship whiz by at the speed of light. What is the time dilation factor k that you observe when you measure the speed of a clock inside that ship and compare it with the speed of a clock that is stationary relative to you?
     (a)   1
     (b)   0
     (c)   Infinity
     (d)   It is not defined.

8.   Some light beams will follow curved paths
     (a)   under no circumstances.
     (b)   when measured inside a space ship that is coasting at high speed.
     (c)   when measured in the presence of an extreme gravitational field.
     (d)   when measured from a reference frame that is not accelerating.

9. Suppose that you get on a space ship and travel toward the star Sirius at 150,000 km/s, which is approximately half the speed of light. If you measure the speed of the light arriving from Sirius, what figure will you obtain?
   (a) 150,000 km/s
   (b) 300,000 km/s
   (c) 450,000 km/s
   (d) It cannot be calculated from this information.

10. Clocks in different locations are impossible to synchronize from every possible reference frame because
   (a) the speed of light is absolute, finite, and constant and is the fastest known speed.
   (b) the speed of light depends on the location of the reference frame from which it is measured.
   (c) the speed of light depends on velocity of the reference frame from which it is measured.
   (d) there is no such thing as a perfect clock.

Test: Part Four

Do not refer to the text when taking this test. A good score is at least 30 correct. Answers are in the back of the book. It is best to have a friend check your score the first time so that you won't memorize the answers if you want to take the test again.

1. The main sequence on the Hertzsprung-Russell diagram is
    (a) where the red giants are found.
    (b) where the white dwarfs are found.
    (c) where the Sun is found.
    (d) where stars are before hydrogen fusion begins.
    (e) where stars end up after they have burned out.

2. Stars seem to twinkle when observed from Earth's surface because
    (a) dispersion of the starlight occurs in outer space.
    (b) turbulence in Earth's atmosphere refracts the starlight.
    (c) the stars actually are changing in brilliance.
    (d) the solar wind refracts the starlight.
    (e) the geomagnetic field bends the starlight.

3. If a neutron star is massive enough so that gravitation overpowers all other forces during the final collapse, the object will in theory become
    (a) a black dwarf.
    (b) a supernova.
    (c) a white dwarf.
    (d) an event horizon.
    (e) a space-time singularity.

4. When Michelson and Morley measured the speed of light in various directions, they discovered that
    (a) the speed of light is slowest in the direction in which Earth travels through space.
    (b) the speed of light is fastest in the direction in which Earth travels through space.
    (c) Earth drags the luminiferous ether along with itself.

    (d) the speed of light is the same in all directions.

    (e) the speed of light cannot be accurately determined.

5. A black dwarf is

    (a) a small planet with low albedo.

    (b) a star in the process of formation, still dark because nuclear fusion has not yet begun.

    (c) any small, dark object in the Cosmos.

    (d) an object whose gravitation is so intense that not even light can escape.

    (e) a white dwarf that has burned out and cooled down.

6. Near the plane of our own Milky Way's spiral disk, it is almost impossible to see distant galaxies and quasars because

    (a) they are too far away.

    (b) the gas and dust in the plane of the Milky Way obscure the view.

    (c) there are few distant objects in the plane of the Milky Way.

    (d) the gravitation of the Milky Way bends light from such objects away from us.

    (e) No! It is easier to see distant objects near the plane of the Milky Way than in other regions of space.

7. Suppose that a space ship whizzes by so fast that clocks on board seem, as seen from our point of view, to be running at half speed. If the rest mass of the ship is 50 metric tons, what will be the mass of the ship from our point of view as it whizzes by?

    (a) 25 metric tons

    (b) 50 metric tons

    (c) 100 metric tons

    (d) 400 metric tons

    (e) It cannot be determined without more information.

8. Gravitational waves

    (a) are like ripples in space and time.

    (b) cannot penetrate solid objects.

    (c) have been proven to be a theoretical fiction and not to exist in reality.

    (d) cause black holes to form.

    (e) travel faster than light.

9. Einstein's principle of equivalence states that

    (a) gravitational force is just like acceleration force.

    (b) force equals mass times acceleration.

    (c) the speed of light is constant no matter what.

    (d) the speed of light is the highest possible speed.

    (e) the shortest distance between two points is a straight line.

10. The term *near IR* refers to

    (a) energy at wavelengths slightly longer than visible red light.

(b) energy at wavelengths slightly shorter than visible red light.

(c) energy at wavelengths slightly longer than visible violet light.

(d) energy at wavelengths slightly shorter than visible violet light.

(e) energy from objects comparatively near the Solar System.

11. Regardless of whether a reference frame is accelerating or not, light rays always

(a) travel in straight lines.

(b) follow the shortest possible path between two points in space.

(c) travel in curved paths.

(d) are repelled by gravitational fields.

(e) travel fastest in the direction of motion.

12. The distances to galaxies closer than about 10 million light years can be inferred by observing

(a) Cepheid variables in the galaxies.

(b) the extent to which the galaxies are tilted as we view them.

(c) the extent to which the spectra of the galaxies are blue-shifted.

(d) the waveforms of the pulses emitted by the galactic nuclei.

(e) the intensity of x-rays emitted by the spiral arms of the galaxies.

13. A teaspoonful of neutrons packed tightly together

(a) would mass many thousands of kilograms.

(b) would fly apart because like charges repel.

(c) would instantly disintegrate.

(d) would mass about the same as a teaspoonful of electrons.

(e) would undergo nuclear fusion.

14. Fill in the blank in the following sentence: "According to astronomer Thomas Gold, the magnetic field in the immediate vicinity of a pulsar can be _____."

(a) as intense as the field at Earth's surface

(b) perfectly uniform

(c) trillions of times as intense as the field at Earth's surface

(d) doughnut-shaped

(e) as weak as a trillionth of the intensity of the field at Earth's surface

15. No quasar has ever been observed that has blue-shifted spectral lines. This lends support to the theory that

(a) quasars are objects that have been ejected from our galaxy.

(b) quasars are objects in the Solar System.

(c) quasars have weak gravitational fields.

(d) quasars are distant and are receding from us.

(e) No! There are plenty of quasars with blue-shifted spectra.

16. One minute of arc is equal to

(a) the distance light travels in 1 minute.

(b) 1/60 of an angular degree.

(c) 1/60 of a full circle.

(d) the angular distance the Sun travels across the sky in 1 minute.

(e) 1/60 of 1 hour of right ascension.

17. When astronomers scrutinized the so-called spiral nebulae, it was eventually discovered that they are

(a) rotating clouds of gas and dust in the Milky Way.

(b) black holes sucking in interstellar gas.

(c) exploding stars.

(d) distant congregations of stars outside the Milky Way.

(e) a mystery to this day; no one yet knows what they are.

18. The "ticks" emitted by a pulsar

(a) occur at irregular and unpredictable intervals.

(b) occur fastest when the pulsar is high in the sky and slowest when the pulsar is low in the sky.

(c) occur fastest when the pulsar is low in the sky and slowest when the pulsar is high in the sky.

(d) generally have rough waveforms, unlike the signals from radio transmitters.

(e) have regular waveforms, just like the signals from radio transmitters.

19. A hypothetical point where matter enters our Universe from another space-time continuum is sometimes called

(a) a quasar.

(b) a white hole.

(c) a black hole.

(d) an event horizon.

(e) a pulsar.

20. Most astronomers believe the Sun will explode as a supernova

(a) 1 million to 2 million years from now.

(b) 500 million to 1 billion years from now.

(c) 1 billion to 2 billion years from now.

(d) 5 billion to 15 billion years from now.

(e) never.

21. If a space ship from Earth landed on a planet made of antimatter,

(a) the space ship would sink to the center of the planet.

(b) the space ship would land normally, but no passengers could get off without being annihilated.

(c) there would be a terrific explosion.

(d) it would be just the same as if the planet were made of matter.

(e) No! The ship could never land because it would be repelled by the antimatter planet.

22. Relativistic spatial distortion occurs

(a) only at speeds faster than the speed of light.

(b) only when objects accelerate.

(c) only along the axis of relative motion.

(d) only for extremely dense or massive objects.

(e) only within black holes.

23. A spectroscope is used for
   (a) enhancing the quality of images seen through a telescope.
   (b) evaluating electromagnetic signals received at radio wavelengths.
   (c) transmitting signals in the hope of contacting extraterrestrial beings.
   (d) scrutinizing visible light by breaking it down by wavelength.
   (e) measuring the parallax of stars.

24. Imagine that a space ship whizzes by so fast that clocks on board seem, as seen
   from our point of view on Earth, to be running at one-third their normal speed.
   Suppose that you mass 60 kg on Earth and have a friend riding on the ship who
   also masses 60 kg on Earth. If your friend measures his mass while traveling
   on the ship, what will he observe it to be?
   (a) 20 kg
   (b) 60 kg
   (c) 180 kg
   (d) 540 kg
   (e) It cannot be determined without more information.

25. Planetary nebulae
   (a) form around massive planets.
   (b) are clouds of gas and dust from which planets form.
   (c) are gas and dust attracted by the gravitational fields of planets.
   (d) have stars at their centers.
   (e) are irregular in shape.

26. Spatial distortion can be caused by all the following *except*
   (a) acceleration.
   (b) gravitation.
   (c) high relative speed.
   (d) black holes.
   (e) the solar wind.

27. The gravitational radius of an object
   (a) is directly proportional to its mass.
   (b) is inversely proportional to its mass.
   (c) is directly proportional to the square of its mass.
   (d) is inversely proportional to the square of its mass.
   (e) does not depend on its mass.

28. When we look at a quasar that is 8 billion light-years away, we see
   (a) the quasar as it appears right now.
   (b) the quasar as it will appear 8 billion years in the future.
   (c) the quasar as it appeared before the Solar System existed.
   (d) an image that has traveled all the way around the known Universe.
   (e) an illusion because astronomers doubt that anything exists that is 8 billion
        light-years distant.

29. The heliopause is
    (a) the region in space where the solar wind gives way to the general circulation of interstellar gas and dust.
    (b) the region in the Sun's corona beyond which the temperature begins to drop.
    (c) the region inside the Sun where radiation gives way to the convection.
    (d) the time in the future at which the Sun's hydrogen fuel will be all used up.
    (e) the time in the future at which the Sun's nuclear fusion reactions will all cease.

30. Time travel into the future might be possible by taking advantage of
    (a) relativistic time dilation.
    (b) relativistic mass distortion.
    (c) relativistic spatial distortion.
    (d) the gravitational pull of the Earth.
    (e) nothing! Time travel into the future is theoretically impossible.

31. Suppose that two superaccurate atomic clocks, called clock A and clock B, are synchronized on Earth so that they agree exactly. Now imagine that clock B is placed aboard a space vessel and sent to Mars and back. The clock readings are compared after the ship returns. What do we find?
    (a) Clocks A and B still agree precisely.
    (b) Clock A is behind clock B.
    (c) Clock A is ahead of clock B.
    (d) Any of the above, depending on the extent to which the ship accelerated during its journey.
    (e) None of the above

32. Which of the following terms does *not* refer to a type of variable star?
    (a) RR Lyrae
    (b) Mira
    (c) White dwarf
    (d) Eclipsing binary
    (e) Cepheid

33. In the Milky Way galaxy, the Solar System is believed to be located
    (a) in one of the spiral arms halfway from the center of the disk to the edge.
    (b) high above the plane of the galaxy's disk.
    (c) in a black hole at the center.
    (d) between spiral arms at the outer edge of the galaxy's disk.
    (e) in the central bulge but not within the central black hole itself.

34. In today's conventional model of the atom,
    (a) electrons orbit in circles and all in the same plane.
    (b) electrons orbit in ellipses with the nucleus at one focus.
    (c) electrons and protons comprise the nucleus.
    (d) there are more electrons than protons.
    (e) electrons exist in spherical shells surrounding the nucleus.

35. When astronomers talk about "dark matter," they are referring to
    (a) clouds of dust in interstellar space.
    (b) planets and moons in the Solar System.
    (c) black-dwarf stars.
    (d) asteroids, meteoroids, and comets that are too far from the Sun to glow.
    (e) hypothetical cosmic "stuff" that has mass but cannot be seen.

36. Fill in the blank in the following sentence: "When a space ship moves at a speed approaching the speed of light relative to an observer, that observer will see a clock on the ship appear to _____."
    (a) run too fast
    (b) stop
    (c) run too slowly
    (d) run infinitely fast
    (e) run at normal speed

37. A distance of 1 Mpc is
    (a) 0.001 parsec.
    (b) 1,000 parsecs.
    (c) 1 million parsecs.
    (d) 1 billion parsecs.
    (e) dependent on the angle at which it is measured.

38. If a neutron star collapses to within its event horizon,
    (a) electromagnetic rays leaving the surface at low angles are trapped.
    (b) electromagnetic rays cannot escape from the surface into outer space.
    (c) it rebounds and explodes, causing a supernova.
    (d) it disappears without a trace.
    (e) No! A neutron star cannot collapse to within its event horizon.

39. Suppose that an astronomer finds an object that looks like a glowing ball of stars, and its spectrum is significantly red-shifted. The astronomer concludes that the object is
    (a) an elliptical galaxy.
    (b) a quasar.
    (c) a black hole.
    (d) a globular cluster.
    (e) an emission nebula.

40. Globular star clusters are believed to be
    (a) galaxies far from the Milky Way.
    (b) comprised of young stars.
    (c) comprised of old stars.
    (d) approaching us at high speed.
    (e) receding from us at high speed.

PART 5

Space Observation and Travel

CHAPTER 17

Optics
and Telescopes

Until a few hundred years ago, the only instrument available for astronomical observation was the human eye. This changed in the 1600s when several experimenters, including such notables as Galileo Galilei and Isaac Newton, combined lenses and mirrors to make distant objects look closer. Since then, *optical telescopes* have become larger and more sophisticated. So have the ways in which the light they gather is scrutinized.

Basic Optics

You have learned that visible light always take the shortest path between two points and that it always travels at the same speed. These are the cornerstones of relativity theory and can be taken as axiomatic as long as the light stays in a vacuum. However, if the medium through which light passes is significantly different from a vacuum, and especially if the medium changes as the light ray travels through it, these principles of relativity do not apply.

Let's focus our attention on what happens when light passes through a medium such as glass or is reflected by mirrors. If a ray of light passes from air into glass or from glass into air, the path of the ray is bent. Light rays change direction when they are reflected from mirrors. This has nothing to do with relativity. It happens all the time, everywhere you look. It even takes place within your own eyes.

LIGHT RAYS

What is a *ray of light*? Definitions vary. Informally, a thin shaft of light, such as that which passes from the Sun through a pinhole in a piece of cardboard, can be called a ray or beam of light. In a more technical sense, a ray can be considered to be the path that an individual photon (light particle) follows through space, air, glass, water, or any other medium.

Light rays have properties of both particles and waves. This duality has long been a topic of interest among physicists. In some situations, the *particle model* or *corpuscular model* explains light behavior very well, and the wave model falls short. In other scenarios, the opposite is true. No one has actually seen a ray of light; all we can see are the effects produced when a ray of light strikes something. Yet there are certain things we can say about the way in which rays of light behave. These things are predictable, both qualitatively and quantitatively. When we know these facts about light, we can build high-quality instruments for observing the Cosmos at visible wavelengths.

REFLECTION

Prehistoric people knew about reflection. It would not take an intelligent creature very long to figure out that the "phantom in the pond" actually was a reflection of himself or herself. Any smooth, shiny surface reflects some of the light that strikes it. If the surface is perfectly flat, perfectly shiny, and reflects all the light that strikes it, then any ray that encounters the surface is reflected away at the same angle at which it hits. You have heard the expression, "The angle of incidence equals the angle of reflection." This principle, known as the *law of reflection*, is illustrated in Fig. 17-1. The angle of incidence and the angle of reflection are both measured relative to a normal line (also called an *orthogonal* or *perpendicular*). In the figure, these angles are denoted q. They can range from as small as 0 degrees, where the light ray strikes at a right angle, to almost 90 degrees, a grazing angle.

If the reflective surface is not perfectly flat, then the law of reflection still applies for each ray of light striking the surface at a specific point. In such a case, the reflection is considered with respect to a flat plane passing through the point tangent to the surface at that point. When many parallel rays of light strike a curved or irregular reflective surface at many different points, each ray obeys the law of reflection, but the reflected rays do not all emerge parallel. In some cases they converge; in other cases they diverge. In still other cases the rays are scattered haphazardly.

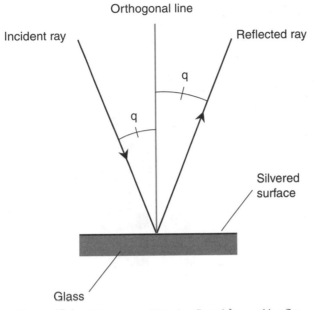

Figure 17-1.   When a ray of light is reflected from a shiny flat surface, the angle of incidence is equal to the angle of reflection. Here, both angles are denoted q.

REFRACTION

Prehistoric people noticed refraction as well as reflection; a clear pond looks shallower than it actually is because of this effect. People who used spears to catch fish learned to compensate for the effects of refraction; the images of fish were displaced more or less depending on the angle at which they were observed. The cause and the far-reaching uses of visible-light refraction were not known or understood until quite recently. There is evidence that ancient Greeks and Romans knew how to make crude lenses for the purpose of focusing light beams, but more refined applications apparently evaded them.

When light rays cross a flat boundary from one clear medium into another having different light-transmission properties, the rays are bent, or *refracted*. An example is shown in Fig. 17-2 when the *refractive index* of the initial medium, called medium X in the figure, is higher than that of the final medium, called Y. (The refractive index, also called the *index of refraction*, is defined in the next section.) A ray striking the boundary at a right angle passes through without changing direction. However, a ray that hits at some other angle is bent; the greater the angle of incidence, the

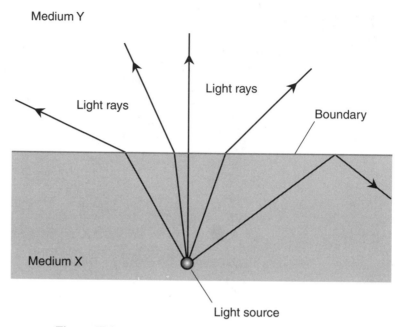

Medium Y

Light rays

Light rays

Boundary

Medium X

Light source

Figure 17-2. Rays of light are bent more or less as they cross a
boundary between media having different properties.

sharper is the turn. When the angle of incidence reaches a *critical angle*,
then the light ray is not refracted at the boundary but is reflected back into
medium *X*. This is *total internal reflection*.

If the directions of the light rays in Fig. 17-2 are reversed, they still follow the same paths. Thus a ray originating in medium *Y* and striking the
boundary at a grazing angle is bent downward at a considerable angle. This
causes significant distortion of images when viewed from underwater. You
have surely seen this effect if you are a scuba diver. The entire landscape
above the water looks as if it is viewed through a wide-angle lens.

If the refracting boundary is not perfectly flat, then the principle shown
by Fig. 17-2 still applies for each ray of light striking the boundary at a specific point. The refraction is considered with respect to a flat plane passing
through the point tangent to the boundary at that point. When many parallel rays of light strike a curved or irregular refractive boundary at many different points, each ray obeys the same principle individually. As a whole,
however, the effect can be much different than is the case for a flat boundary. In some cases parallel rays converge after crossing the boundary; in
other cases they diverge. In still other cases the rays are scattered.

REFRACTIVE INDEX

Different media transmit light at different speeds. This does not violate the fundamental principle of relativity theory. The speed of light is absolute in a vacuum, where it travels at 299,792 km/s or 186,282 mi/s expressed to six significant digits. However, light travels more slowly than this in other media because the relativistic principle only applies for a vacuum.

In air, the difference in the speed of light is slight, although it can be significant enough to produce refractive effects at near-grazing angles between air masses having different densities. In water, glass, quartz, diamond, and other transparent media, light travels quite a lot more slowly than it does in a vacuum. The refractive index of a particular medium is the ratio of the speed of light in a vacuum to the speed of light in that medium. If c is the speed of light in a vacuum and c_x is the speed of light in medium X, then the index of refraction for medium X, call it r_x, can be calculated simply:

$$r_x = c/c_x$$

Always use the same units when expressing c and c_x. According to this definition, the index of refraction of any transparent material is always greater than or equal to 1.

The greater the index of refraction for a transparent substance, the more a ray of light is bent when it passes the boundary between that substance and air. Different types of glass have different refractive indices. Quartz refracts more than glass, and diamond refracts more than quartz. The high refractive index of diamond is responsible for the multicolored shine of diamond stones.

DISPERSION

The index of refraction for a particular substance depends on the wavelength of the light passing through it. Glass slows down light the most at the shortest wavelengths (blue and violet) and the least at the longest wavelengths (red and orange). This variation of the refractive index with wavelength is known as *dispersion*. It is the principle by which a prism works (Fig. 17-3). The more the light is slowed down by the glass, the more its path is deflected when it passes through the prism. This is why prisms cast rainbows when white light shines through them.

Dispersion is important in optical astronomy for two reasons. First, a prism can be used to make a *spectrometer*, which is a device for examining

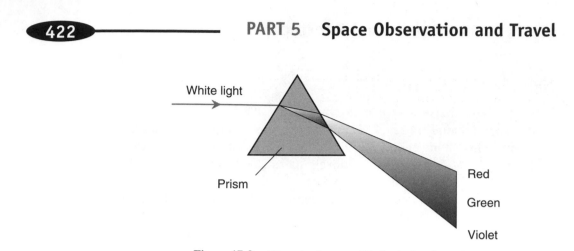

Figure 17-3. Dispersion is responsible for the fact that a glass prism "splits" white light into its constituent colors.

the intensity of visible light at specific wavelengths. (Gratings are also used for this.) Second, dispersion degrades the quality of white-light images viewed through lenses unless those lenses are specially made to cancel out the effect.

Lenses and Mirrors

The ways in which visible light is reflected and refracted can be used to advantage. This was first discovered when experimenters noticed that specially shaped pieces of glass could make objects look larger or smaller. The refractive properties of glass have been used for centuries to help correct nearsightedness and farsightedness. Lenses work because they refract light more or less depending on where and at what angle the light strikes their surfaces. Curved mirrors have much the same effect when they reflect light.

THE CONVEX LENS

You can buy a *convex lens* in almost any novelty store or department store. In a good hobby store you should be able to find a magnifying glass up to 10 cm (4 in) or even 15 cm (6 in) in diameter. The term *convex* arises from the fact that one or both faces of the glass bulge outward at the center. A convex lens is sometimes called a *converging lens*. It brings parallel light rays to a sharp *focus* or *focal point*, as shown in Fig. 17-4A, when those rays are parallel to the axis of the lens. It also can *collimate* (make parallel) the light from a point source, as shown in Fig. 17-4*B*.

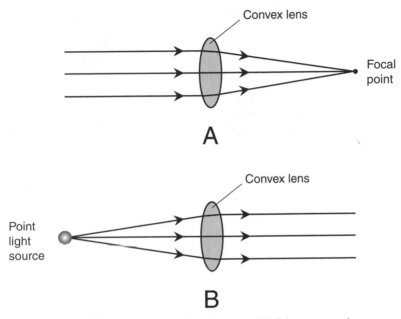

Figure 17-4.   At *A* a convex lens focuses parallel light rays to a point. At *B* the same lens collimates light from a point source at the focus.

The properties of a convex lens depend on the diameter of the lens, as well as on the difference in thickness between the edges and the center. The larger the diameter, the greater is the light-gathering power. The greater the difference in thickness between the center and the edges, the shorter is the distance between the lens and the point at which it brings parallel light rays to a focus. The effective area of the lens, measured in a plane perpendicular to the axis, is known as the *light-gathering area*. The distance between the center of the lens and the focal point (as shown in Fig. 17-4*A* or *B*) is called the *focal length*. If you look through a convex lens at a close-up object such as a coin, the features are magnified; they appear larger than they look with the unaided eye.

The surfaces of convex lenses generally are spherical. This means that if you could find a large ball having just the right diameter, the curve of the lens face would fit neatly inside the ball. Some convex lenses have the same radius of curvature on each face; others have different radii of curvature on their two faces. Some converging lenses have one flat face; these are called *planoconvex lenses*.

THE CONCAVE LENS

You will have some trouble finding a *concave lens* in a department store, but you should be able to order them from specialty catalogs or Web sites. The term *concave* refers to the fact that one or both faces of the glass bulge inward at the center. This type of lens is also called a *diverging lens*. It spreads parallel light rays outward (Fig. 17-5A). It can collimate converging rays if the convergence angle is correct (see Fig. 17-5B).

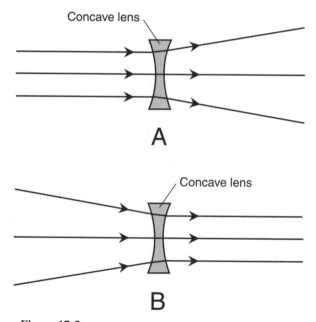

Figure 17-5.   At *A* a concave lens spreads parallel light rays.
At *B* the same lens collimates converging light rays.

As with convex lenses, the properties of a concave lens depend on the diameter and the extent to which the surface(s) depart from flat. The greater the difference in thickness between the edges and the center of the lens, the more the lens will cause parallel rays of light to diverge. If you look through a concave lens at a close-in object such as a coin, the features are reduced; they appear smaller than they look with the unaided eye.

The surfaces of concave lenses, like those of their convex counterparts, generally are spherical. Some concave lenses have the same radius of curvature on each face; others have different radii of curvature on their two faces. Some diverging lenses have one flat face; these are called *planoconcave lenses*.

THE CONVEX MIRROR

A *convex mirror* reflects light rays in such a way that the effect is similar to that of a concave lens. Incident rays, when parallel, are spread out (Fig. 17-6A) after they are reflected from the surface. Converging incident rays,

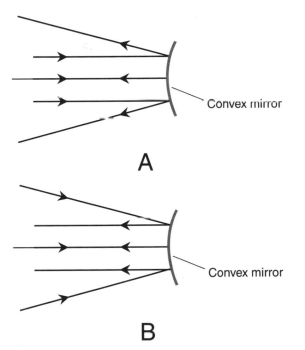

Figure 17-6.   At *A* a convex mirror spreads parallel incident light rays. At *B* the same mirror collimates converging incident light rays.

if the angle of convergence is just right, are collimated by a convex mirror (see Fig. 17-6B). When you look at the reflection of a scene in a convex mirror, the objects all appear reduced. The field of vision is enlarged, a fact that is used to advantage in some automotive rear-view mirrors.

The extent to which a convex lens spreads light rays depends on the radius of curvature. The smaller the radius of curvature, the greater is the extent to which parallel incident rays diverge after reflection.

THE CONCAVE MIRROR

A *concave mirror* reflects light rays in a manner similar to the way a convex lens refracts them. When incident rays are parallel to each other and to the axis of the mirror, they are reflected so that they converge at a focal point (Fig. 17-7A). When a point source of light is placed at the focal point, the concave mirror reflects the rays so that they emerge parallel (see Fig. 17-7B).

The properties of a concave mirror depend on the size of the reflecting surface, as well as on the radius of curvature. The larger the light-gathering

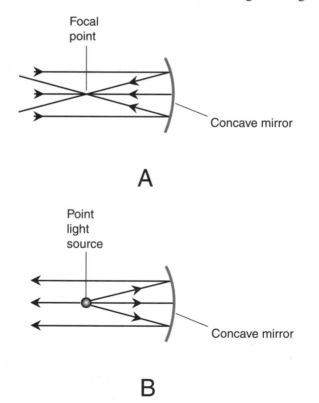

Figure 17-7.   At *A* a concave mirror focuses parallel light rays to a point.
At *B* the same mirror collimates light from a point source at the focus.

area, the greater is the light-gathering power. The smaller the radius of cur-vature, the shorter is the focal length. If you look at your reflection in a con-vex mirror, you will see the same effect that you would observe if you placed a convex lens up against a flat mirror.

Concave mirrors can have spherical surfaces, but the finest mirrors have surfaces that follow the contour of an idealized three-dimensional figure called a *paraboloid*. A paraboloid results from the rotation of a parabola, such as that having the equation $y = x^2$ in rectangular coordinates, around its axis. When the radius of curvature is large compared with the size of the reflecting surface, the difference between a *spherical mirror* and a *parab-oloidal mirror* (more commonly called a *parabolic mirror*) is not notice-able to the casual observer. However, it makes a big difference when the mirror is used in a telescope.

Refracting Telescopes

The first telescopes were developed in the 1600s and used lenses. Any tel-escope that enlarges distant images with lenses alone is called a *refracting telescope*.

GALILEAN REFRACTOR

Galileo devised a telescope consisting of a convex-lens *objective* and a con-cave-lens *eyepiece*. His first telescope magnified the apparent diameters of distant objects by a factor of only a few times. Some of his later telescopes magnified up to 30 times. The *Galilean refractor* (Fig. 17-8A) produces an *erect image*, that is, a right-side up view of things. In addition to appearing right-side-up, images are also true in the left-to-right sense. The *magnifi-cation factor*, defined as the number of times the angular diameters of dis-tant objects are increased, depends on the focal length of the objective, as well as on the distance between the objective and the eyepiece.

Galilean refractors are still available today, mainly as novelties for ter-restrial viewing. Galileo's original refractors had objective lenses only 2 or 3 cm (about 1 in) across; the same is true of most Galilean telescopes found today. Some of these telescopes have sliding, concentric tubes, providing variable magnification. When the inner tube is pushed all the way into the

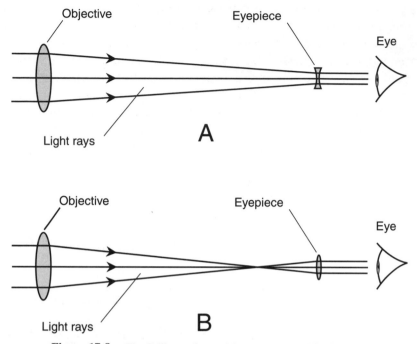

Figure 17-8. The Galilean refractor (*A*) uses a convex objective and a concave eyepiece. The Keplerian refractor (*B*) has a convex objective and a convex eyepiece.

outer one, the magnification factor is the lowest; when the inner tube is pulled all the way out, the magnification is highest. The image remains fairly clear over the entire magnification-adjustment range. These instruments are sometimes called *spy glasses*.

KEPLERIAN REFRACTOR

Johannes Kepler's refracting telescope employed a convex-lens objective with a long focal length and a smaller convex-lens eyepiece with a short focal length. Unlike the Galilean telescope, the *Keplerian refractor* (see Fig. 17-8*B*) produces an *inverted image*; it is upside-down and backwards. The distance between the objective and the eyepiece must be exactly equal to the sum of the focal lengths of the two lenses in order for the image to be clear. The magnification factor depends on the ratio between the focal lengths of the objective and the eyepiece.

The Keplerian telescope is preferred over the Galilean type mainly because the Keplerian design provides a larger *apparent field of view*. This is the angular diameter, as seen directly by the eyes, of the circular region in which objects appear through the telescope. Galilean telescopes have apparent fields of view so narrow that looking through them is an uncomfortable experience.

The magnification factor of a Keplerian telescope can be changed by using eyepieces with longer or shorter focal lengths. The shorter the focal length of the eyepiece, the greater is the magnification factor, informally known as *power*, assuming that the focal length of the objective lens remains constant.

The largest refracting telescope in the world is a Keplerian refractor, located at the Yerkes Observatory in Wisconsin. Its objective lens has a diameter of 40 in, or slightly more than 1 m. Keplerian refractors are used by thousands of amateur astronomers worldwide.

LIMITATIONS OF REFRACTORS

A well-designed refracting telescope is a pleasure to use. Nevertheless, there are certain problems inherent in their design. These are known as *spherical aberration*, *chromatic aberration*, and *lens sag*.

Spherical aberration results from the fact that spherical convex lenses don't bring parallel light rays to a perfect focus. Thus a refracting telescope with a spherical objective will focus a ray passing through its edge a little differently than a ray passing closer to the center. The actual focus of the objective is not a point but a very short line along the lens axis. This effect causes slight blurring of images of objects that have relatively large angular diameters, such as nebulae and galaxies. The problem can be corrected by grinding the objective lens so that it has paraboloidal rather than spherical surfaces.

Chromatic aberration occurs because the glass in an objective lens refracts the shortest wavelengths of light slightly more than the longest wavelengths. The focal length of any given convex lens therefore is shorter for violet light than for blue light, shorter for blue light than for yellow light, and shorter for yellow light than for red light. This produces rainbow-colored halos around star images, as well as along sharply defined edges in objects with large angular diameters. Chromatic aberration can be corrected by the use of *compound lenses*. These lenses have two or more sections made of different types of glass; the sections are glued together with a special transparent adhesive. Such objectives are called *achromatic lenses* and are standard issue in refracting telescopes these days.

Lens sag occurs in the largest refracting telescopes. When an objective is larger than approximately 1 m (about 40 in) across, it becomes so massive that its own weight distorts its shape. Glass is not perfectly rigid, as you have noticed if you have seen the reflection of the landscape in a large window on a windy day. There is no way to get rid of this problem with a refractor except to take the telescope out of Earth's gravitational field.

Reflecting Telescopes

The problems inherent in refracting telescopes, particularly chromatic aberration and lens sag, can be largely overcome by using mirrors instead of lenses as objectives. A *first-surface mirror*, with the silvering on the outside so that the light never passes through glass, can be ground so that it brings light to a focus that does not vary with wavelength. Mirrors can be supported from behind, so it is possible to make them larger than lenses without running into the sag problem.

NEWTONIAN REFLECTOR

Isaac Newton designed a *reflecting telescope* that was free of chromatic aberration. His design is still used in many reflecting telescopes today. The *Newtonian reflector* employs a concave objective mounted at one end of a long tube. The other end of the tube is open to admit incoming light. A small, flat mirror is mounted at a 45-degree angle near the open end of the tube to reflect the focused light through an opening in the side of the tube containing the eyepiece (Fig. 17-9A).

The flat mirror obstructs some of the incoming light, slightly reducing the effective surface area of the objective mirror. As a typical example, suppose that a Newtonian reflector has an objective mirror 20 cm in diameter. The total surface area of this mirror is approximately 314 centimeters squared (cm^2). If the eyepiece mirror is 3 cm square, its total area is 9 cm^2, which is about 3 percent of the total surface area of the objective.

Newtonian reflectors have limitations. Some people find it unnatural to "look sideways" at objects. If the telescope has a long tube, it is necessary to use a ladder to view objects at high elevations. These annoyances can be overcome by using a different way to get the light to the eyepiece.

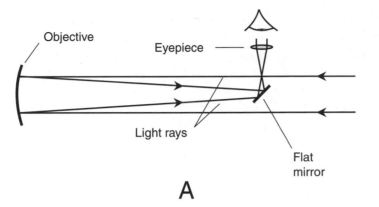

A

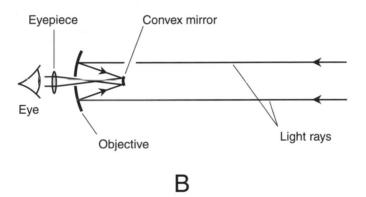

B

Figure 17-9.   The Newtonian reflector (*A*) has an eyepiece set into the side of the tube. In the Cassegrain reflector (*B*), the eyepiece is in the center of the objective mirror.

CASSEGRAIN REFLECTOR

Figure 17-9*B* shows the design of the *Cassegrain reflector*. The eyepiece mirror is closer to the objective than in the Newtonian design. It is not angled, but it is convex. The convexity of this mirror increases the effective focal length of the objective mirror. Light reflects from the convex mirror and passes through a small hole in the center of the objective containing the eyepiece.

The Cassegrain reflector can be made with a physically short tube and an objective mirror having a smaller radius of curvature than that of a Newtonian telescope having the same diameter. Thus the Cassegrain telescope is less massive and less bulky. Cassegrain reflectors with heavy-duty mountings are physically stable, and they can be used at low magnification to obtain wide views of the sky.

Telescope Specifications

Several parameters are significant when determining the effectiveness of a telescope for various applications. Here are the most important ones.

MAGNIFICATION

The *magnification*, also called *power* and symbolized ×, is the extent to which a telescope makes objects look closer. Actually, telescopes increase the observed sizes of distant objects, but they do not look closer in terms of perspective. The magnification is a measure of the factor by which the apparent angular diameter of an object is increased. A 20× telescope makes the Moon, whose disk subtends about 0.5 degrees of arc as observed with the unaided eye, appear 10 degrees of arc in diameter. A 180× telescope makes a crater on the Moon with an angular diameter of only 1 minute of arc (1/60 of a degree) appear 3 degrees across.

Magnification is calculated in terms of the focal lengths of the objective and the eyepiece. If f_o is the effective focal length of the objective and f_e is the focal length of the eyepiece (in the same units as f_o), then the magnification factor m is given by this formula:

$$m = f_o/f_e$$

For a given eyepiece, as the effective focal length of the objective increases, the magnification of the whole telescope also increases. For a given objective, as the effective focal length of the eyepiece increases, the magnification of the telescope decreases.

RESOLVING POWER

The *resolution*, also called *resolving power*, is the ability of a telescope to separate two objects that are not in exactly the same place in the sky. It is measured in an angular sense, usually in seconds of arc (units of 1/3,600 of a degree). The smaller the number, the better the resolving power.

The best way to measure a telescope's resolving power is to scan the sky for known pairs of stars that appear close to each other in the angular sense. Astronomical data charts can determine which pairs of stars to use for this purpose. Another method is to examine the Moon and use a detailed map of the lunar surface to ascertain how much detail the telescope can render.

Resolving power increases with magnification, but only up to a certain point. The greatest image resolution a telescope can provide is directly proportional to the diameter of the objective lens or mirror, up to a certain maximum dictated by atmospheric turbulence. In addition, the resolving power depends on the acuity of the observer's eyesight (if direct viewing is contemplated) or the coarseness of the grain of the photographic or detecting surface (if an analog or digital camera is used).

LIGHT-GATHERING AREA

The light-gathering area of a telescope is a quantitative measure of its ability to collect light for viewing. It can be defined in centimeters squared (cm^2) or meters squared (m^2), that is, in terms of the effective surface area of the objective lens or mirror as measured in a plane perpendicular to its axis. Sometimes it is expressed in inches squared (in^2).

For a refracting telescope, given an objective radius of r, the light-gathering area A can be calculated according to this formula:

$$A = \pi r^2$$

where π is approximately equal to 3.14159. If r is expressed in centimeters, then A is in centimeters squared; if r is in meters, then A is in meters squared.

For a reflecting telescope, given an objective radius of r, the light-gathering area A can be calculated according to this formula:

$$A = \pi r^2 - B$$

where B is the area obstructed by the secondary mirror assembly. If r is expressed in centimeters and B is expressed in centimeters squared, then A is in centimeters squared; if r is in meters and B is in meters squared, then A is in meters squared.

ABSOLUTE FIELD OF VIEW

When you look through the eyepiece of a telescope, you see a circular patch of sky. Actually, you can see anything within a cone-shaped region whose apex is at the telescope (Fig. 17-10). The *absolute field of view* is the angular diameter q of this cone; q can be specified in degrees, minutes, and/or seconds of arc. Sometimes the angular radius is specified instead of the angular diameter.

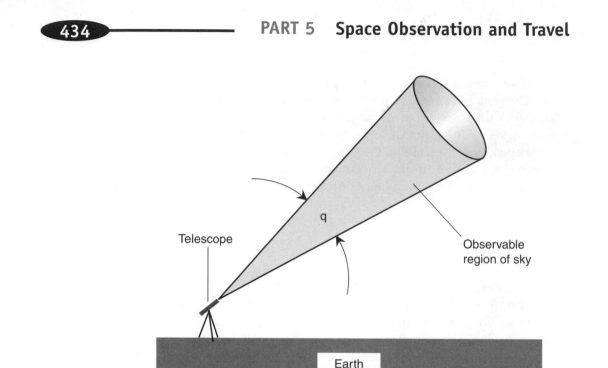

Figure 17-10.   A telescope's absolute field of view, q, is measured
in angular degrees, minutes, and/or seconds of arc.

The absolute field of view depends on several factors. The magnification of the telescope is important. When all other factors are held constant, the absolute field of view is inversely proportional to the magnification. If you double the magnification, you cut the absolute field of view in half. If you reduce the magnification by a factor of 4, you increase the absolute field of view by a factor of 4.

The viewing angle provided by the eyepiece—the apparent field of view—is important. Some types of eyepieces have a wide apparent field, such as 60 degrees or even 90 degrees. Others have narrower apparent fields, in some cases less than 30 degrees. Galileo's first refracting telescope had an apparent field of view only a few degrees wide.

Another factor that affects the absolute field of view is the *focal ratio* or *f-ratio*: the objective's focal length divided by its actual diameter as measured in the same units. A telescope's *f*-ratio is denoted by writing *f*, followed by a forward slash, followed by the ratio expressed as a number. Thus, for example, if the focal length of the objective is 200 cm and its actual diameter is 20 cm, the *f*-ratio is *f*/10. In general, the larger the *f*-ratio, the smaller is the maximum apparent field of view that can be obtained with the telescope. Long, narrow telescopes have the smallest maximum apparent fields of view; short, fat ones have the widest maximum fields.

Observation Limitations

The quality of the image obtainable with a telescope depends on some things the astronomer can control, such as the objective diameter, the quality of the optics, and the location of the observatory. Other problems are unavoidable. Here are some of the factors that affect the performance of telescopes.

OBJECTIVE DIAMETER

The greatest useful magnification a telescope can provide is approximately 20× per centimeter of objective diameter, or 50× per inch. Visible light has properties of a wave disturbance, and the waves have physical length in space. The objective lens or mirror must measure at least a certain number of wavelengths in diameter to provide a given image resolution.

If you have a telescope that is 10 cm (4 in) across, you can obtain up to approximately 200× of magnification with it. Beyond this, you can get things to "look bigger," but they also become more blurred, so the resolution of the images you see does not improve.

QUALITY OF OPTICS

The importance of good *optics* (that is, high-quality lenses and mirrors) is obvious. The best optics can be expensive, and the cheapest optics are usually not so good. However, the quality of the optics is not an absolute and direct function of expense.

It is important to keep the objective lens or mirror, the secondary mirror (if any), and the eyepieces clean and free of condensation. In some locations, the formation of dew can be a constant problem. Frost can form if the temperature is low enough and the humidity is relatively high. Dew and frost not only cloud the optics, but they also accelerate the accumulation of dirt.

PHYSICAL STABILITY

When a telescope is aimed at an object, you expect it to stay there. Vibration transmitted from the ground can cause blurring of the image. Swaying produced by the wind can make viewing difficult. Either of these effects, when they occur in conjunction with the use of a time-exposure camera attached to a telescope, injure the image quality.

The mounting that secures the telescope to the base must be physically rugged so that the telescope will not waver once it is aimed at an object. The platform on which the entire observatory is placed likewise must be solid. The effects of vibration and swaying are exaggerated by high magnification.

EARTH'S ROTATION

When a telescope is aimed at a certain spot in the sky, celestial objects move across the field of view because Earth rotates on its axis. For most regions of the sky, this apparent motion is from east to west. Depending on the type of mounting used, image rotation also can take place. This rotation does not occur with objects on the celestial equator, but it does happen with all other objects unless an *equatorial mounting* and motor drive are used to compensate for Earth's rotation.

AIR TURBULENCE

The atmosphere interferes with observation of celestial objects. The main trouble is the fact that the air is not uniformly dense at every point above the surface. Pockets of warmer and cooler air are constantly swirling and tumbling, and these have the effect of slightly refracting the light passing from space to the surface. This refraction is responsible for the twinkling of stars, whose angular diameters are virtually zero even when viewed through a telescope. It blurs the images of planets, their moons, and the details within distant galaxies.

Air turbulence limits the practical magnification of any telescope to about $500\times$ no matter how large the objective. It can be somewhat greater than $500\times$ for observatories atop high mountains in regions where the atmosphere is relatively stable, such as Mauna Kea in Hawaii. In many locations, however, it is less than $500\times$. The only way to completely get around this limitation is to put the telescope in outer space.

AIRGLOW AND SKYGLOW

The atmosphere scatters visible light, particularly at the shorter wave-lengths (blue and violet). This produces an effect called *airglow*. It is responsible for the blue color of the daytime sky. Daytime airglow makes it impossible to see any stars or planets, except Venus and, of course, the Sun. Airglow occurs at night to a much lesser extent. It washes out the dimmest stars and limits the faintness of objects that can be observed through large telescopes. Airglow at night is worst at the time of the full Moon, and it is aggravated by the proximity of large cities because of myriad human-made lights. Dust and air pollution aggravate airglow problems.

In outer space, there is some scattering of light even though there is no airglow because space itself is not a perfect vacuum. One example of this is the *gegenschein* (counterglow) that occurs in a direction opposite the Sun. Atoms in space scatter light from the Sun and also from the distant stars and galaxies. This *skyglow* places a theoretical limit on the faintness of objects that can be observed, no matter how large the telescope and regardless of where it is located. In general, however, it is possible to see much fainter objects from space than from Earth's surface.

Telescope Peripherals

At the world's large observatories, the astronomers make use of instruments to record, analyze, and enhance the images. Rarely does a telescope operator simply peer through an eyepiece.

PHOTOMETER

Sometimes it is necessary to measure the brightness of a celestial object rather than merely making a good guess or a subjective comparison. A *photometer* is an instrument that does this. An astronomical photometer is a sophisticated version of the lightmeter, which is used by photographers to determine camera exposure time. The device consists of a *photosensor* placed at the focal point of the telescope. An amplification circuit multiplies the sensitivity of the sensor. The output of the amplifier can be connected to a circuit that plots the light intensity as a function of time.

Many celestial objects, such as variable stars and visible pulsars, change in visual magnitude with time. Variable stars fluctuate slowly, but some pulsars blink so fast that they look like ordinary stars until a graph is plotted using a photometer capable of resolving into brief intervals of time. Photometers can be made sensitive in the infrared (IR) and ultraviolet (UV) ranges as well as in the visible spectrum.

SPECTROMETER

Visible light can be broken down into the colors of the rainbow, each hue representing a specific wavelength. This can be done by a prism with a triangular or trapezoidal cross section. It also can be done by passing the light through or reflecting it from a *diffraction grating*, which is a clear plate or mirror with thousands of tiny parallel opaque bands etched on it. The grating works because of the *interference patterns* produced by light waves passing through the gaps between the dark bands. This is an entirely different phenomenon from the refraction that occurs in a prism, but the practical effect is similar.

A *spectrometer*, also known as a *spectroscope*, is a device intended for analyzing visible light at all its constituent wavelengths. Some spectrometers also work at IR or UV wavelengths. Stars, galaxies, quasars, and some nebulae have spectra that contain dark absorption lines at certain wavelengths. Other nebulae are dark except at specific *emission wavelengths* that manifest themselves as bright lines in a spectrum. The patterns of lines allow scientists to determine the amounts of various chemicals that comprise celestial objects after corrections are made for the absorption effects of Earth's atmosphere.

Figure 17-11A is a functional diagram of a simple spectrometer. The light-sensitive surface can be photographic film or a matrix of *optoelectronic sensors*. The maximum image resolution obtainable by a spectrometer of this type is limited by the grain of the film or the number of pixels per centimeter in the sensor. Higher resolution can be achieved by a scheme such as that diagrammed in Fig. 17-11B. The rotating prism causes the spectrum to sweep across the objective of a viewing scope. A light sensor connected to the scope measures the intensity of the rays. The angle of the prism at any given moment in time is fed to a computer along with the sensor output. This produces a graph of the spectrum that is called, not surprisingly, a *spectrograph*.

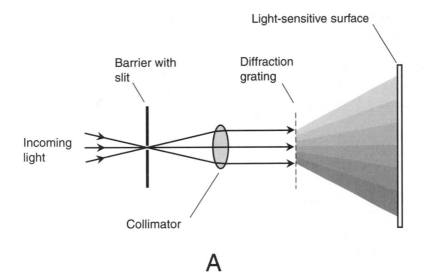

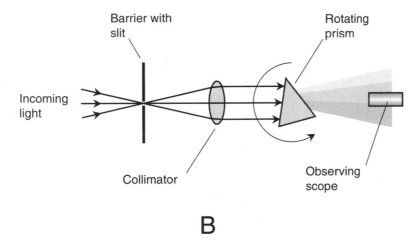

Figure 17-11.   Functional diagrams of a simple spectrometer
(*A*) and a high-resolution spectrometer (*B*).

CAMERAS AND FILM

Astrophotography is the art of recording the images from a telescope on photographic film. Long exposures make it possible for astronomers to "see" objects much dimmer than they could by looking directly through the telescope.

The sensitivity of an astrophotographic system depends on the light-gathering area of the telescope, on the speed of the film, and on the length of time the film is exposed to the image. The image resolution depends on the size of the emulsion particles in the film, as well as on the telescope's magnification and light-gathering area. Some cameras and films can record images at near-IR or near-UV wavelengths as well as in the visible range.

Film behaves in strange ways when exposed to visible light over long periods of time. This is one of the reasons digital imaging systems are gradually replacing film cameras in astrophotography.

CHARGE-COUPLED DEVICE (CCD)

A *charge-coupled device* (CCD) is a camera that converts visible-light images into digital signals. Some CCDs also work with IR or UV. Astronomers use CCDs to record and enhance images of all kinds of celestial objects. Common digital cameras work on a principle similar to that of the CCD.

The image focused on the retina of your eye or on the film of a camera is an *analog image*. It can have infinitely many configurations and infinitely many variations in hue, brightness, contrast, and saturation. However, a digital computer needs a *digital image* to make sense of and enhance what it "sees." Binary digital signals have only two possible states: on and off. These are also called *high* and *low* or 1 and 0. It is possible to get an excellent approximation of an analog image in the form of high and low digital signals. This allows a computer program to process the image, bringing out details and features that otherwise would be impossible to detect.

A simplified block diagram of a CCD is shown in Fig. 17-12. The image falls on a matrix containing thousands or millions of tiny sensors. Each sensor produces one *pixel* (picture element). The computer (not shown) can employ all the tricks characteristic of any good graphics program. In addition to rendering high-contrast or false-color images, the CCD and computer together can detect and resolve images much fainter than is possible with conventional camera film.

The Space Telescope

Once astronomers realized the extent to which Earth's atmosphere limits the resolution and faintness of objects that can be seen or photographed

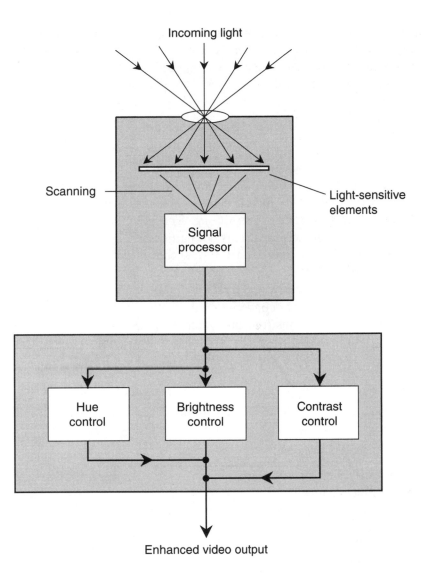

Figure 17-12.
Functional diagram of a charge-coupled device (CCD) and basic image-processing system. A computer can provide additional image enhancement.

with telescopes, they began to dream about putting a telescope in space. The Hubble Space Telescope (HST) was the first major optical instrument to be placed in Earth orbit and used for intensive observation within the visible-light range.

ASSETS

Space telescopes get above the "dirty window" of gases that absorb, disperse, and scatter the light. They are also freed from the constraints imposed by gravity. This makes it possible to build gigantic instruments without concern for lens or mirror sag. The HST has a mirror approximately 2.4 m (94 in) across. This is not particularly large as telescopes go, but we should expect that some future space telescopes will have much larger mirrors.

Problems caused by dew and frost are eliminated in outer space because there is no water vapor to condense on the optics. There is no atmospheric dust to accumulate on instruments either. The HST can be covered when not in use, so meteoric dust can be kept out. The telescope contains onboard computers and steering devices and is powered by batteries recharged with solar panels.

The HST uses a Cassegrain design. The f-ratio is optimized. Several instruments are available to obtain, process, and analyze images. These include:

- High-speed photometer
- Wide-field camera
- Faint-object camera
- Faint-object spectrometer
- Multiobject spectrometer
- IR camera

CHALLENGES

Operating a telescope in space presents certain problems that users of Earth-based telescopes don't have to deal with. The most obvious constraint, at least with the HST, is the fact that the observer is not physically present to attend to the apparatus. The telescope must be aimed and the instruments operated by remote control.

Pointing and tracking of the HST are somewhat more complicated than is the case with an Earth-based telescope. This is so because the instrument is not fixed with respect to Earth but is in orbit. The telescope is also subject to instability because of the *action-reaction* principle of physics, also known as *Newton's third law*. If anything moves in the telescope assembly, its momentum produces a reaction that causes the whole machine to slowly tumble unless corrective action is taken. The telescope is kept "on target" using *star trackers* that follow the images of certain reference stars in fixed positions relative to the telescope axis. Gyroscopes add stability.

Temperature fluctuations affect all large telescopes. A change of even a few degrees Celsius can cause the optics to expand and contract enough to affect the focal length and sharpness of the image. Even in the relatively temperature-stable environment of Earth's atmosphere, the cooling that takes place after sundown can cause problems unless measures are taken to minimize the effect. In an orbiting telescope such as HST, the difference in temperature between "day" (telescope in sunlight) and "night" (telescope in Earth's shadow) is much greater. Sophisticated temperature-compensation equipment is necessary to prevent this from severely degrading the performance of the telescope.

A space telescope can be placed in an orbit directly above the *terminator*, also known as the *gray line* or *twilight line*, on Earth's surface. In this way, the telescope is kept in sunlight all the time. This evens out the temperature fluctuations, although some variation still occurs as the Sun shines on different parts of the instrument when the telescope is aimed at different parts of the sky. This type of orbit also ensures that the solar panels are constantly exposed to plenty of light so that there is solar power available all the time.

Refer to the text if necessary. A good score is 8 correct. Answers are in the back of the book.

1. A simple convex lens has a focal length that varies slightly depending on the wavelength of the light passing through it. When such a lens is used as the objective of a telescope, this effect results in
   (a) dispersion.
   (b) spherical aberration.

(c) chromatic aberration.

    (d) nothing! The premise is wrong. A convex lens has the same focal length for all wavelengths of light passing through it.

2. At the surface of Earth, the maximum useful magnification of a telescope, no matter how large the objective, is approximately 500× because of

    (a) skyglow.

    (b) air turbulence.

    (c) dispersion.

    (d) diffraction.

3. A spectrometer can record its image on

    (a) photographic film.

    (b) an eyepiece.

    (c) a secondary mirror.

    (d) a gegenschein device.

4. The use of a CCD with a telescope can enhance the

    (a) focal length.

    (b) magnification.

    (c) image contrast.

    (d) spectral output.

5. According to the law of reflection,

    (a) a ray of light traveling from a medium having a low refractive index to a medium having a higher refractive index is reflected at the boundary.

    (b) a ray of light traveling from a medium having a high refractive index to a medium having a lower refractive index is reflected at the boundary.

    (c) a ray of light always reflects from a shiny surface in a direction exactly opposite the direction from which it arrives.

    (d) None of the above

6. A Cassegrain type reflecting telescope has an objective mirror with a diameter of 300 mm and an eyepiece with a focal length of 30 mm. The magnification is

    (a) 100×.

    (b) 10×.

    (c) 9,000×.

    (d) impossible to calculate from this information.

7. A diverging lens

    (a) can collimate converging rays of light.

    (b) can focus the Sun's rays to a brilliant point.

    (c) is also known as a convex lens.

    (d) is ideal for use as the objective in a refracting telescope.

8. Suppose that the speed of red visible light in a transparent medium is 270,000 km/s. What, approximately, is the index of refraction for this substance with respect to red light?
   (a) 0.900
   (b) 1.11
   (c) 0.810
   (d) It cannot be calculated from this information.

9. As the magnification of a telescope is increased,
   (a) the image resolution decreases in direct proportion.
   (b) physical stability becomes more and more important.
   (c) the light-gathering area increases in direct proportion.
   (d) dimmer and dimmer objects can be seen.

10. An advantage of space telescopes over Earthbound telescopes is the fact that
   (a) space telescopes have less mass.
   (b) space telescopes do not need power supplies.
   (c) space telescopes are not subject to temperature changes.
   (d) space telescopes are not subject to airglow.

CHAPTER 18

Observing the Invisible

The optical telescope was invented long before scientists knew that visible light represents only a tiny part of a continuum of energy wavelengths. Isaac Newton believed that visible light was composed of tiny particles or *corpuscles*. Today we recognize these particles as *photons*. However, light is more complex than can be represented by the simple *corpuscular theory*. The same is true of all forms of radiant energy.

Electromagnetic Fields

The wave nature of visible light, and of other forms of radiant energy, is the result of synergistic interaction of electrical and magnetic forces. Charged particles, such as electrons and protons, are surrounded by *electrical (E) fields*. Magnetic poles produce *magnetic (M) fields*. The fields extend into the space surrounding the charged particles or magnetic poles, and when the fields are strong enough, their effects can be noticed at a considerable distance. When the E and M fields vary in intensity, the result is an *electromagnetic (EM) field*.

Orderly, well-defined EM fields are generated by voltages or currents that vary in a rhythmic way. Conversely, an EM field can give rise to alternating voltages or currents. These effects can occur over vast distances in space.

STATIC E AND M FIELDS

If you've ever played with permanent magnets, you have noticed the attraction between opposite poles and the repulsion between like poles. Similar effects take place with electrically charged objects. These forces seem to operate only over short distances under laboratory conditions. This is so because static (steady, unchanging) E and M fields weaken rapidly, as the distance between poles increases to less than the smallest intensity we can detect. In theory, the fields extend into space indefinitely.

Physicists have known for a long time that a constant electric current in a wire produces an M field around the wire. The *lines of magnetic flux* are perpendicular to the direction of the current. It is also known that the existence of a constant voltage difference between two nearby objects produces an E field; the *lines of electrical flux* are parallel to the direction in which the voltage varies most rapidly with distance. When the intensity of a current or voltage changes with time, things get more interesting.

FLUCTUATING FIELDS

A fluctuating current in a wire or a variable voltage between two nearby objects gives rise to both an M field and an E field. These fields regenerate each other, so they can travel for long distances with less attenuation than either type of field all by itself. The E and M lines of flux in such a situation are perpendicular to each other everywhere in space. The direction of travel of the attendant EM field is perpendicular to both the E and M lines of flux, as shown in Fig. 18-1.

In order for an EM field to exist, electrons or other charge carriers not only must be moving, but also must be accelerating. That is, their velocity must be constantly changing. The most common method of creating this sort of situation is the introduction of an alternating current (ac) in an electrical conductor. It also can result from the bending of charged-particle beams by E or M fields.

FREQUENCY

The frequency of an EM field can be very low, such as a few cycles per second, or hertz (abbreviated Hz), and can range upward into the thousands, millions, billions, or trillions of hertz. A frequency of 1,000 Hz is a *kilohertz* (kHz); a frequency of 1,000 kHz is a *megahertz* (MHz); a frequency

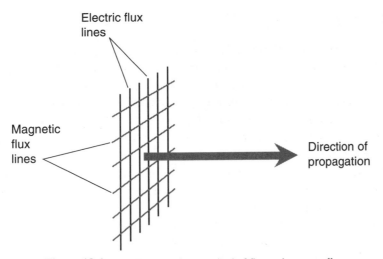

Electric flux lines

Magnetic flux lines

Direction of propagation

Figure 18-1. An EM wave is comprised of fluctuating, mutually perpendicular electric and magnetic lines of flux. The field travels perpendicular to both sets of flux lines.

of 1,000 MHz is a *gigahertz* (GHz); and a frequency of 1,000 GHz is a *terahertz* (THz).

WAVELENGTH

Electromagnetic waves travel through space at the speed of light, which is approximately 2.99792×10^8 m/s. This is often rounded up to 3.00×10^8 m/s, expressed to three significant figures. The wavelength of an EM field in free space gets shorter as the frequency becomes higher. At 1 kHz, the wavelength is about 300 km. At 1 MHz, the wavelength is about 300 m. At 1 GHz, the wavelength is about 0.300 m or 300 mm. At 1 THz, an EM signal has a wavelength of 0.3 mm—so small that you would need a magnifying glass to see the waves, if they were in fact directly visible.

The frequency of an EM wave can get much higher than 1 THz; some of the most energetic rays known have wavelengths of 0.000001 *nanometer* (10^{-6} nm). The nanometer equivalent to 10^{-9} m and is used commonly by scientists to measure the wavelengths of EM disturbances at visible wavelengths and shorter. A microscope of great magnifying power would be needed to see an object with a length of 1 nm. Less commonly, the *Ångström* (Å) is used to denote extremely short wavelengths. This unit is 1/10 of a nanometer; that is, 1 Å = 10^{-10} m = 0.1 nm.

MANY FORMS

The discovery of EM fields led to the "wireless" radio and ultimately to the sophisticated and complex variety of communications systems we know today. Radio waves are not the only form of EM radiation. As the frequency increases above that of conventional radio, we encounter new forms. First come the *microwaves*. Then comes *infrared* (IR), or "heat rays." After that comes visible light, ultraviolet (UV) radiation, x-ray energy, and gamma-ray energy.

In the opposite, and less commonly imagined, sense, EM fields can exist at frequencies far below those of radio signals. Some *extremely-low-frequency* (ELF) fields have frequencies less than the 50 or 60 Hz of ac utility electricity. In theory, an EM wave can go through one complete cycle every hour, day, year, thousand years, or million years. Some astronomers suspect that stars and galaxies generate EM fields with periods of years, centuries, or millennia.

The EM Spectrum

The wavelengths of the lowest-frequency EM fields can extend, at least theoretically, for light-years. The shortest gamma rays have, as we have already mentioned, wavelengths that measure only a tiny fraction of a nanometer. In between these extremes lie all the forms of EM energy.

THE EM WAVELENGTH SCALE

To illustrate the range of EM wavelengths, scientists often use a logarithmic scale. It is necessary to use a logarithmic scale because the range is so great that a linear scale is impractical. The left-hand portion of Fig. 18-2 is such a logarithmic scale and shows wavelengths from 10^8 down to 10^{-12} m. Each division, in the direction of shorter wavelength, represents a tenfold decrease, known as a mathematical *order of magnitude* (not to be confused with star magnitude). Utility ac is near the top of this scale; the wavelength of 60-Hz ac in free space is quite long. Gamma rays are at the bottom; their EM wavelengths are tiny.

From this example, it is easy to see that visible light takes up only a tiny sliver of the EM spectrum. However, this diagram makes the visible por-

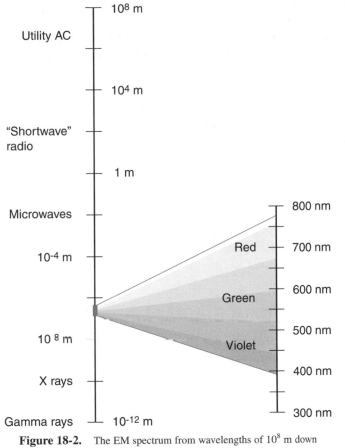

Figure 18-2.   The EM spectrum from wavelengths of 10^8 m down to 10^{-12} m, and an exploded view of the visible-light spectrum within.

tion of the realm look much larger than it is in linear proportion. If the scale were linear in this illustration, the visible slice would be thinner than the diameter of an atom. Along the right-hand scale, visible wavelengths are denoted in nanometers.

HOW LITTLE WE SEE!

The next time you get a chance, look through a red or blue colored piece of glass or cellophane. Such a color filter greatly restricts the view you get of the world because only a narrow range of visible wavelengths can pass through it. Different colors cannot be ascertained through the filter. For example, when a scene is viewed through a red filter, blue appears the same as black, and crimson appears the same as gray or white. Other colors look red with varying degrees of saturation, but there is little or no variation in the hue. If our eyes had built-in red color filters, our view of the world would be much different; we would essentially be color-blind.

When considered with respect to the entire EM spectrum, all optical instruments suffer from the same sort of handicap we would have if the lenses in our eyeballs were tinted red. The range of wavelengths we can detect with our eyes is approximately 770 nm at the longest and 390 nm at the shortest. Energy at the longest visible wavelength appears red to our eyes, and energy at the shortest visible wavelength appears violet. The intervening wavelengths show up as orange, yellow, green, blue, and indigo.

In order to "see" EM energy waves above and below the visible spectrum, we need special apparatus. Astronomers have devised an amazing variety of instruments that can detect energy all the way from the radio spectrum through microwaves, IR, visible, UV, x-rays, and even gamma rays.

ATMOSPHERIC EM TRANSPARENCY

The observation of the Universe at the visible wavelengths is not greatly hindered by the atmosphere of Earth. Some diffusion and refraction take place, but the air is essentially transparent at the range of 770 to 390 nm. At some other wavelengths, however, energy from outer space can't make it to the ground, so we can't detect it unless we put instruments high up in the atmosphere or in outer space.

At radio wavelengths greater than a few meters, Earth's *ionosphere*, at altitudes ranging from about 60 km (40 mi) to 400 km (250 mi), causes absorption, refraction, and reflection of EM waves. This keeps signals from space away from us at certain radio frequencies. This same effect keeps *shortwave* radio waves near the Earth and makes long-distance communication possible.

At UV wavelengths, the ozone in the air is highly absorptive. In the x-ray and gamma-ray parts of the EM spectrum, our atmosphere completely obscures our view of the heavens. However, some portions of the invisible spectrum are observable with ground-based equipment, especially the wavelengths between microwaves and visible light.

One of the most significant developments in the exploration of space at invisible wavelengths has been the *radio telescope*. *Radio astronomy* has become a refined science since the development of wireless communications equipment and in particular since the end of World War II in the 1940s. High-altitude rockets, orbiting satellites, and the advent of the space age have opened up the realms of UV, x-ray, and gamma-ray astronomy.

Radio Astronomy

The science and art of radio astronomy began as the result of an accident. Karl Jansky was conducting investigations at a wavelength of 15 m (a frequency of about 21 MHz) in the shortwave radio band to determine the directional characteristics of *sferics*, or radio noise that originates from natural sources in Earth's atmosphere, particularly thunderstorms. The antenna was not particularly large. However, Jansky found, in addition to the radio noise caused by thunderstorms, a weaker and steady noise of unknown origin.

THE MYSTERY NOISE

Human-made noise was ruled out when Jansky noticed that the source of the faint noise seemed to change with the time of day. It was found to have a rotational period of 23 hours and 56 minutes, exactly the same as the sidereal rotation period of the Earth. Jansky concluded that the radio noise was of extraterrestrial origin, and he found that it was coming from the direction of the constellation Sagittarius, which lies in the same direction as the center of our Milky Way galaxy. Other parts of the galaxy also produced radio noise, Jansky found, but none of it was up to the amplitude of the noise coming from Sagittarius.

Jansky was interested in the phenomenon and wanted to continue the research in the field with equipment designed specifically for receiving signals from space, but his superiors and the people who funded his work weren't impressed by his "mystery noise." As a result, he did not pursue radio astronomy any further. However, Jansky's discovery of the noise coming from the Milky Way did not pass entirely unnoticed. A radio engineer named Grote Reber began to get interested in radio astronomy as a hobby, in conjunction with his activities as an amateur radio operator. Radio amateurs, also called *ham operators*, have been known to make radical communications discoveries. Reber built a large parabolic dish antenna in his back yard. His neighbors were amazed (and fortunately, tolerant) as the assembly of the 10-m (31-ft) bowl-shaped reflector progressed.

Reber's antenna was not fully steerable but could be moved only up and down along the celestial meridian from the southern horizon through the zenith to the northern horizon. As Earth rotated on its axis during the course of a day, different parts of the observable sky passed across the focal axis of the antenna. Many radio telescopes use this kind of steering system. By tilting the antenna from horizontally south through the zenith to horizontally north, the entire *radio sky* can be mapped if the astronomer is willing to take the necessary time.

Reber's first tests were conducted at the fairly short wavelength of 9 cm, corresponding to a radio frequency of 3.3 GHz, or 3,300 MHz. Reber checked the most familiar objects in the sky, such as the Sun, the Moon, and the planets. No signals were detected. At a wavelength of 1.87 m, or about 160 MHz, however, Reber did find noise coming from the Milky Way.

Astronomers took notice of the work of Jansky and Reber, and plans were made to construct large radio antennas to receive signals from the Cosmos.

THE HARDWARE

The most important part of any radio or wireless receiver is the antenna. This is especially true of a radio telescope. Radio signals from space are much fainter than standard broadcast or microwave signals. To determine the location in the sky from which a signal is arriving, it is necessary that a radio telescope antenna have exceptional resolving power, also known as *directivity*. It must be sensitive only to signals in the direction in which it is aimed, and it must be able to reject signals coming from other directions. The old-fashioned television (TV) receiving antenna, which you still see occasionally on home and business rooftops, is a directional antenna, but the radio telescope requires a much more precise antenna than this. Radio telescope antennas more closely resemble outsized satellite TV antennas.

The *gain* (a logarithmic measure of the sensitivity) of an antenna is also important in the design of a radio telescope. The gain and directivity both depend on the physical size of the antenna. For a given amount of gain and directivity, a dish antenna must have at least a certain diameter, measured in wavelengths. For a given fixed antenna size, the sensitivity and directivity increase as the wavelength decreases.

Of course, even the most sensitive and directional antenna is useless without a good receiver. Most of the radio noise that comes from space sounds like the noise generated inside the electronic circuits of a radio receiver, and this compounds the problem of radio reception from the Cosmos. (Tune an old AM radio receiver to a frequency where there is no station. The faint hiss is internal noise; this is what radio astronomers generally hear from space.) The most advanced receiver designs must be used in a radio telescope to obtain the greatest possible amplification and sensitivity.

The location of the radio telescope antenna is important, just as is the site for any optical observatory. Human-made interference can ruin the operation of a radio telescope. Such interference comes from all kinds of electrical appliances, such as hair dryers, light dimmers, electric blankets, and thermostats. Automobile ignition systems are a severe problem for those who attempt radio reception of faint signals. A rural location is therefore superior to an urban site for a radio telescope.

INTERFEROMETRY

With all these factors in mind, scientists set out to build sophisticated radio telescopes. One of the most famous early instruments employed a 250-ft steerable dish and was located at Jodrell Bank in Cheshire, England. This project, completed in the 1950s, was proposed and overseen by the physicist A. C. B. Lovell. He went through great personal difficulties in arranging the construction of this radio telescope.

Not all radio telescopes use single-dish antennas. There are schemes for obtaining exceptional directivity that are more physically workable than the construction and operation of one huge parabolic reflector. The *interferometer*, pioneered by Martin Ryle of Cambridge University and J. L. Pawsey of Australia, provides superior resolving power using two separate antennas. When two antennas, spaced many wavelengths away from each other, are connected to the same receiver, an *interference pattern* occurs. There are many *lobes*, or directions in which the signals arriving at the two antennas add together. There are also many *nodes*, or directions from which the signals cancel each other out. The farther apart the antennas, the more numerous are the lobes and nodes, and the narrower they become. Each lobe covers a smaller part of the sky than the main lobe of any single antenna.

Figure 18-3 shows *horizontal-plane directional patterns* of the sort used by antenna engineers for a hypothetical single antenna (*A*) and a pair of antennas in an interferometer arrangement (*B*). Imagine that you are high above Earth, looking straight down on the antennas and at such an altitude that the pair of antennas (at *B*) looks like a single point. Also imagine that

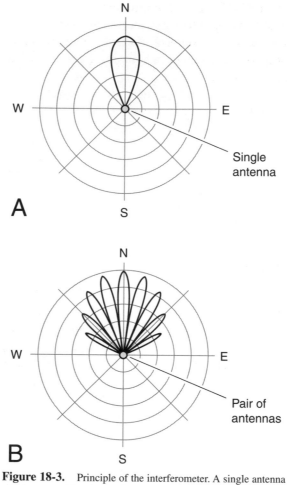

Figure 18-3. Principle of the interferometer. A single antenna
has one wide response lobe (*A*). Two antennas, properly
interconnected, have numerous narrow response lobes (*B*).

both radio telescopes are aimed at the northern horizon. The curves show the
relative sensitivity of the radio telescope as a function of the azimuth. These
are two-dimensional slices of the true pattern, which is three-dimensional.
In three-space, the lobes are shaped like tapered cigars.

Interferometry cannot provide the sensitivity of a huge dish measuring
many kilometers in diameter, but it does provide the equivalent directivity
at a far lower cost and inconvenience. In some cases, the radio image resolution can be on the order of a few seconds of arc.

Today, there are radio telescopes in many countries throughout the world. These radio telescopes have proven worth the trouble and the expense of their construction. The mysterious, fascinating quasars and pulsars were found using radio telescopes; only later did astronomers start analyzing these objects with optical telescopes.

THE RADIO SKY

When a radiotelescope with sufficient resolving power is used to map the sky, certain regions of greater and lesser radio emission are found. The center of our galaxy, located in the direction of the constellation Sagittarius, is a powerful radio source. The Sun is a fairly strong emitter of radio waves, as is the planet Jupiter.

A strong source of radio waves is found in the constellation Cygnus, and it has been named *Cygnus A*. The Australian radio astronomers J. Bolton and G. Stanley determined that Cygnus A has a very tiny angular diameter, and they also found many other localized radio sources. This led to the development of a system for naming radio sources. A significant celestial source of radio waves is designated according to the constellation in which it is found, followed by a letter that indicates its relative radio intensity within that constellation. The letter *A* is given to the strongest source in a given constellation, the letter *B* to the second strongest, and so on. Cygnus A is the strongest source of radio emissions in Cygnus and also, it so happens, in the entire sky. It is so small in diameter that its output fluctuates because of effects of Earth's ionosphere as the signals pass through on their way to the surface. Cygnus A is a *radio galaxy*.

Using radio telescopes, maps of the sky have been made, in the same way that optical astronomers make star and galactic maps. Radio maps do not look like optical maps. Instead, they appear like topographic maps used in geologic surveys or like computerized abstract art. Regions of constant radio emission are plotted along lines, which tend to be curved. Or they can be rendered as pixelated images in color or grayscale, as shown in Fig. 18-4, an image of a hypothetical radio galaxy viewed edgewise. (The smaller objects are hypothetical foreground stars within our own galaxy.) The better the directivity of the radio telescope, the greater is the number of discrete radio objects that can be defined on such a map.

In radio maps of the entire sky, the Milky Way shows up as a group of lines or colored regions with their widest breadth (representing the greatest intensity) in the constellation Sagittarius. Other galaxies have been found

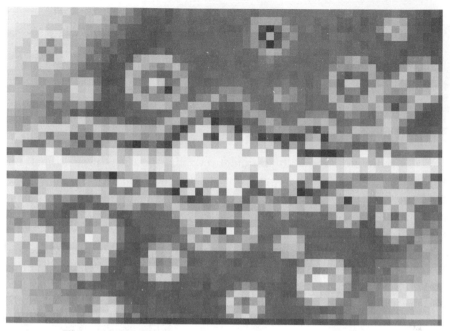

Figure 18-4. Radio map of a celestial object, in this case a hypothetical galaxy viewed edgewise. Shades of gray indicate relative radio brightness.

that emit radio frequency (rf) energy. Scientists at Cambridge University, in the early days of radio astronomy, identified four different external galaxies as radio sources. One of these is the Great Nebula in Andromeda, approximately 2.2 million light-years from our own galaxy.

RECEPTION FROM THE SOLAR SYSTEM

As radio astronomy evolved, scientists turned their attention to several objects in our own Solar System. One of these is the Sun. The *radio Sun* is somewhat larger than the visible solar disk, and it appears oblate or flattened along the plane of the equator. This is to say, the apparent diameter of the radio Sun is smallest through the poles and largest through the equator.

Visible solar flares are also observed with radio telescopes. Such flares have long been associated with disruption of the ionosphere of our planet, a phenomenon that wreaks havoc with radio broadcasting and communications at some frequencies. There are several different kinds of solar flares at radio wavelengths. Radio outbursts from the Sun usually portend a disturbance in the Earth's magnetic field a few hours afterward as the high-energy particles arrive and are focused toward the Earth's north and south magnetic poles. Then, at night, we see the *aurora* (northern lights and southern lights). We also observe an abrupt change in radio wave propagation at some frequencies.

Radio observations of the Moon and the planets have enabled astronomers to more accurately ascertain the surface temperatures, especially of planets with thick atmospheres such as the "gas giants" Jupiter, Saturn, Uranus, and Neptune.

Jupiter produces exceptionally strong radio emissions and has a fairly high temperature deep within its shroud of gas. At a wavelength of about 15 m, the EM radiation from Jupiter is almost as strong as that from the Sun. Jupiter is also a strong radio source at shorter wavelengths. Some of this radiation can be attributed to the fact that Jupiter generates considerable heat of its own, in addition to reflecting energy from the Sun. However, the internal heat of Jupiter cannot account for all the radio emissions coming from the giant planet. Several theories have been formulated in an attempt to explain the unusual levels of EM radiation coming from Jupiter. According to one idea, numerous heavy thunderstorms rage through the thick atmosphere, and the radio noise is caused by lightning. However, the noise is too intense for this idea to fully explain it. A more plausible theory is that electrons, trapped by the intense magnetic field of Jupiter and accelerated by the high rotational speed of the planet, cause a form of EM emission called *synchrotron radiation*.

Radar Astronomy

The huge parabolic antennas used in some radio telescopes, as well as other forms of high gain antenna systems, can work for transmitting in the same way as they function for receiving. The large power gain developed in the interception of a faint signal from space also can increase the effective transmitted power of a signal greatly. By sending out short pulses of radio energy, generated by a transmitter in the laboratory, and then listening for possible echoes, astronomers can make accurate determinations of the distances to other objects in the Solar System. The use of radio telescopes for distance determination, motion analysis, and surface mapping of extraterrestrial objects is called *radar astronomy*.

THE MOON AND VENUS

The distance to Earth's Moon has been refined to within a few centimeters, and it is possible to tell how fast the Moon is moving toward or away from Earth as it revolves around our planet. The surface of the Moon has been mapped by radar, and the quality of detail rivals that of optical photographs. Radar astronomy has proven to be a valuable tool in the study of the surface of Venus because that planet's thick clouds make it impossible to see the surface with optical apparatus.

The main challenge to pursuing radar investigations of the planets has been the development of radar sets with enough power and sensitivity. The planets are very distant in terms of human-made radio transmissions. In addition, they do not reflect radio signals the way a mirror reflects visible light. Only a minuscule part of the rf signal reaching a planet is reflected back in the general direction of the Earth; of this reflected wave, only a tiny portion ever gets back to the antenna. Radar technology eventually was perfected to overcome the path loss between Earth and Venus. The radar telescope enabled scientists to discover that Venus has a retrograde motion on its axis. Before that time, astronomers had made educated guesses concerning the planet's rotation based on the motion of the cloud tops, and these guesses had all been wrong.

Radar mapping of the surface of Venus was attempted because rf signals easily penetrate the visually opaque clouds. Astronomers found that the surface is solid but irregular. High-resolution maps were obtained when the planet was scrutinized using radar sets aboard Venus-orbiting spacecraft. The solidity of the surface was not known with certainty until it was verified by radar.

MERCURY

Observations of Mercury have been made by radar. Surface details of Mercury are difficult to resolve with optical telescopes based on the Earth's surface because that planet is always near the Sun in the sky. The rotation rate of Mercury was once believed to be 88 Earth days, identical with its period of revolution around the Sun. The radar telescope revealed that one rotation is completed in 59 Earth days. Mercury has a day of its own, but it is long and strange by our standards.

Echoes from the planet Mercury have been used to verify one of the predictions of Albert Einstein's general theory of relativity. According to

Einstein's equations, radio waves passing close to a massive object, such as the Sun, should appear to "slow down" because of the curvature of space in the gravitational field. All radiant energy, according to general relativity, is affected in this way near massive celestial objects.

Experimenters bounced radar signals off Mercury as it passed on the far side of the Sun (Fig. 18-5). According to general relativity, an illusion should occur in which Mercury seems to deviate about 65 km outside its orbital path as it passes behind the Sun. This deviation was found, and it takes place to the exact extent predicted by Einstein. The echoes are slightly delayed as the signals pass near the Sun on their way to and from Mercury.

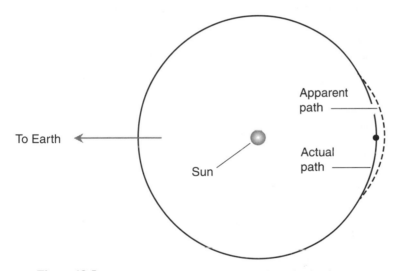

Figure 18-5.   Mercury appears to deviate outside its orbit when it passes behind the Sun. (The extent of the deviation is exaggerated in this drawing.)

METEOR ORBITS

Radar astronomy is useful in the study of meteors. Meteors apparently come, for the most part, from inside the Solar System; they are not interstellar wanderers. This can be deduced by determining the velocities of large numbers of meteors relative to Earth. The velocities of meteors as they enter the atmosphere can be measured accurately using radar. Such measurements are difficult or impossible to do by visual means. From the radar information, an astronomer can figure out the original paths of meteors through space. These paths always turn out to be orbits around the Sun.

Meteors arrive during the day as well as at night, and the radar telescope can "see" them in visual daylight as well as in visual darkness. This gives the radar-equipped observer another advantage over the visual observer.

SOLAR OBSERVATION

The Sun has been observed by radar. Even though the surface is not solid, the outward motion of subatomic particles has been detected: the *solar wind*. The "surface" of the Sun is different in the radio part of the EM spectrum as compared with the visible-light portion.

The Sun has, of course, no solid surface as does the Earth, Venus, or Mercury. The apparent diameter of the solar globe depends on the EM wavelength at which the Sun is observed. This makes it possible to examine the motion of the gases at different levels. Great turbulence exists there; we know this because Doppler shifts are constantly observed. Radar telescopes allow astronomers to calculate how fast the gases rise and fall as the Sun's surface boils in an endless storm.

EFFECTIVE RANGE

The maximum range of a radar telescope is limited by two factors. First, there is *path loss*, caused by the sheer physical distances over which radar signals must travel on their way from the antenna to the target and back. Second, the free-space speed of EM-wave propagation is finite. Although 299,792 km/s (186,282 mi/s) seems fast on a terrestrial scale, it is sluggish with respect to the Cosmos.

Path loss increases with distance and mandates more sensitive receivers and more powerful transmitters as the distance gets greater. There is a practical limit to how sensitive any rf receiver can be made. There is also a limit to how much power can be generated in a radar transmitter and a limit to how much gain can be realized with an antenna of manageable size. There is yet room for engineers to design and build larger antennas, more sensitive receivers, and more powerful transmitters. Eventually, however, economic considerations must prevail over scientific curiosity.

The propagation-speed constraint is insurmountable no matter what the size of the hardware budget. An echo from Pluto returns to Earth approximately 10 hours after the signal is sent. An echo from the nearest star system would not return for almost 9 years. Most of the star systems in our galaxy are so far away that the echoes from a radar set will not come back until many human lifetimes have passed. Earth-based radar astronomy will never be useful in the study of objects outside the Solar System.

Infrared Astronomy

EM energy at radio frequency has much greater wavelengths than energy in the visible part of the spectrum. The shortest radio microwaves measure approximately 1 mm in length; the reddest visible light has a wavelength of a little less than 0.001 mm. This is a span of a thousandfold, or three mathematical orders of magnitude, and it is called the *infrared (IR) spectrum*. In terms of frequency, IR lies below red visible light. It is from this fact that IR gets its name: the prefix *infra-* means "below" or "under." Our bodies sense infrared as radiant warmth or heat. The IR rays are not literally heat, but they produce heat when they strike an absorptive surface such as human skin.

OBSERVATIONAL EQUIPMENT

Stars, galaxies, planets, and other things in the Cosmos radiate at all wavelengths, not only at wavelengths convenient for humans to observe. In some portions of the IR spectrum, the atmosphere of our planet is opaque. Between about 770 nm (the longest visible red wavelength) and 2 micrometers (μm), our atmosphere is reasonably clear, and it is possible to observe IR energy in this wavelength range from surface-based locations. To see celestial images at longer IR wavelengths, the observations must be made from high in the atmosphere or from space.

The Moon, the Sun, and the planets all have been observed in IR, as have some stars and galaxies. IR observing equipment resembles optical apparatus. This is true of telescopes as well as cameras. Similar lenses, films, and sensors are used, and excellent resolution can be obtained. Special kinds of film have been developed recently, making observation possible at longer and longer IR wavelengths.

STARS IN IR

IR astronomy has helped scientists to discover certain peculiar dim stars that seem to radiate most of their energy in the IR range. Visually, such stars appear red and dim. However, like a faintly glowing electric-stove burner, they are powerful sources of IR. These stars have relatively low surface temperatures compared with other stars. At first thought, this seems to be a paradox, but the peak wavelength at which an object radiates is a direct function of the temperature. "Cool" stars produce radiation at predominantly longer wavelengths than "hot" stars. The hottest stars are comparatively weak radiators of IR. When astronomers talk about temperatures of celestial objects, they usually refer to the *spectral temperature*, which is determined by examining the EM radiation intensity from the object at various wavelengths.

IR astronomy is important in the study of evolving stars and star systems. As a cloud of interstellar dust and gas contracts, it begins to heat up, and its peak radiation wavelength becomes shorter and shorter. A cool, diffuse cloud radiates most of its energy in the radio part of the EM spectrum. Hot stars radiate largely in the UV and x-ray regions. Sometime between the initial contraction of the nebula and the birth of the star, the peak emission wavelength passes through the IR. The observation of IR is also important in the analysis of dying stars. As a white dwarf cools down and becomes a black dwarf, its peak radiation wavelength decreases. On its way toward ultimate cold demise, the star emits, for a certain period of time, most of its energy in the IR.

MEASURING TEMPERATURE

The characteristic EM radiation from a star is a form of emission known as *blackbody radiation*. A *blackbody* is a theoretically perfect absorber and radiator of EM energy at all wavelengths. Any object having a temperature above absolute zero ($-273°C$ or $-459°F$) has a characteristic pattern of wavelength emissions that depends directly on the temperature. For any EM-radiating object (and this includes everything in the Universe), the emission strength is maximum at a certain defined wavelength and decreases at longer and shorter wavelengths. If EM intensity is graphed as a function of wavelength or frequency, the result is a curve that resembles a statistical distribution (Fig. 18-6).

By observing an object at many different wavelengths including the radio region, the IR, the visible range, and the UV and x-ray spectra, the point of maximum emission can be found. Sometimes it can be inferred even if observations are not actually made at that wavelength by plotting points on an intensity-versus-wavelength graph and connecting the points

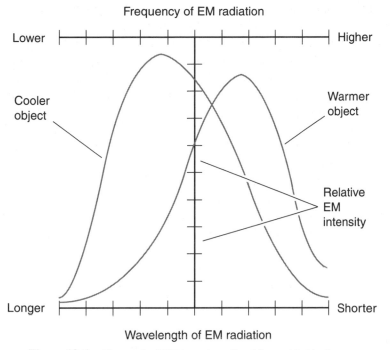

Frequency of EM radiation

Lower

Cooler
object

Warmer
object

Relative
EM
intensity

Longer

Shorter

Wavelength of EM radiation

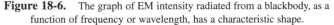

Figure 18-6.   The graph of EM intensity radiated from a blackbody, as a
function of frequency or wavelength, has a characteristic shape.

with a smooth curve. From the maximum-emission wavelength, the temperature of the object can be estimated based on the assumption that it behaves as a blackbody. Figure 18-7 is a rough graph, on a logarithmic scale, of the function employed by scientists for this purpose.

Ultraviolet and Beyond

As the wavelength of an EM disturbance becomes shorter than that of visible violet light, the energy contained in each individual photon increases. The UV range of wavelengths starts at about 390 nm and extends down to approximately 1 nm. The x-ray range extends from roughly 1 nm down to 0.01 nm. (The precise dividing line between the shortest UV and the longest x-ray wavelengths depends on whom you ask.) The gamma-ray spectrum consists of EM waves shorter than 0.01 nm. *Cosmic radiation* is

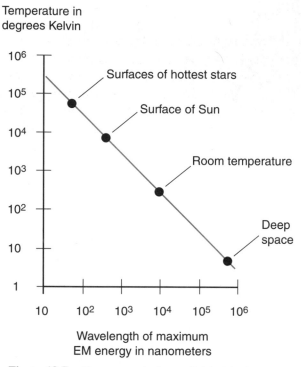

Temperature in
degrees Kelvin

Figure 18-7.   Temperature, in degrees Kelvin (absolute), as a
function of the maximum-amplitude wavelength of a blackbody.

entirely different in origin; it arises from high-speed subatomic particles
thrown off by the most energetic celestial objects.

THE UV ATMOSPHERIC "WINDOW"

As the wavelength decreases, the atmosphere of Earth becomes highly
absorptive at about 290 nm. At still shorter wavelengths, the air is essentially opaque. (This is a good thing because it protects the environment against
damaging UV radiation from the Sun.) The atmosphere scatters some EM
radiation even in the visible blue and violet parts of the spectrum. This is why
the sky appears blue to our eyes. Ground-based observatories can see something of space at wavelengths somewhat shorter than the visible violet, but
when the wavelength gets down to 290 nm, nothing more can be seen. At the
shortest UV wavelengths (*hard UV* and *extreme UV*), as in the case of the far
IR, it is necessary to place observation apparatus above the atmosphere.

"SEEING" UV

Glass is virtually opaque to UV, so ordinary cameras with glass lenses cannot be used to take conventional photographs in this part of the spectrum. Instead, a pinhole-type device is used, and this severely limits the amount of energy that passes into the detector. While a camera lens has a diameter of several centimeters, a pinhole is less than a millimeter across. This does not present a problem for photographing the Sun or the Moon, but for other celestial objects it is not satisfactory.

For analysis of fainter celestial objects in UV, an instrument called a *spectrophotometer* is used. This device is a sophisticated extended-range spectrometer in which a diffraction grating (not a glass prism) is used to disperse EM energy into its constituent wavelengths. By moving the sensing device back and forth, any desired wavelength can be singled out for observation, even those in the IR or UV spectrum. The principle of the spectrophotometer is shown in Fig. 18-8. In the long-wavelength or *soft-UV* range, a photoelectric cell can serve as a sensor. In the hard- and extreme-UV spectra, radiation counters are sometimes used, similar to the apparatus employed for the detection and measurement of x-rays and

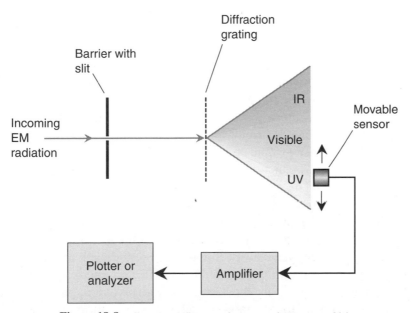

Figure 18-8.   Functional diagram of a spectrophotometer, which can be used to sense and measure UV radiation.

gamma rays. For photographic purposes, ordinary camera film works in the soft-UV range, but special film, rather like x-ray film, is used to photograph hard-UV and extreme-UV images.

SOURCES OF UV

The hot type O and B stars are strong sources of UV radiation. These stars evidently radiate more energy in the UV than in any other part of the EM spectrum. Type O and B stars generally are young stars. Within the visible part of the spectrum, their energy tends to be concentrated at the shortest wavelengths, so such stars look blue to us. The surface temperatures of these stars are much greater than the temperature at the surface of our Sun, which is a type G star. Type O and B stars have surface temperatures ranging from about 15,000 to 25,000°C (27,000 to 45,000°F).

Type A and F stars radiate smaller amounts of UV energy than type O and B stars. Most type A and F stars look white. Still, these stars are hotter than the Sun. Temperatures at their surfaces range from about 8,000 to 15,000°C (14,000 to 27,000°F).

Type K and M stars have the coolest surfaces of any stars, ranging down to about 1,500°C (2,700°F). The greatest amount of visible radiation from such stars falls into the wavelengths corresponding to red and orange. These stars produce comparatively little UV radiation.

Supernovae produce fantastic amounts of visible light, but they also emit large amounts of UV. The explosion of a supernova within a few light-years of the Solar System would be a spectacular sight; the brilliance of the star would exceed that of the full Moon. However, the UV radiation, despite the distance, would compare with that from our own Sun. You could get a "starburn" from the supernova, even at night, and certainly without realizing it, because the IR intensity would be relatively low. The shortest UV rays, which would cause the most damage to life on our planet, would be largely kept at bay by the atmosphere.

A supernova, after having thrown off a cloud of gas in the process of exploding, causes ionization of the cloud because of UV radiation. This causes the gas to fluoresce, or glow visibly. Such a glowing cloud can be seen from many light-years away. Sometimes nearby stars, of the hot types O and B, cause ionization of interstellar gas. This produces such spectacular astronomical objects as the Great Nebula in Orion, the Horsehead Nebula, and others. Bright emission nebulae betray the presence of UV sources in their vicinity. A fluorescent lightbulb operates on the same principle as the emission nebulae; the coating on the inside of such a bulb is set to glow by UV radiation from the gases within.

THE SUN IN UV

The Sun's surface temperature is in the neighborhood of 6,000°C (11,000°F). Our parent star emits some UV, but not a great deal of it as stars go. This is fortunate for the kind of life that has evolved on this planet. If the Sun were a hotter star, life on any Earthlike planet in its system would have developed in a different way, if at all.

The Sun's UV radiation has been investigated using equipment aboard rockets and satellites. The UV surface of the Sun is somewhat above the visual surface. This tells us that as the altitude above the photosphere (visible surface) increases the temperature rises. If our eyes suddenly became responsive to a range of wavelengths only half as long as they actually are—say, a continuum of 200 to 400 nm—the disk of the Sun would seem a little larger than it appears in the visible range. We would ascribe to the Sun a different photosphere.

If our eyes suddenly became UV eyes, the Sun not only would look slightly larger in size but it also would appear less bright. The atmosphere of Earth transmits EM rays poorly in the short-wavelength part of the range 200 to 400 nm. Assuming that the longest detectable wavelengths appeared "red" to us, we would consider the Sun a ruddy star. Vision is a subjective thing. We would be equally impressed with the unnatural blueness and brilliance of the Sun if our eyes suddenly became responsive to, say, a wavelength range of 800 to 1,600 nm.

X-RAYS

The x-ray spectrum consists of EM energy at wavelengths from approximately 1 to 0.01 nm. This is 2 mathematical orders of magnitude. Proportionately, the x-ray spectrum is vast compared with the visible range.

As the wavelength of x-rays become shorter, it becomes increasingly difficult to direct and focus them. This is so because of the penetrating power of the short-wavelength rays. A piece of paper with a tiny hole can work for UV photography; in the x-ray spectrum, the radiation passes right through the paper. However, if x-rays encounter a reflecting surface at a nearly grazing angle, and if the reflecting surface is made of suitable material, some degree of focusing can be realized. The shorter the wavelength of the EM energy, the smaller the angle relative to the surface must be if reflection is to take place. At the shortest x-ray wavelengths, the angle must be smaller than 1 degree of arc. This grazing reflection effect is shown in Fig. 18-9A. The focusing mirror is tapered in the shape of an elongated paraboloid. Figure 18-9B is a rough illustration of how an *x-ray telescope* achieves its focusing. As parallel x-rays enter the aperture of the reflector, they strike its inner surface at a grazing angle. The x-rays are brought to a focal point, where a radiation counter or detector is placed.

The resolving power of an x-ray telescope, such as the one shown in Fig. 18-9B, is not as good as that obtainable with optical apparatus, but it does allow the observation of some celestial x-ray sources. As is the case with

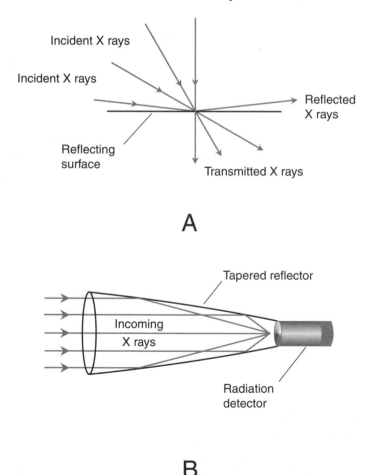

Figure 18-9. At *A*, x-rays are reflected from a surface only when they strike at a grazing angle. At *B*, a functional diagram of an x-ray focusing and observing device.

UV radiation, x-rays from space must be viewed from above the atmosphere of our planet; x-ray telescopes aboard rockets and satellites send their information back to Earth by radio.

SOURCES OF X-RAYS

After the development of high-altitude rockets and space vehicles, it became possible to look at the x-ray sky. Powerful x-ray sources were found, but there at first appeared to be no explanation for some of them. Even type O and B stars do not produce large amounts of radiation in the x-ray spectrum; they are not hot enough. The most interesting x-ray objects appeared to produce more energy in the x-ray region than at longer wavelengths.

Some tentative hypotheses have been brought forward to explain intense, pointlike x-ray objects, which have, because of their apparent location within our galaxy, been called *x-ray stars*. They are not supernovae; their radiation wavelengths are too short even for that. Besides, they are not visually bright enough to be supernovae. X-ray stars are found more commonly than supernovae. Some astronomers theorize that the x-ray objects are binary stars in very close mutual orbits. Matter from one of the stars in a binary system could be torn away from the other member by gravitational forces. The stars might even be in mutual contact. The exchange of matter between two stars in such close association could account for the production of large amounts of x-rays.

Another suggestion concerning the nature of x-ray stars has been given: They are binary star systems in which one member is a neutron star or a very dense black dwarf. Still another theory holds that the strange stars are binary systems containing black holes. The gravitational influence of a neutron star or black hole is sufficient to account for the x-rays; as matter is torn from the visible member of such a binary system, the hotter interior layers are exposed, and this can produce radiation at very short wavelengths. The idea that matter is being ripped out of a star is supported by the existence of Doppler shifts in the x-rays.

Some x-ray objects seem to be outside of our galaxy. Certain quasars and radio galaxies have been associated with strong sources of x-rays. Some astronomers have hypothesized that interaction among the photons of radiant energy at different wavelengths, as they collide with each other, could be responsible for the x-ray emissions from extragalactic objects. Some galaxies and quasars apparently have regions of tremendously high temperature—hotter than anything we know in our Milky Way—and this state of affairs can generate highly energetic EM waves that peak in the x-ray portion of the EM spectrum.

GAMMA RAYS

As the wavelength of EM energy becomes shorter than the hardest x-rays, it becomes more and more difficult to obtain an image. The cutoff point where the x-ray region ends and the gamma-ray region begins is approximately 0.01 nm. Gamma rays can get shorter than this without limit. The gamma classification represents the most energetic of all EM fields. Short-wavelength gamma rays can penetrate several centimeters of solid lead or more than a meter of concrete. They are even more damaging to living tissue than x-rays. Gamma rays come from radioactive materials, both natural (such as radon) and human-made (such as plutonium).

Radiation counters are the primary means of detecting and observing sources of gamma rays. Gamma rays can dislodge particles from the nuclei of atoms they strike. These subatomic particles can be detected by a counter. One type of radiation counter consists of a thin wire strung within a sealed, cylindrical metal tube filled with certain gases. When a high-speed subatomic particle enters the tube, the gas is ionized for a moment, and conduction occurs between the inner wire and the cylinder. A voltage is applied between the wire and the outer cylinder so that a pulse of current occurs whenever the gas is ionized. This pulse produces a click in the output of an amplifier connected to the device.

A simplified diagram of a radiation counter is shown in Fig. 18-10. A glass window with a metal sliding door is cut in the cylinder. The door can be opened to let in particles of lower energy and closed to allow only the fastest particles to get inside. High-speed particles, which are tiny yet massive for their size, have no trouble penetrating the window glass if they are moving fast enough. Yet gamma rays can penetrate into the tube with ease, even when the door is closed.

COSMIC PARTICLES

If you sit in a room with no radioactive materials present and switch on a radiation counter with the window of a tube closed, you'll notice an occasional click from the device. Some of the particles come from the Earth; there are radioactive elements in the ground almost everywhere (usually in small quantities). Some of the radiation comes indirectly from space. These

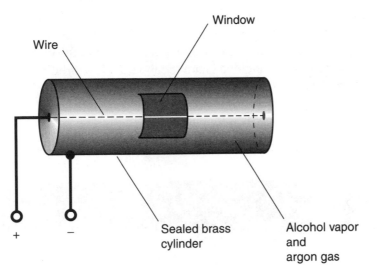

Wire

Window

Sealed brass
cylinder

Alcohol vapor
and
argon gas

+   −

Figure 18-10.   Simplified diagram of a radiation counter.

particles strike atoms in the atmosphere, and these atoms in turn eject other subatomic particles that arrive at the counter tube.

The direction of arrival of high-speed atomic particles can be determined, to a certain extent, by means of a device called a *cloud chamber*. The air in a small enclosure can be treated especially to produce condensation when a subatomic particle enters, and the path of the particle will show up as a vapor trail.

In the early 1900s, physicists noticed radiation apparently coming from space. They found that the strange background radiation increased in intensity when observations were made at high altitude; the radiation level decreased when observations were taken from underground or underwater. This space radiation has been called *secondary cosmic radiation* or *secondary cosmic particles*. The actual particles from space, called *primary cosmic particles*, usually do not penetrate far into the atmosphere before they collide with and break up the nuclei of atoms. To observe primary cosmic particles, it is necessary to ascend to great heights, and as with the UV and x-ray investigations, this was not possible until the advent of the space rocket.

While the radiation in the EM spectrum—the radio waves, IR, visible light, UV, x-rays, and gamma rays—consists of photons traveling at the speed of light, cosmic particles are matter, traveling at speeds almost, but not quite, as fast as light. At such high speeds, the protons, neutrons, and other heavy particles gain mass because of relativistic effects, and this renders them almost immune to Earth's magnetosphere. Such particles, arriving in the upper atmosphere, come to us in a nearly perfect straight-line path despite the magnetic field of our planet. By carefully observing the trails of the particles in a cloud chamber aboard a low-orbiting space ship, it is possible to ascertain the direction from which they have come. Over time, cosmic-particle maps of the heavens can be generated and compared with maps at various EM wavelengths.

Quiz

Refer to the text if necessary. A good score is 8 correct. Answers are in the back of the book.

1. Radio frequency noise caused by lightning in the atmosphere of our planet is called
    (a) sferics.
    (b) resolution.
    (c) aperture.
    (d) ground noise.

2. The resolving power of a dish-antenna radio telescope depends largely on
    (a) the focal length of the reflector in meters.
    (b) the diameter of the reflector in wavelengths.
    (c) the distance to the source of the signal.
    (d) the gain of the amplifier.

3. When observed at UV compared with visible wavelengths, the Sun
    (a) appears brighter.
    (b) appears to rotate much more rapidly.
    (c) is invisible.
    (d) has a larger angular diameter.

4. In order for an EM field to be produced, a charge carrier must be
    (a) positive.
    (b) negative.
    (c) moving.
    (d) accelerating.

5. A binary star system, of which one member is a bright visible star and the other member is a black hole orbiting close to the visible star,
    (a) radiates virtually all its energy at radio wavelengths.
    (b) appears to us as a visible pulsar.

    (c) can emit x-rays.

    (d) is known as a blackbody.

6. Some radio waves from space cannot reach Earth's surface because of

    (a) the solar wind.

    (b) sferics.

    (c) the ionosphere.

    (d) lack of resolution.

7. The Sun emits most of its EM energy at

    (a) radio and IR wavelengths.

    (b) IR and visible wavelengths.

    (c) visible and UV wavelengths.

    (d) UV and x-ray wavelengths.

8. An EM field whose wavelength is 550 nm in free space would appear to us as

    (a) a radio wave.

    (b) infrared radiation.

    (c) visible light.

    (d) ultraviolet radiation.

9. The wavelength of greatest EM energy intensity from a blackbody depends on

    (a) the distance of the object from Earth.

    (b) the angular diameter of the object.

    (c) the mass of the object.

    (d) the temperature of the object.

10. The radar telescope has been a valuable tool in

    (a) the search for life on other worlds.

    (b) finding new comets.

    (c) observing the surface of Venus.

    (d) mapping the radio sky.

CHAPTER 19

Traveling and Living in Space

Ever since human beings first realized that the nighttime stars are distant suns, adventurous humans have dreamed about traveling among them. During the late 1800s and early 1900s, the true vastness of the Universe became apparent, and the sheer number of other galaxies and stars led astronomers to imagine that myriad Earthlike planets exist, with climates ideal for human beings. Like the New World before the Vikings, these places beckon to dreamers. Science-fiction writers and moviemakers have a field day when it comes to "space stories." Starting with the launch of the first Earth-orbiting satellites in the 1950s, fiction has been evolving into reality.

Why Venture into Space?

The proponents of space travel and colonization use several arguments to support their contention that the human species (that is, *us*) should reach for the stars. However, there are plenty of people who think we have no need to leave Earth and no business trying. Here are some arguments for and against human efforts to venture into the Cosmos.

DEATH OF THE SUN

When we observe it casually, our Sun seems like a steady, unchanging, eternal source of energy. When compared with a single human lifespan, this is true. Even if we look back thousands of years to the earliest known human civilizations, we have little or no evidence that the Sun's output was ever different than it is today. Astronomers know that the Sun will not shine forever, though. In a few billion years (where a billion is considered to be 1 billion or 10^9), the Sun will bloat into a red giant, and the inner planets will be incinerated. All the stars we now see will die eventually too. From the vantage point of our little speck of dust, heaven and Earth will pass away.

In the years following World War II, we developed weapons capable of annihilating much of the life on this planet. Even today, despite the end of the so-called cold war, there remain enough nuclear bombs to turn back the clock of civilization hundreds of years. As our species finds new ways to create disaster for itself, rational people wonder whether we will survive another 4,000 years, let alone another 4 billion. However, even the most optimistic folks must face the fact that our lease on this Earth is for a finite term.

Even if we evolve into the most enduring and cooperative species imaginable, the time will come when we must leave this Solar System and look for someplace else to live. We have a lot of work to do before we will be able to roam among the stars. If we prepare for the Sun's demise, we will be ready when it sets on the *last perfect day*. If we do not prepare, we will face extinction. Therefore, say the proponents of space travel, we might as well set our goals and start working toward them now because the stakes could not be higher, and the road will be difficult.

SEEKING EXTRATERRESTRIAL LIFE

Another reason for traveling among the stars is the opportunity to discover new life forms. Adherents of this philosophy argue as follows: "Are there other intelligent civilizations out there? If so, why should we wait for them to come to us? Let's go to them!"

If we decide to travel to other stars and galaxies in the hope of meeting extraterrestrial beings, will we do it with the intention of learning from them and joining with them to create an enlightened interstellar civilization? Or will we go on missions of conquest, intending to exploit and subjugate creatures from other worlds, take over their resources, and set up outposts from which to conduct a campaign to conquer the Universe? It's reasonable to suppose that if we survive as a species long enough to develop the technology to travel freely among the stars, we will have learned to cooperate with each other and thus will have benevolent intentions.

CELESTIAL RESOURCES

Even if there is no intelligent life in the Universe other than us, there are plenty of resources out there. The asteroids, the Moon, and the satellites of the outer planets have been suggested as sources of substances that could be mined without disturbing the environment here on Earth. If we could find a way to get the metals and minerals we need from extraterrestrial deposits, we could plant forests and expand wildlife preserves on our own planet. We could make Earth a more pleasant place to live and never have to worry about running out of the "stuff" we need to carry on.

The main problem with extraterrestrial resources is the fact that they are far away. Getting to them and then transporting them back to Earth would be expensive and would require vast amounts of fuel. Imagine a transport vessel designed to carry 100 billion kilograms of minerals from Io, the volcano-covered satellite of Jupiter. Think of the awkward task of navigating a spacecraft the size of Manhattan filled to the brim with liquid methane from Titan, Saturn's largest moon! The inertia of such a load would be formidable. Accelerating and decelerating a transport ship having that much mass and navigating a craft of such a size would require technology and resources the likes of which we can hardly comprehend. Yet, say some scientists, the reward would be worth the effort.

SOLVE EARTH'S PROBLEMS FIRST

The people who argue against the quest to travel to other planets, star systems, and galaxies point out that we have plenty of problems that need attention right here on Earth. Why spend huge sums of money, translating into countless human hours of work, on a program that is not guaranteed to bring tangible rewards? How can we justify a space program of any kind when most of the world's people lack decent housing, food, and medical care?

It makes sense that we should get our own planet in order before we go looking for other planets to visit or colonize and before we run off to distant stars without knowing what we will find. It makes sense to conclude that unless we first learn to take better care of ourselves on Earth, we will never survive the rigors of space. The vastness of deep space and the lengths of time that interstellar journeys will take present unforeseen dangers and extreme boredom. This will require cooperation, courage, and patience of a sort that, historically, has been exhibited rarely, even by the best human leaders and societies.

Proponents of space travel counter this argument as follows: If we wait until our species is perfect before we reach for the stars, we'll never make the leap. At some point we must decide whether or not we want to confine ourselves to the third planet in orbit around a mortal G2 star in one of the spiral arms of a typical disk-shaped galaxy. The more adventurous among us have already made this decision.

Robot Astronauts

The American space program reached a turning point in 1969 when *Apollo 11* landed on the Moon, and for the first time, a creature from Earth walked on another world. Some people think the visitor from Earth should have been a robot. Some scientists argue that there is no need to risk people's lives by sending them into space. Robotic space probes have been sent near planets in the Solar System. Robotic spacecraft have landed, as of this writing, on Venus and Mars, and comets have been explored up close. Why not have robotic spacecraft do our cosmic wanderings?

ALMOST LIKE BEING THERE

Robots could, in theory, be used to explore outer space while people stay safely back on Earth and work the robots via remote control. A human being can wear a control suit and have a distant humanoid robot, called a *telechir*, mimic all movements. The robot can be some distance away. The remote-control operation of a robot is called *teleoperation*. When the remote control has feedback that gives the operator a sense of being where the robot is, the system is called *telepresence* (Fig. 19-1).

With high-end *virtual reality*, it is possible to duplicate the feeling of being in a place to such an extent that the person can imagine that he or she is really there. Stereoscopic vision, binaural hearing, and a crude sense of touch can be duplicated. Imagine stepping into a gossamer-thin suit, walking into a chamber, and existing, in effect, on the Moon or Mars, free of danger from extreme temperatures or deadly radiation! With remote control, virtual reality can have its basis in actual reality.

Despite the assets of robotic space travel, some people say that it defeats the ultimate reason for having a space program: the romantic adventure of living beings roaming the Cosmos. There is another problem too:

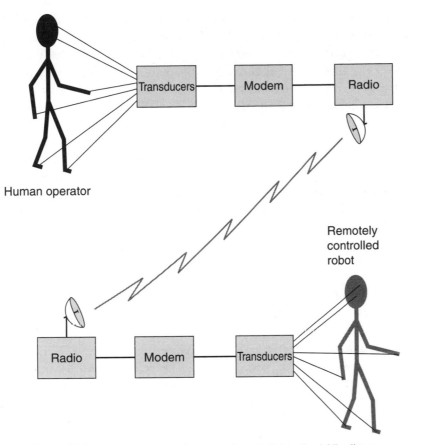

Figure 19-1.   In telepresence, a human operator can "virtually visit" a distant place using a remotely controlled humanoid robot.

Communications signals don't travel very fast on an interstellar or intergalactic scale.

THE LIGHT-SPEED PROBLEM

If robots are used in space travel with the intention of having the machines replace human astronauts, then the distance between a robot and its operator cannot be very great. The reason is that the control and response signals propagate through space at only 299,792 km/s (186,282 mi/s).

The Moon is 1.3 light-seconds from Earth. If a remotely controlled robot, rather than Neil Armstrong, had stepped onto the Moon on that summer day in 1969, its Earthbound operator would have had to deal with a delay of 2.6 seconds between command and response. It would take each command 1.3 seconds to get to the Moon and each response 1.3 seconds to get back to Earth. True telepresence is impossible with a delay like this. Experts say that the maximum delay for realistic telepresence is a tenth of a second (0.1 s). Therefore, the distance between the robot and its controller cannot be more than 0.5, or 1/20, light-second. This is about 15,000 km or 9,300 mi, slightly more than the diameter of the Earth.

Suppose that astronauts are in orbit around a planet whose environment is too hostile to allow an in-person visit. Then a robot might be sent down. An example of such a planet is Venus, whose crushing surface pressures would kill an astronaut clad in even the most advanced pressure suit. It would be easy to sustain an orbit of less than 15,000 km above Venus, so telepresence would be feasible.

Propulsion Systems

All space vehicles require some means of attaining high speed. The most obvious schemes work according to the *action-reaction* principle. When matter is ejected from the rear of a space ship, the ship moves forward. The spacecraft can be rotated 180 degrees, and the matter thereby ejected from the front when it is necessary to decelerate.

THE CHEMICAL ROCKET

During World War II, the Germans demonstrated that rockets could send small payloads rapidly from one point to another on Earth's surface. This technology led to larger and larger rockets, culminating in the American *Saturn* booster and the Space Shuttle and in the Russian *Soyuz* booster. These rockets can put people and satellites into Earth orbit and can even hurl them beyond the influence of the Earth's gravitation.

Chemical rockets are still used by all spacecraft today. They literally burn materials to produce thrust. If you've had any experience with model rockets, you know the basics of how a rocket works. Some rockets use solid fuel (model rockets, for example), whereas others employ liquid fuel. The larger the rocket, and the more fuel it carries, the farther it can send a payload.

It is tempting to think that in order to freely roam among the stars, we need only build a huge rocket. Unfortunately, things aren't this simple. The rocket necessary to send a manned mission to one of the Sun's nearby neighbor stars and back within the span of a human lifetime would require many times more fuel than a practical rocket could ever carry. If we are to become an interstellar species, we'll have to come up with something a lot more efficient than the chemical rocket engine.

Put your mind in "futurist mode." Think about chemical rockets for awhile. Soon you'll begin to see that chemical combustion is a ridiculous means by which to attempt to explore the vast tracts of outer space. It's even sillier than burning the remains of eons-old plants and animals to propel vehicles over the Earth's surface and through the lower atmosphere. Someday people will look back on these technologies and wonder why humanity stuck with them for so long.

THE ION ENGINE

Hot gases produced by the combustion of flammable fuels are not the only way to produce thrust. Another method that has been suggested as a means to get spacecraft starbound makes use of powerful linear particle accelerators. Instead of using the high-speed subatomic particles to smash atoms, however, the particles are ejected out the rear of the device, resulting in a forward impulse.

Figure 19-2 is a simplified functional diagram showing how an *ion engine* can work. The source ejects large quantities of ionized gas. Hydrogen is a logical choice; it is plentiful in the Universe. It could be

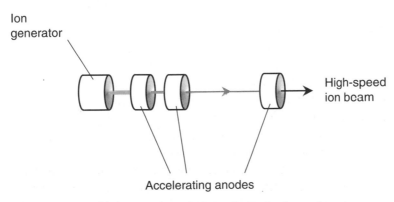

Figure 19-2.   Conceptual diagram of an engine that produces
thrust by accelerating atomic nuclei (ions) to high speed.

"scooped up" from interstellar space to supply the fuel for the engine. The positive ions of hydrogen are protons. They can be accelerated by negatively charged *anodes*, through which the particles pass. As the protons go through each succeeding anode, they gain more speed until finally, when they pass through the last anode, they are moving so fast that the reaction force pushes the spacecraft forward.

Ion engines do not produce much thrust, but they are efficient. They can keep operating for a long time. Once a spacecraft has gone beyond the gravitational influence of the Earth, an ion engine eventually could bring the ship up to enough speed to make an interstellar journey within a human lifespan.

Unfortunately, linear particle accelerators require enormous amounts of power. The only known system that can provide enough power is a nuclear reactor. If a hydrogen fusion reactor is ever developed, it could be used to power the ion engine. But then the fusion reaction itself would be a better source of thrust than accelerated ions.

FUSION ENGINES

Several types of hydrogen-fusion-powered spacecraft have been proposed by scientists and aerospace engineers as alternatives for obtaining the speeds necessary for long-distance space journeys using fuel that would be of reasonable mass. These include designs known as the *Orion*, the *Daedalus*, and the *Bussard Ramjet*.

In the Orion space ship, hydrogen-fusion bombs would be exploded at regular intervals to drive the vessel forward. The force of each blast, properly deflected, would accelerate the ship. The blast deflector would be strong enough to withstand the violence of the bomb explosions, and it would be made of material that would not melt, vaporize, deform, or erode because of the explosions. The blast deflector also would serve as a radiation shield to protect the astronauts in the living quarters. The bombs used in the Orion space ship could not be too large, lest each explosion cause the occupants to be injured or killed by the resulting acceleration force (also known as *g-force*).

A smoother ride would be provided by the Daedalus design. This would replace the bombs with a nuclear fusion reactor that would, in effect, produce a miniature Sun. This vessel would need a blast deflector similar to that used in the Orion design. The advantage of the Daedalus ship would be that the acceleration would be steady rather than intermittent. Thus Daedalus could attain greater speeds in less time than Orion without subjecting the astronauts to excessive *g*-force. Either the Orion or the Daedalus ships could reach

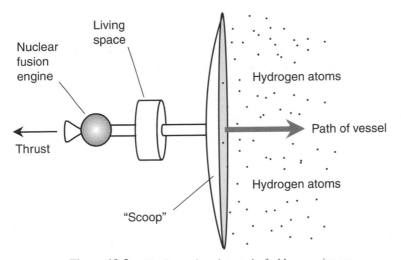

Figure 19-3.   The Bussard ramjet gets its fuel by scooping up
hydrogen atoms from space, which is not a perfect vacuum.

approximately 10 percent of the speed of light. This would not be sufficient
to cause significant relativistic time dilation, but it would make it possible to
reach the nearest star, Proxima Centauri, in about 45 years.

The most intriguing nuclear fusion design is known as the Bussard
Ramjet. This is similar to the Daedalus, but it would not have to carry nearly as much fuel. Once the ship got up to a certain speed, a huge scoop in the
front could gather up enough hydrogen atoms from interstellar space to provide the necessary fuel for hydrogen fusion reactions (Fig. 19-3). The greater
the speed attained by this vessel, the more hydrogen it could sweep up, thus
helping it go even faster. The Bussard ramjet would work especially well in
gaseous nebulae, provided those regions were not too peppered with meteoroids and asteroids. Some scientists think that this type of space ship could
attain speeds great enough to take advantage of relativistic time dilation, so
that the occupants would age more slowly than the Universe around them.
This would shorten the time they would have to spend en route to and from
distant stars, at least according to their point of view.

MATTER-ANTIMATTER ENGINES

Every particle of matter has an "evil twin." The *antiparticle* corresponding
to the electron is the positron; the counterpart of the neutron is the *anti-*

neutron; the antiparticle for the proton is the *antiproton*. Physicists and science-fiction writers alike call this stuff *antimatter*. Small amounts of antimatter have been isolated in laboratories.

Antimatter does exactly what the science-fiction writers say it will when it combines with an equal amount of matter: Both the matter and the antimatter are turned completely into energy. This happens so perfectly that it makes hydrogen fusion seem inefficient by comparison. Because of this fact, matter-antimatter reactors, should they ever be designed and put into production, will solve the world's energy problems. They will make it possible to build space ships that can accelerate to nearly the speed of light. They also will make it possible to build a bomb that can blow our whole planet to smithereens.

One big problem with antimatter is containment. What do we put it in? How can we keep it from coming into contact with the walls of the ordinary-matter chamber in which it is stored? One suggested scheme would employ powerful magnetic fields. Antiprotons (with a negative electrical charge) could be kept in defined "clouds" within the chamber and allowed to escape at a controlled rate, combining with ionized hydrogen gas (protons) to produce the energy necessary for thrust. Such a system, while theoretically plausible, would require a level of precision that engineers have not yet attained. One power failure, one accident, or one stray meteoroid could cause the antimatter to get out of control, resulting in instant destruction of the vessel and annihilation of the crew and passengers.

Another problem with any matter-antimatter reactor will be the intense radiation it produces. The EM rays from this type of engine can be expected to consist largely of short-wavelength gamma rays with tremendous penetrating power. Without some method of shielding, the occupants of a matter-antimatter–powered space ship will be killed by this radiation.

THE STELLAR SAIL

There is an entirely different way to get a space ship to move, at least in the vicinity of the Sun or another star. All stars emit energy in the form of high-speed particles. Some of these particles are atomic nuclei; others are photons of electromagnetic (EM) energy. The Sun, and presumably any star, produces a constant stream of these particles—a *stellar wind*—that rushes radially outward from the star. If we are near enough to the Sun or another star, it ought to be possible to sail on this stellar wind, in much the same way as sailing ocean vessels functioned before the advent of steam, fuel combustion, or nuclear engines.

A space ship with a *stellar sail* (also known as a *light sail*) would require an enormous sheet of reflective fabric. This sheet would be attached to the living quarters (Fig. 19-4). The range of travel would be limited to the Solar System (or whatever other star system we might happen to be visiting). It would be easier to travel outward away from the star than inward toward it, for obvious reasons. However, just as sailors of olden times managed to make progress into the wind by tacking (taking a zigzag path), we would be able to navigate in any direction given sufficient stellar wind speed. To move in closer to the star, we would follow a spiraling path inward, our direction of travel subtending an angle of slightly less than 90 degrees with respect to the stellar wind.

The stellar sail requires no on-board fuel, at least in the ideal case. However, navigation would be tricky. Sudden stellar flares would produce a dramatic increase in the stellar wind because of the large number of rel-

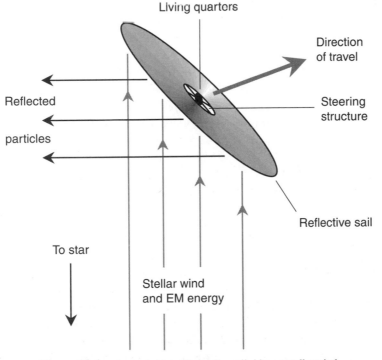

Figure 19-4.   A space ship with a stellar sail rides on stellar wind, just as a sailboat rides on atmospheric wind.

atively massive particles ejected during such an event. Near any planet with a magnetic field, including Earth, the particles are deflected, and the stellar wind does not necessarily "blow" away from the star.

The classic ocean sailing ships did not come in all the way to the beach but dropped anchor in deep water and sent small boats to shore. During space voyages, the main ship could draw in its stellar sail and "drop anchor" by falling into a planetary orbit. Landings could be made with small shuttles that would be sent out by and return to the main ship.

Interplanetary Travel

As of this writing, humans have paid live visits to one other world in the Universe besides the Earth: our own Moon. We have sent robotic probes to the surfaces of Venus and Mars. We have flown remotely controlled vessels past the outer planets.

It is one thing to send a machine to, say, Titan, the largest moon of Saturn. It is another matter to send people there. Machines can survive without air, water (H_2O), or food. Machines can tolerate far more radiation than can human beings, and they can put up with long periods of zero gravity. Humans are more fragile. In order to travel in interplanetary space, astronauts will need vessels equipped to keep them alive and safe so that they can reach their destinations and return to live normal lives thereafter. They also will have to realize that there are certain perils they cannot completely avoid.

GETTING HELP

An interplanetary journey will require months or years, given the propulsion technology we have right now. Even if nuclear-fusion engines are developed, a trip to the outermost planets will take weeks or months. Can we design a space vessel that will keep its occupants alive and well for this length of time? Tests conducted to date suggest that we can. Both the Americans and the Russians have sent people to Earth-orbiting space stations and let them stay there for periods comparable with those required for interplanetary travel.

There's one big difference between spending a long time in Earth orbit and spending the same amount of time en route to and from another planet. This difference is the ease of supply renewal. If something goes wrong with the International Space Station, we can send some technicians in a Space Shuttle to fix the problem. If the food goes bad or the H_2O-recycling system goes awry, help is just a shuttle away. This won't be the case for vessels halfway between Earth and Saturn.

How will future astronauts and cosmonauts bail themselves out of trouble when Earth-based help is not immediately available? One idea involves redundancy, also known as the *buddy system*. Rather than sending only one vessel on an interplanetary quest, we could send two identical vessels with two crews. Each vessel would carry sufficient supplies for twice the number of travelers on board (plus a little extra to allow for unexpected delays). In this way, if something bad happens to one ship, all the astronauts could get into the other one while the affected vessel is repaired or, at worst, abandoned.

ARTIFICIAL GRAVITY

Human beings can tolerate fairly long periods of weightlessness, known as *zero-g*, although it is not known if a person could survive in that environment indefinitely. There are serious health consequences. One problem is loss of mineral matter from the bones. Even with heavy exercise, an astronaut loses calcium from the bones when the bones are not required to support the body against the force of gravity. This weakens the skeleton so that when the astronaut returns to Earth, bone fractures can occur easily. In addition, the calcium, which is excreted in the urine, can cause kidney stones. Muscle wasting also takes place. Cardiopulmonary (heart and lung) functions decline. An astronaut in zero-*g* gets out of shape fast.

Artificial gravity will be a necessity for the well-being of astronauts who travel among the planets. This could be done by rotating a large, wheel-shaped vessel around the axis that points in its direction of travel. You saw this in Chapter 5 (see Fig. 5-6). Another method, which takes advantage of the buddy system, would involve tethering two identical vessels together with a strong cable and spinning the whole assembly like a huge baton (Fig. 19-5). If the cable is a few hundred meters long, such a tethered assembly does not have to revolve very fast to provide an acceleration force of 1 *g*, equivalent to the gravity at the surface of the Earth.

Using small rockets in each ship, the spin rate would be adjusted to get astronauts accustomed to whatever gravitational force they would encounter at their destination. Computers would take care of the navigation. It would take a while for the travelers to get used to looking out the windows without getting dizzy. Casual stargazing would be done only for amusement, if at all.

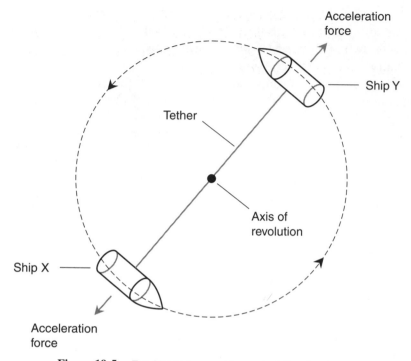

Figure 19-5.　Two identical space ships can be tethered and spun, producing artificial gravity for the occupants.

METEOROIDS

Interplanetary travel won't require relativistic speeds, but meteoroids will present a hazard nevertheless. A space ship must accelerate to about 11 km/s (7 mi/s) to escape the gravitational field of Earth. Higher speeds are necessary to reach planets in the outer part of the Solar System. Although most of the debris that existed in the primordial Solar System has been swept up by the planets and their moons, there are plenty of "space rocks" left. Have you ever been driving on a highway, following too closely behind a big truck, and had a piece of gravel strike your windshield? That little pebble was moving at about 20 m/s. Imagine what would happen if it were flying 1,000 times that fast!

A meteoroid the size of a pea, should it strike a space ship at interplanetary-travel speeds, would have an effect similar to that of a bullet fired from a high-powered rifle. A direct hit by a meteoroid the size of a soccer ball would, at the very least, cripple a space vessel, requiring evacuation.

How will interplanetary travelers see these objects coming? One possible way to see swarms of meteoroids is to use powerful radar, but this will not resolve individual objects. If we make enough interplanetary journeys, occasional impacts are bound to take place.

SOLAR FLARES

As you have learned, solar flares taking place on the visible side of the Sun are followed by a barrage of high-speed subatomic particles. These particles do not reach Earth's surface because the atmosphere blocks them. In addition, Earth's magnetic field deflects the charged particles toward the polar regions. In interplanetary space, however, there is nothing to prevent the particles from bombarding space ships. When such particles hit the metal shell of a spacecraft, deadly radiation is produced, in much the same way as they produce radiation when they strike the target of a particle accelerator.

Solar flares are impossible to predict. But they can be seen using telescopes, especially at certain wavelengths. The subatomic particles from a solar flare travel fast, but more slowly than light. When a solar flare is observed by watchful interplanetary travelers, they'll have some time to take cover. An interplanetary space vessel will need a radiation shelter, such as a small room lined with lead. Astronauts will have to enter the shelter as soon as a flare is seen and wait out the radiation storm. Such a shelter will be massive, and this will increase the amount of fuel the vessel consumes. However, it will be part of the price we'll have to pay if we want to be sure we reach our destination alive.

PLANETARY MAGNETOSPHERES

Subatomic particles are always emitted from the Sun. The intensity is much greater during a solar flare than at other times, but even on an ordinary solar day, there are plenty of charged atomic nuclei flying out from our parent star. These charged particles are accelerated near planets that have strong *magnetospheres* (magnetic fields). This produces radiation belts around such planets.

The radiation regions surrounding Earth are known as the *Van Allen belts*. Before the first *Apollo* spacecraft went to the Moon, scientists weren't sure if astronauts would survive passage through these regions or, if they did survive, whether or not they would develop *radiation sickness*. Apollo astronauts reported seeing tiny, pointlike flashes of light as a result of energetic particles striking the retinas of their eyes. This effect can be produced by a subatomic particle coming from any direction—even through the back of the head. Fortunately, serious radiation illnesses did not occur; the symptoms can range from nausea to hair loss and internal bleeding.

The radiation belts surrounding Jupiter are many times larger and more intense than the Earth's Van Allen belts. Astronauts planning to land on any of Jupiter's moons will have to know whether or not they will be within a high-radiation zone. If a vessel orbits Jupiter at or near the altitude of any of the radiation belts, the astronauts will sicken and die. Radiation detectors will be standard-issue on interplanetary ships.

Saturn also has a strong magnetic field, but not as powerful as that of Jupiter. Saturn is farther from the Sun, so the subatomic-particle bombardment is less intense. Even so, astronauts visiting that planet's moon system will have to be careful. Only time will tell if or to what extent space travelers will be able to spend time in the Jupiter and Saturn systems without ill effects from the planetary radiation belts.

WATER AND AIR

Human astronauts require life-support systems. In the twentieth and early twenty-first centuries, H_2O and air were carried along for the trip. Air, and especially oxygen (O_2), can be compressed by liquefaction. Water cannot be compressed. Supplies of O_2 and H_2O in early spacecraft had to be adequate at the start of the journey to ensure that, with proper recycling, they would last until the end. In interplanetary missions, it is not practical to load up a space ship with all the necessary air and H_2O beforehand.

Fortunately, H_2O exists on other worlds in the Solar System. The polar caps of Mars contain H_2O ice. Some of the moons of Jupiter and Saturn also have H_2O. Evidence suggests that traces of H_2O ice exist on crater floors at the poles of our own Moon, in regions that never receive direct sunlight. Even with these supplies, H_2O will be a precious thing on any interplanetary craft.

Oxygen can be obtained by dissolving certain minerals, such as ordinary table salt or baking soda, in H_2O and then passing an electric current through the solution. This is called *electrolysis of water*. The electricity can come from solar panels or from nuclear reactors on the vessel. To some extent, O_2 can be recirculated. When human beings breathe air at sea level, they take in a gas that is 21 percent O_2; when they exhale there is still 16 percent O_2. The 5 percent difference consists of carbon dioxide (CO_2). If this CO_2 is filtered out, the air can be breathed again. By using pressure regulators in addition to filters, a given parcel of air may be used three or four times.

FOOD

Human food must consist of living matter. We can't live off of minerals alone. We have no choice in this regard: Astronauts have to carry all their food along with them, and it must all come from supplies on Earth.

If life is ever discovered on another planet, will it be edible, and will it provide sustenance? Will anyone be daring enough to boil and consume extraterrestrial organisms, if they are ever found, and find out? The answer is, of course, yes. Someone will try it. Even if such a meal doesn't injure or kill the person who eats it, there is no guarantee that it will have much nutritional value. Until we find a source of safe, nutritious "ET food," dehydrated food, along with some plants grown in artificial environs, will be the norm on interplanetary journeys. We can be certain that astronauts will get tired of this bland stuff. By the time they return to Earth, they will have some strange cravings. A peanut-butter sandwich will be a gourmet meal.

Food can be freeze-dried, and nutritional supplements can be added. By carefully regulating the number of calories each person gets daily, staying well fed on interplanetary voyages should not be a problem. Unmanned spacecraft, loaded with food supplies, can be placed in orbit around destination worlds or even parachuted down to their surfaces to provide additional food supplies without the need for carrying them on the main vessels.

Interstellar and Intergalactic Travel

All the challenges of interplanetary travel will exist in interstellar space on an exaggerated scale. In addition, there will be new problems and inspirations.

WATER, AIR, AND FOOD—AGAIN

The essentials of life will be harder to come by on interstellar missions than on trips to other worlds in the Solar System. This will open up new avenues of technology. There should be no lack of work for people who want to design star-wandering ships.

Hydrogen exists in the voids between the stars, although it is at extremely low pressure, comparable with the best laboratory vacuum ever obtained on Earth. This hydrogen (H_2) can be combined with O_2 to provide H_2O. But where will the O_2, necessary not only for H_2O but for a breathable atmosphere, come from? Some scientists think it exists bound up in icy rocks floating among the stars. The distant Oort Cloud, the belt of comets that surrounds our Solar System and hopefully other star systems, can be expected to provide a source of H_2O, and therefore of O_2 as well. The challenge will be snaring the comets while traveling through the cloud at speeds of many kilometers per second!

Another way of getting breathable air is to split apart the CO_2 exhaled by the astronauts, mixing the O_2 with the ever-present nitrogen gas, and setting the carbon residue aside. No one has yet figured out an easy way to do this, but optimists believe that this problem eventually will be solved.

Food on interstellar voyages will be grown in the form of plant life, as well as carried along in the form of protein, vitamin, and mineral supplements. Exhaled human CO_2 will be a blessing here. Plants convert CO_2 into O_2, which is released into the air; the residual carbon is used by the plants to build their own living matter. Special gardens will be provided for the dual purpose of supplementing the O_2 stores and obtaining food. The gardens also will serve an aesthetic purpose. Astronauts will find psychological and emotional respite from the rigors of their artificial environment by sitting or strolling on the "garden deck" among the plants.

COSMIC RADIATION

The Sun is not the only source of high-speed subatomic particles. All the stars in the Universe emit them. The cores of galaxies are intense sources, some more than others. Supernovae can produce great quantities of radiation. There are x-ray and gamma-ray objects scattered throughout the galaxies. During a long interstellar or intergalactic voyage, astronauts will be exposed to unknown quantities of this radiation.

In the long term, we should not be surprised if long-distance space travelers have an above-average incidence of cancer. If multigeneration space voyages are carried out, the later generations will be subject to more-often-than-usual occurrences of birth defects. In the extreme, high-intensity cosmic radiation will shorten the life spans of space travelers, and they will be sick for much of their lives.

Hopefully, some scheme will be found to protect interstellar astronauts from cosmic radiation. Radiation shelters, mentioned earlier in this chapter as a means of staying safe from the perils of solar flares, may serve as sleeping quarters to minimize long-term exposure. The deadly cosmic particles may be deflected away from interstellar spacecraft by devices yet to be invented and perfected.

UNKNOWNS

It would be naive and arrogant of us to suppose that we are aware of all the dangers and challenges interstellar and intergalactic travelers will face. What about dark matter? We know it exists, but in what form? Billions upon billions of tiny black holes? Quadrillions of meteoroids and dormant comets? These are not the sorts of things a starship captain will want to encounter at 99 percent of the speed of light (or at any speed).

The large-scale structure—the "shape"—of the Universe is not known with certainty. One theory holds that the Universe is a gigantic four-dimensional *hypersphere*, with our space continuum comprising the three-dimensional curved surface. Other theories give the Universe different shapes such as a hypersaddle, or a hypertunnel. Curvature of space, according to Einstein's general theory of relativity, is the inevitable companion of gravitation. What if gravitation, working over distances of billions of light-years, has different effects than those with which we are familiar on a local scale?

Historically, scientists have assumed that the laws of physics are the same everywhere in the Universe. By observing the structures and spectral emissions of distant galaxies, we can see that they appear to behave in a manner similar to those closer to us. But are the physical constants the same? We would like to think so, but we do not know with absolute certainty. If the distant history of our Universe, according to the Big Bang theory, is any indicator, we have reason to suspect that physical constants change with time. Because time and distance are inextricably linked on a cosmic scale, the very foundations of reality may not be the same at the end of a long intergalactic journey as they were at the beginning. How might this affect human beings making such a trip?

TIME DILATION

If a spacecraft could be accelerated to high enough speed, time would flow more slowly for the occupants of the vessel than for everything in

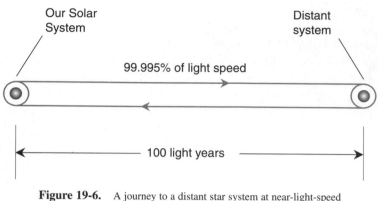

Figure 19-6. A journey to a distant star system at near-light-speed would dilate time by a large factor.

the surrounding Universe. This is the result of relativistic time dilation, which you learned about in Chapter 16. This could work to the advantage of interstellar and intergalactic travelers in some ways, although it would be a mixed blessing.

Suppose that we have a spacecraft with powerful engines that can boost it to almost the speed of light. We're going to visit a star system 100 light-years away (Fig. 19-6). We won't spend any time at the distant system once we get there; this is just a trial run to see if we can do it. Our space ship has a propulsion system so advanced that it averages 99.995 percent of the speed of light during the outbound and return journeys! This produces a time dilation factor of 100. For every second that passes in the outside Cosmos, only 0.01 s passes for us in the ship.

The round-trip distance is 200 light-years. We arrive back in our Solar System, according to Earth-based timekeepers, slightly more than 200 years after we departed. This is longer than a lifetime. When we return, we do not know a single person on our home planet. Even our children are deceased (assuming medical technology has not extended the average human lifespan severalfold). However, according to our own sense of time, the whole round trip has taken only a little more than 2 years.

This time dilation effect is no illusion. It is real. We only need to provide for a 2-year journey, even though it takes 200 years according to timekeepers in the greater Cosmos. If we could get even closer to the speed of light, we could travel to stars 1,000 light-years, 10,000 light-years, or even 100,000 light-years distant in the same 2 years. Our Milky Way galaxy is 100,000 light-years across.

NO RETURN

Extrapolating still further, there is no limit to the distance we might travel within the span of 2 years according to our sense of time, provided that we could accelerate to a speed arbitrarily close to the speed of light in free space. However, if we dare to travel over intergalactic distances, we must realize that we are leaving home forever. Even if we return to Earth, millions or billions of years will have passed. Will human beings even exist any more? If so, what will they be like? What about the climate? If we are gone more than about 4 billion years, the Sun will be in its red-giant phase when we return, and Earth will not be the sort of planet we would want to visit, if it exists at all.

Will we ever make journeys like this if the technology becomes available? In his book *Journey to the Stars*, Robert Jastrow expresses the belief that we will. Journeys of this sort have been undertaken before. South Pacific natives searched for, found, and settled mysterious islands north of the equator (now known as Hawai'i) by embarking into the unknown in outrigger canoes. European adventurers discovered lands in a new hemisphere (the Americas) after sailing across stormy seas. These people placed little importance on "returning home." To them, "home" was wherever they were bound. The most daring of our descendants, if and when we build ships that can travel among the stars, will have the same attitude. They will venture into the Cosmos because the alternative—*not* to—will be unthinkable.

 **Quiz**

Refer to the text if necessary. A good score is 8 correct. Answers are in the back of the book.

1. In one proposed spacecraft design, the hydrogen necessary for a nuclear-fusion-powered space vessel would be obtained from
   (a) the surface of the Sun.
   (b) interstellar space.
   (c) moon rocks.
   (d) the atmosphere of Jupiter or Saturn.

2. Which of the following propulsion systems, assuming that they all can be developed and used in space ships, will provide the most power for the least fuel mass?

(a) The ion engine
(b) The conventional rocket
(c) The matter-antimatter engine
(d) The nuclear fusion engine

3. Suppose that you are on a space ship bound for Mars and you observe a bright flare on the surface of the Sun. The most important thing you must do is
(a) take advantage of the extra energy by turning the ship's solar panels to directly face the Sun.
(b) take cover in the ship's radiation shelter.
(c) set the communication system to a frequency that will not be interfered with by the approaching magnetic storm.
(d) nothing special; carry on as usual.

4. Realistic telepresence spanning extremely long distances is
(a) not feasible because radio receivers and transmitters cannot be made sensitive or powerful enough.
(b) impractical because the speed of electromagnetic signals through space is only 299,792 km/s.
(c) unworkable because the necessary signal bandwidth is impossible to obtain.
(d) quite practical; there is no limit to the distance over which realistic telepresence can be accomplished.

5. Visitors to the moons of Jupiter will have to be especially aware of the peril presented by
(a) the planet's gravitation.
(b) the planet's poisonous atmosphere.
(c) radiation in the planet's magnetosphere.
(d) none of the above; the Jupiter system will pose no special dangers.

6. One of the most common, and most powerful, arguments that some people make against space exploration cites the fact that
(a) the extraterrestrial planets and the stars cannot be reached because they are too far away.
(b) Earth is a perfectly fine place and it will last forever, so there will never be any need to venture into space.
(c) we ought to spend our limited resources to solve problems here on Earth before we spend money or energy on space programs.
(d) there is not likely to be life elsewhere in the Universe, so there is no point in wasting our time searching for it.

7. Overexposure to ionizing radiation such as gamma rays or cosmic particles is known to cause all the following *except*
(a) overeating, leading to obesity.
(b) genetic mutations.

(c) increased incidence of cancer.
(d) pointlike flashes in the field of vision.

8. Scientists are certain that if our species survives long enough we eventually
will have to leave Earth and seek another place to live because
(a) the human population will become too great for Earth to support.
(b) Earth's atmosphere will run out of oxygen.
(c) humans will develop an insatiable desire to travel in space.
(d) the Sun will not remain a life-sustaining star forever.

9. Ion engines are efficient, but they
(a) cause acceleration in bursts that would make space travel uncomfortable
or dangerous.
(b) operate at extreme temperatures, requiring elaborate means to prevent the
hardware from melting or vaporizing.
(c) produce extreme amounts of gamma rays, posing a hazard to the occupants of space ships in which they are used.
(d) do not produce very much thrust.

10. The Moon, the asteroids, and other planets' moons are believed by some
astronomers to contain plenty of
(a) natural resources such as minerals and metals.
(b) ozone to replenish the protective layer in Earth's upper atmosphere.
(c) life forms to provide food for Earth's population.
(d) land suitable for farming.

CHAPTER 20

Your Home Observatory

Now that you have some knowledge of what goes on beyond the reaches of Earth's atmosphere, you can more fully appreciate what you see when you look up after the Sun goes down. You don't have to spend your life's savings on hardware, but a few instruments can help you see a lot of interesting celestial objects.

Location, Location, Location!

As our cities swell, good places for astronomical viewing are becoming hard to find. We light up the darkness so that our streets are safe for driving powered vehicles, even as the exhaust from those machines thickens the veil between us and the Sun, Moon, planets, stars, nebulae, and galaxies. Tall buildings turn fields into canyons. Some children grow up without learning to recognize any celestial objects other than the Sun and Moon. It doesn't have to be this way.

   If you happen to live in a rural area, especially in one of the less populated parts of the country, consider yourself blessed.

BEFORE YOU VENTURE OUT

Read this chapter before you start to shop for astronomical viewing aids. Then check out several stores; hobby shops are excellent. If there is a local astronomy club in your area, find out where and when it meets, and get some input from experienced amateur astronomers before spending any money. Your needs will depend on what you want to see "up there" and how important amateur astronomy really is to you.

Sky and Telescope magazine online has information about astronomy clubs all over the world. Go to the following Web site:

http://www.skypub.com

Click on "Site Map," and then click on "Astronomical Directory." As you know if you have used a computer online lately, the Web is always changing, and by the time you read this, the links may be different. In that case, go to this site:

http://www.google.com

Click on "Advanced Search," and input the words *astronomy clubs* in the "exact phrase" box. Then take it from there!

ESCAPING

Wherever you live, you need not travel far to get to a place where latter-day contrivances don't interfere with your view of the nighttime sky. There are plenty of places, even near Boston, London, or Sydney, where the stars twinkle and the planets stand out like beacons. In this respect, ironically, some of the world's poorest people are well-to-do. Have you ever wondered what folks in remote Afghanistan or Tibet see when they look at the sky on a clear and moonless night?

The next time the weather is favorable for sky watching, get out in a rural area, out on a big country lake, or offshore in the ocean in a small boat. Find a quiet place, a safe place, where human and animal pests will not disturb you. Bring along some insect repellent unless it's winter. If it is winter, wear plenty of warm clothing! You're not going to be jogging around or doing aerobics. Don't trespass or put yourself in danger. Put at least 75 miles between yourself and the nearest big town.

Don't expect to find a spot entirely without any human-made lights in view, but if they're few in number and more than a city block away, it should be good enough. Give your eyes at least 15 minutes to adjust to the darkness. Then gaze upward. Better yet, lie flat on your back with an unobstructed half-sphere of sky above you.

NIGHT VISION

Astronomers have always had a problem with night vision. Now you'll find out first hand how they deal with it. On one side of the visibility equation, your eyes must adjust to the darkness, especially when the Moon is not above the horizon. On the other side of the equation, you'll want to read star maps or consult other reference materials from time to time. You might have to check eyepiece specifications, make adjustments to a telescope, or otherwise fiddle around with "stuff." You'll need some sort of lamp to do this.

Get a flashlight and some red cloth or thin red tissue paper. This will serve as a color filter. Cover the business end of the flashlight with the filter. Secure it with a rubber band. The resulting light should be dim; you'll have to experiment with various coverings to find out what works best. Use a flashlight with size D cells or, better yet, a lantern with one of those bulky 6-volt batteries. Be sure the cells or batteries are fresh, and carry a spare bulb. The light from the lamp should be bright enough so that you can read your star charts, eyepiece numbers, and other information after your eyes have fully adjusted to the darkness. But it shouldn't be any brighter than that.

Red light has some special properties. It docs not desensitize your eyes to the extent white light does. If you keep the filtered light source just bright enough so that you can read by it (but no brighter), it won't interfere with your stargazing. Another plus: Red light attracts fewer insects than white light.

GETTING YOUR BEARINGS

Once your eyes have adjusted to the darkness, it's time to locate some stars, constellations, or planets. These vary depending on the time of year, the hour of the evening, and the latitude on the Earth at which you happen to live. You can refer back to Chapters 1, 2, and 3 to locate some of the major constellations and to figure out what point(s) of reference to use. The positions of the Moon and the planets, as you know, vary among the background of stars.

Current maps of the heavens can be viewed by going to Weather Underground at the following Web site:

http://www.wunderground.com

Click on the "Astronomy" link. You can input your location anywhere in the world, as well as the hour of the evening or night, and get a complete map of the sky. If this link isn't available for some reason, *Sky and Telescope* online has excellent printable star maps. Go to:

http://www.skypub.com

If you can afford it, bring a notebook computer along on your stargazing expedition and have it equipped with wireless Internet access. In this way, you can check out the star maps on the fly. Turn down the display brightness to a low level so that it won't degrade your night vision.

The circumpolar constellations are the best reference to begin with. This is so because they're always above the horizon regardless of the time of year, unless you happen to live in the tropics (between approximately 20°N lat and 20°S latitude).

Your eyes alone can see a lot of interesting things in the sky once you know where to look. Mysterious fuzzy spots appear. Certain dim objects seem to pop out when you look slightly away from them, only to maddeningly vanish when you look straight at them. This is normal; it is a result of the anatomy of human eyes. The center of your field of vision is known as the *fovea*, representing the point on your retina where your gaze is directed precisely. This is where your eyes' *image resolution*, also called *resolving power*, is greatest. However, this comes at the expense of *sensitivity*, which is better slightly off-center in your field of vision. Sensitivity and resolving power both can be improved dramatically, of course, with binoculars and telescopes. However, then you sacrifice absolute field of view.

Binoculars

Some amateur astronomers recommend that you obtain a pair of binoculars before you spend any money on a telescope. This is an individual choice. Binoculars are good for general star viewing at low magnification. Telescopes are a requirement for resolving detail in the planets, observing lunar terrain up close, or examining Sunspots.

BASIC STRUCTURE

Figure 20-1 is a simplified functional diagram of a pair of binoculars. You can think of the assembly as two identical telescopes placed beside each other. The eyepieces are spaced to match the distance between the pupils of the observer's eyes. This spacing is adjustable. In most types of binoculars,

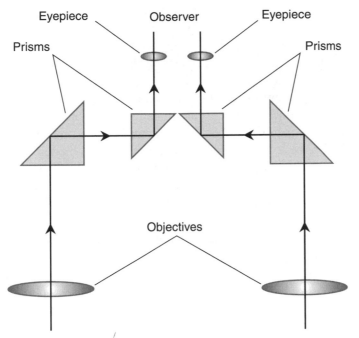

Figure 20-1. Functional diagram of a pair of binoculars.

when the eyepiece spacing is adjusted, the spacing between the objectives also varies.

The objectives are farther apart than the eyepieces. This exaggerates perspective for scenes within a few hundred meters but does not affect perspective for celestial objects, which are too far away for parallax to exist relative to any single observation point. The light enters the objectives, passes through a pair of prisms that bring the light beams closer together by means of internal reflection, and finally leaves the eyepieces to enter the observer's eyes. The principle of operation of each half of a pair of binoculars is identical to that of a Keplerian refracting telescope. The prisms turn the image right-side up and also orient the view properly left to right.

SIZE SPECIFICATIONS

Binoculars are rated in terms of the magnification (the number of times the apparent diameters of distant objects are increased), as well as in terms of the objective-lens diameter in millimeters (mm). You'll see a pair of numbers separated by a multiplication symbol, for example 7 × 50, printed somewhere on the assembly. The first number is the magnification, and the second number is the objective-lens diameter.

In general, the light-gathering power of binoculars is proportional to the square of the objective-lens diameter. However, this holds true only when the binoculars are optimized for a particular observer. If you divide the objective-lens diameter by the magnification, you should get a number between approximately 4 and 8. This number is called the *exit pupil* of the instrument. For best viewing, the exit pupil of a pair of binoculars should be the same as the diameter of the pupils of the observer's eyes (in millimeters) when adjusted to the darkness. In general, larger exit pupils (6 to 8 mm) are a good match for younger observers, and smaller exit pupils (4 to 6 mm) are better for older observers.

In terms of physical bulk and mass, binoculars range from tiny to huge. At least, this is the impression you'll get. Some binoculars can fit in your pocket. (But always keep them in a carrying case when you're not using them). Others are so large that you'll want a tripod to support them. The most massive binoculars will make your arms tired if you have to hold them up for a long time. High-magnification binoculars, especially those greater than 8×, need the extrasteady support that a tripod can provide.

The biggest binoculars are more appropriately called *binocular telescopes* or *stereoscopic telescopes*. These are fabulous for viewing star clusters, galaxies, and nebulae. They also can deplete the average person's bank account.

OPTICS

The prisms inside binoculars serve as mirrors to reflect the incoming light between the widely spaced objectives and the narrowly spaced eyepieces. Prisms provide better image resolution and contrast than mirrors. The best prisms are called *porro prisms* (Fig. 20-2A). They reflect the light entirely by total internal reflection, which you learned about in Chapter 17. Less effective but still superior to mirrors are *roof prisms* (see Fig. 20-2B), which have aluminized back surfaces that help reflect the light rays. Porro prisms are more expensive than roof prisms because a higher grade of glass must be used to get the highest amount of total internal reflection to occur without aluminized surfaces.

Another factor to consider in binoculars is whether or not the lenses are specially coated to minimize the amount of light they reflect. As you have

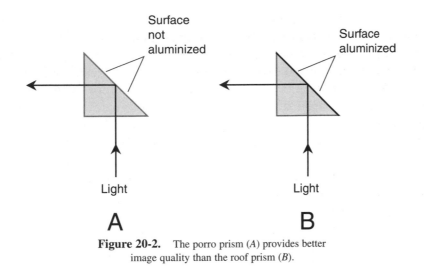

Surface
not
aluminized

Surface
aluminized

Light

Light

A

B

Figure 20-2. The porro prism (*A*) provides better
image quality than the roof prism (*B*).

seen if you've looked at the window of a darkened house from the outside
during the daytime, all glass reflects light as well as transmitting it. Any
light reflected is light that doesn't pass through the glass. In binoculars or
a telescope, you'll want as much of the light as possible to reach your eyes
and not be reflected back into space from the objective(s) or into the internal chamber of the instrument from the eyepiece(s). The best lenses have
multiple coatings on both the inside surfaces and the outside surfaces.
These binoculars will be specified as having *fully multicoated optics*.

FIELD OF VIEW

When you look through a properly adjusted pair of binoculars, with the
focus and barrel spacing optimized for your vision, you should see a single, large circular region. The absolute field of view is the angular diameter, in degrees of arc, of this region as measured against the background of
distant stars. In some sets of binoculars, the absolute field of view is
expressed in terms of feet at 1,000 yards instead of degrees.

The absolute field of view, which is defined in the same way as it is with
telescopes (see Chapter 17), depends on the magnification and also on the
apparent diameter of the circular region—the apparent field of view—as
seen through the eyepiece. Binoculars with large apparent fields of view
offer a more pleasant viewing experience than those with narrower apparent fields regardless of the absolute field of view.

Even if a pair of binoculars has a wide apparent field of view, the image quality won't necessarily be good. How well do objects near the outer periphery of the field stay in focus compared with objects near the center? Are stars near the edge distorted or blurred? To what extent do "little rainbows" appear around stars, especially near the edge of the field? You can't easily test these things in a hobby store when you are deciding which pair of binoculars to buy. Therefore, it's a good idea to check out the return policy of any store with which you do business. If the binoculars prove unsatisfactory, you should be able to return them to exchange for a better pair or to get a full refund within a few days of purchase. Keep the sales receipt!

Choosing a Telescope

If you want to see planetary detail or intricate features on the Moon, you will need a telescope. The theory of telescope operation is discussed in Chapter 17. Here we'll take a pragmatic view. If reading this chapter tempts you to go out and buy a telescope, that's fine, but don't spend more money than you can afford. Sleep on the idea before you act on it.

There are dozens of telescopes available for hobbyists. Some are inexpensive; others cost as much as a car. This chapter isn't meant to be a shopping guide, but you should be aware of certain assets and limitations of the most popular hobby telescope designs. These include the Keplerian refractor, the Newtonian reflector, and a specialized form of Cassegrain reflector known as a *Schmidt-Cassegrain telescope (SCT)*.

F-RATIO

Before we compare the virtues and vices of the various telescope designs, there's a specification you should know about. It is called the *focal ratio*, or *f-ratio*. This specification will appear as the letter *f*, followed by a slash, followed by a number.

If you've done much photography, you know about the *f*-ratio of a camera. In telescopes, the meaning is the same, but the dimensions are larger. The *f*-ratio is equal to the focal length of the objective divided by its diameter or *aperture* (Fig. 20-3). Thus, for example, an objective whose diameter is 20 cm with a focal length of 200 cm is an *f*/10 objective (as shown at *A*). If the focal length is cut to 100 cm, it is an *f*/5 objective (as shown at *B*).

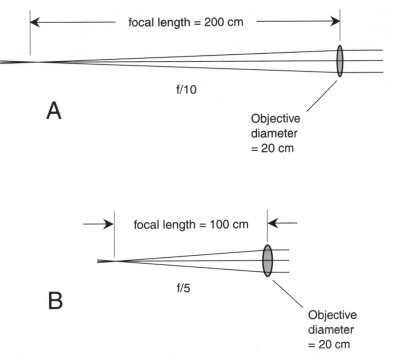

Figure 20-3.   The *f*-ratio of an objective is equal to the focal length divided by the diameter of the lens or mirror. At *A*, an *f*/10 objective; at *B*, an *f*/5 objective.

In general, objectives with smaller *f*-ratios make it possible to get larger absolute fields of view than is the case with large-*f*-ratio objectives. However, there is a tradeoff: The smaller the *f*-ratio, the more difficult it becomes to engineer the optics to provide quality images. This is especially true of refracting telescopes.

KEPLERIAN REFRACTOR

The Keplerian refractor, as you will remember from Chapter 17, uses a convex lens as the objective and a convex lens (actually, a set of two or more convex lenses) as the eyepiece. A well-made Keplerian refractor, mounted on a sturdy base, is a joy to use. The images are sharp, the viewing is stable, and the positioning of the telescope is intuitive. You can get great results without a lot of hassle. A poorly made refractor, however, can be, like anything shoddy, a source of frustration.

Refracting telescopes for amateur use range in diameter from about 50 mm (2 in) to 150 mm (6 in). The cost increases dramatically as the objective diameter increases. Another factor to consider is the manner in which the objective is made. *Achromatic objectives* are the most common; they consist of two different lenses, having different refractive indexes, glued together. This helps to reduce chromatic aberration, the tendency for focal length to vary with color, producing "rainbows" around stars and blurring planetary and lunar images. The greater the *f*-ratio, the less likely you are to have trouble with chromatic aberration in a refracting telescope if all other factors are constant. A typical value is *f*/10.

Apochromatic objectives provide the highest quality in refracting telescopes. As you should expect, telescopes that use this type of objective are expensive. Serious refracting-telescope lovers, to whom price is no object, seek apochromatic refractors. The *f*-ratios are generally smaller than those of the achromatic refractors, sometimes as low as *f*/5, and the image contrast is superior.

NEWTONIAN REFLECTOR

Overall, the Newtonian reflector can provide more light-gathering power for the money than any other hobby telescope. This is its chief advantage. The largest Newtonian reflectors in the amateur market have objective mirrors about 60 cm (2 ft) in diameter. Telescopes this large are heavy and bulky and are inconvenient to transport. But they're fun to use, especially when looking at diffuse objects such as nebulae and galaxies.

Newtonian reflectors come in two types: *normal-field* and *rich-field*. Normal-field telescopes have large *f*-ratios; rich-field telescopes have small *f*-ratios. The cutoff between the two classifications is considered to be *f*/6. Normal-field telescopes produce crisper images for a given aperture, and they are also easier to collimate (adjust) than rich-field Newtonians. However, because the focal length is greater, the normal-field Newtonian's tube is longer, and this makes it more difficult to carry around than a rich-field telescope of the same aperture. The rich-field telescope, as its name implies, can provide a larger absolute field of view than a normal-field telescope.

All Newtonian reflectors have part of their aperture blocked by the secondary mirror and its support. If you take a "star's eye" view and look down the tube of a Newtonian reflector, you'll see the obstruction (Fig. 20-4). This reduces the effective aperture slightly, although this is not significant in most

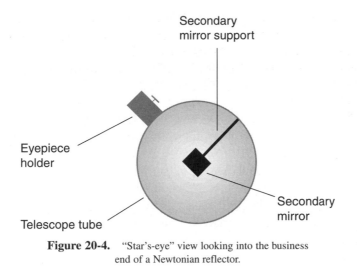

Figure 20-4.   "Star's-eye" view looking into the business
end of a Newtonian reflector.

designs. The bigger problem is that it reduces the image contrast. The light-gathering "hole" is not shaped like a perfect disk; the secondary mirror and its support put an irregular barrier there. Although you cannot notice it when you look through the telescope during normal use, this irregular barrier exacts a toll that refracting telescopes do not have to pay.

Some people dislike the fact that the Newtonian reflector's eyepiece holder is in the side of the tube near the front. However, this is an asset in many, if not most, viewing situations, except with the largest normal-field designs. When looking at objects near the zenith using a modest-sized Newtonian, you don't have to crouch or crane your neck. However, with larger Newtonians, you will need a stepladder to look through the eyepiece. A tall base support pedestal or tripod is also required for large-aperture, normal-field Newtonian reflectors. Otherwise, the mirror end will strike the ground when the instrument is aimed at objects high in the sky. Some people do not feel comfortable standing several rungs up on a stepladder in the dark.

All Newtonian reflectors, especially the rich-field types, exhibit a phenomenon called *coma*, in which objects near the edge of the field of view do not come to a perfect focus. Visually, this can be seen as a radial stretching out of stars near the periphery; in the most extreme cases, the stars near the edge of the apparent view field look like tiny, short-tailed comets. Coma can be overcome by a lens assembly called a *coma corrector* placed immediately on the objective side of the eyepiece.

SCHMIDT-CASSEGRAIN TELESCOPE

In recent years, Schmidt-Cassegrain telescopes (SCTs) have become popular with amateur astronomers. Commercially manufactured SCTs are available with objectives ranging from about 13 cm (5 in) to more than 50 cm (20 in) in diameter.

The SCT has a physically short tube, even though its effective focal length is long. This is possible because of the convex secondary mirror. As a result, SCTs are more portable than Newtonian reflectors of comparable diameter. The eyepiece is located in the center of the objective mirror; this makes viewing more convenient for most people and eliminates the need for a tall tripod or pedestal to support large-diameter instruments. The secondary mirror is supported by a transparent *corrector plate* at or near the front of the telescope tube. This eliminates the need for side supports, so the secondary mirror produces less contrast reduction than is the case with Newtonian reflectors. The corrector plate seals the tube so that dust will not collect on the mirrors inside. Figure 20-5 is a "star's-eye" view into a typical SCT.

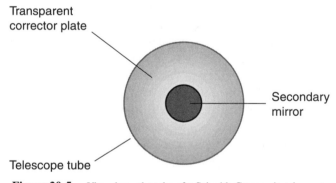

Figure 20-5. View down the tube of a Schmidt-Cassegrain telescope. The transparent corrector plate serves to support the secondary mirror.

On the minus side, the SCT secondary mirror must be quite large to gather all the light from the primary mirror and direct it to the eyepiece. This means a greater reduction in the light-gathering area compared with the Newtonian design. An SCT is somewhat more expensive than a Newtonian of the same diameter. The corrector plate in an SCT has an annoying habit of developing condensation in humid environments.

The SCT is ideal for looking at nebulae and galaxies. If used in conjunction with a solid, heavy-duty, clock-driven equatorial mount (described later in this chapter), and if placed on level ground or pavement away from roads where trucks roll by and cause vibration, this type of telescope is exceptionally stable. It is preferred by many amateur astrophotographers for this reason.

Telescope Accessories

You'll need certain accessories with your telescope. You'll want at least two good eyepieces. You will need some sort of *finder scope* or sighting device. A *Barlow lens* can provide extra magnification for your eyepieces. And, of course, there are *optical filters* of all kinds, some for looking at the Sun, others for the Moon, some for the planets, and others for more sophisticated purposes.

THE EYEPIECE

There are many different designs for telescope eyepieces. All make use of two or more lenses to optimize the apparent field of view, to provide good focus from the center of the view field to the edge, and to make it easy to look through the device. Most eyepieces have focal lengths between 4 and 40 mm.

In general, the longer the focal length of an eyepiece taken as a whole, the lower is the telescope magnification, all other things being equal. Remember how to calculate telescope magnification: Divide the focal length of the objective by the focal length of the eyepiece in the same units. If a telescope has a focal length of 2 m (or 2,000 mm), then a 4-mm eyepiece provides 500× and a 40-mm eyepiece provides 50×. The overall focal length of a telescope eyepiece is not necessarily the same as the focal length of any of its individual lenses.

Eye relief is an important specification of any telescope eyepiece. This is the maximum distance, in millimeters, that the surface of the eye can be away from the surface of the eyepiece lens on the observer side while still letting the observer see the entire apparent field of view. In general, the longer the focal length of an eyepiece, the greater is the eye relief. Larger eye relief numbers translate into easier viewing.

Some people find it difficult and unpleasant to look through short-focal-length eyepieces (6 mm or less) because the lens diameter is more or less proportional to the focal length. A few eyepieces have observer-side lens diameters smaller than the diameter of the pupil of the eye itself. This makes it necessary to bring the eye very close to the eyepiece. If the observer wears glasses, viewing through such eyepieces is compromised. People who don't wear glasses will flinch away from the eyepiece if its surface comes into direct contact with the eyeball.

An eyepiece's *outside barrel diameter* always should match the inside diameter of the telescope's *focusing mount*. This is 31.75 mm (1$^1/_4$ in) in most telescopes, but some instruments have focusing mounts that are 50.80 mm (2 in) across as measured through the inside. Adapters can be found to get small-diameter eyepieces into large-diameter mounts. However, if you want to use a large-diameter eyepiece in a small-diameter mount, you'll have to improvise.

EYEPIECE DESIGNS

Here are four different types of eyepieces you are likely to find on the amateur market. Of these, the *Ramsden* and the *Kellner* are the simplest and therefore the cheapest. The *orthoscopic* and the *Plossl* are more sophisticated and expensive.

Figure 20-6*A* is a cross-sectional diagram of a Ramsden eyepiece. It consists of two planoconvex elements that have the same focal length. The larger lens is toward the telescope objective, and the shorter element is toward the observer. (This is true of virtually all telescope eyepieces.) The convex surfaces of the lenses face inward toward each other, and the flat surfaces face outward. This is an old design, dating all the way back to the 1700s. The Ramsden eyepiece is difficult to optimize because the spacing between the lenses is always a tradeoff between eye relief and the effects of lens aberration.

Figure 20-6*B* shows the Kellner design. It is similar to the Ramsden, except that the observer-side lens is a compound element consisting of a convex lens glued to a planoconcave lens. The compound element, when designed properly, eliminates the chromatic aberration inherent in the Ramsden design. The lenses must be coated to minimize reflection of light inside the eyepiece. Kellner eyepieces work best at the longer focal lengths, providing low to medium telescope magnification.

The orthoscopic eyepiece (see Fig. 20-6*C*) is among the most popular designs in use today. Image distortion and chromatic aberration are eliminated by the three-element compound lens on the objective side. This type of lens is noted for its excellent contrast and its ability to maintain focus

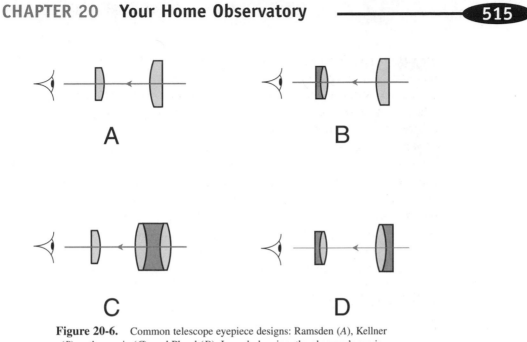

Figure 20-6. Common telescope eyepiece designs: Ramsden (*A*), Kellner
(*B*), orthoscopic (*C*), and Plossl (*D*). In each drawing, the observer's eye is
at left and the incoming light arrives from the right.

from the center of the view field to the edge. In addition, the view field
appears relatively flat as compared with some eyepiece designs that give
the view field a concave (bowl-shaped) appearance. Orthoscopic lenses
work well at all focal lengths.

Figure 20-6*D* is a cross-sectional diagram of a Plossl eyepiece. This
design first gained widespread acceptance among amateur astronomers in
the 1980s. It has all the assets of the orthoscopic eyepiece. The eye relief
of a well-made Plossl is adequate even at the shortest focal lengths. The
observer-side lens has a relatively wide diameter. Plossls with long focal
lengths (25 to 40 mm) are physically bulky, projecting some distance out
from the telescope's eyepiece tube, but they offer the ultimate in viewing
comfort. Some have rubber eye guards to keep out external light.

Other eyepiece designs you might encounter are the *Erfle*, the *zoom*, the
RKE, and the *Huygens*. The sheer variety of eyepieces can confuse the
novice amateur astronomer. You can get on the Internet, enter eyepiece
designs as keyword phrases (for example, *Kellner eyepiece*), and see what
various folks have to say about the different designs. A salesperson at a
hobby shop sometimes can help, but beware. A salesperson may be more
motivated to get you to spend a lot of money than to sell you the best eyepieces for your needs.

STAR DIAGONAL

With refractors and SCTs, the eyepieces are normally in line with the telescope tube. Viewing can be uncomfortable when such a telescope is aimed at objects high in the sky; you have to crouch down and crane your neck. However, there's a simple and common solution to this problem: the *star diagonal*. This device bends the light path without introducing distortion, although it flips the image laterally, as a mirror flips your reflection.

A simple star diagonal employs a prism that causes the light to turn a 90-degree corner because of total internal reflection. The principle is the same as in binoculars. A cutaway view of a basic 90-degree star diagonal is shown in Fig. 20-7. More sophisticated star diagonals provide smaller angles, such as 45 degrees. Some star diagonals use two prisms rather than one, so the image is not laterally reversed.

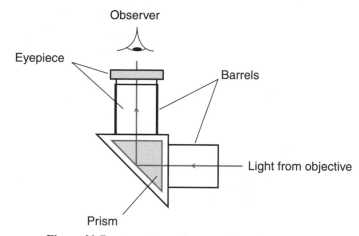

Figure 20-7. A star diagonal uses a prism to bend the light, making it easier to view some celestial objects.

The lateral-reversal feature of basic star diagonals makes it rather inconvenient when you try to find objects in the sky using a star map. You have to imagine everything on the map backwards. With a little practice, however, most people can overcome this mental obstacle.

SIGHTING DEVICE

If you've ever used a telescope in an attempt to locate a planet or star and you didn't have some sort of aiming or sighting device, you know how frustrating such an exercise can be. Except at the very lowest magnifications, you can end up searching for a long time. The simplest sighting devices are similar to gunsights. You aim the telescope as if it were a high-powered rifle. The more advanced type of sighting device has a small laser diode inside; it shines on a slanted glass to produce a variable-brightness red dot in the center of the view field. This dot is used to align the telescope with the object you want to observe.

Before you use it for celestial observations, the sighting device first must be aligned on a terrestrial target that is at least a couple of kilometers away. Find some object on the horizon that is large enough to see through the sighting device (that is, at 1×) yet small enough to fit into the view field of the telescope. Get the object centered in the view field of the telescope, fix the telescope in position, and then adjust the sighting device until the object lines up in it. Then check the view through the telescope again to be sure the object is still centered there. For good measure, go back and check the sighting device again too.

FINDER SCOPE

A more precise device for telescope aiming is a *finder scope*, often called simply a *finder*. This is a small Keplerian refractor. Most finders have objective diameters of 40 to 60 mm and magnify several times. The eyepiece has a pair of fine threads or wires, called *cross hairs*, placed at its focus. These produce a + or × pattern in the view field. The intersection point of the cross hairs is at the center of the view field. The finder position is adjusted until a star that falls at the cross-hair intersection point also shows up in the center of the view field of the main telescope at high magnification.

A finder can be aligned using the same technique as is used for a simple sighting device. The best finders are mounted in a pair of rings, both of which are attached to the main tube of the telescope near the eyepiece. Each ring has three or four adjustment screws. These should be fairly tight (but not so tight that the finder is damaged or the screw threads are stripped). A few finders have single-ring mountings. These are unstable. It is best to stay away from them.

BARLOW LENS

A concave or planoconcave lens can be inserted in any telescope between the eyepiece and the objective, and the effect is to increase the apparent focal length of the objective. This type of lens is called a *Barlow lens*. It is placed close to the eyepiece. The lens is mounted inside a cylinder designed to fit into the eyepiece barrel of the telescope at one end and around the barrel of the eyepiece at the other end (Fig. 20-8).

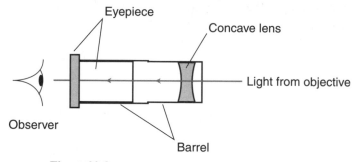

Figure 20-8. A Barlow lens increases the magnification obtainable with a given eyepiece.

Because the Barlow lens increases the effective focal length of the objective, it provides increased magnification when a given eyepiece is used. Most Barlow lenses are rated at 2×. This means that they double the magnification for each eyepiece used. Some Barlow lenses are rated at 3×; these triple the magnification.

A Barlow lens can be useful in two situations. First, it eliminates the need for using eyepieces with extremely short focal lengths when high magnification is desired. An 8-mm eyepiece can be used in place of a 4-mm eyepiece, for example, when a 2× Barlow is inserted in the light path. Most people find an 8-mm eyepiece more viewer-friendly than a 4-mm eyepiece. Another asset of the Barlow lens is that it can double the number of obtainable levels of magnification, provided that you have chosen your eyepieces wisely. Suppose, for example, that you have a telescope whose objective has a focal length 1,000 mm and you have eyepieces whose focal lengths are 20 and 28 mm. This provides magnifications of 50× and 36×, respectively. If you obtain a 2× Barlow lens, you can obtain magnifications of 100× and 72× with the same two eyepieces. This gives you four well-spaced degrees of magnification.

A Barlow lens should not be used in an attempt to get extreme magnification. For example, if you have a telescope whose objective has a focal length of 2,000 mm and you use a 4-mm eyepiece with a 3× Barlow, you can theoretically obtain 1,500×. However, Earth's atmosphere generally makes it futile to try for anything more than 500✕, even with the largest telescopes. The slightest vibration will cause terrible wobbling of the image. In addition, the brightness of an observed image in any particular telescope decreases as the magnification increases. Remember the formula for the highest useful power you can get out of a telescope: approximately 20✕ per centimeter of objective diameter, or 50✕ per inch, with a maximum of 500✕ at sea level and most land-based locations.

FOCAL REDUCER/CORRECTOR

Most hobby SCTs have *f*-ratios of around 10. This is all right for viewing planets, lunar surface features, and some star clusters. However, when looking at nebulae or galaxies, often you will want to reduce the magnification as much as possible. By so doing, you can concentrate the light so that dim, diffuse features show up more clearly against the background of the sky. Reducing the magnification with a given eyepiece also increases the absolute field of view.

A *focal reducer/corrector* is a convex lens that shortens the effective focal length of the SCT objective by a certain amount, usually 37 percent. This means that the effective focal length and the *f*-ratio are both cut to 63 percent of their values without the device installed. You might think of it as the opposite of a Barlow lens. With a 37 percent focal reducer/corrector, an *f*/10 telescope becomes an *f*/6.3 instrument. A focal reducer/corrector is larger in diameter than a Barlow lens and is equipped with a threaded mount that can be screwed into the opening in the objective mirror that passes light into the eyepiece holder.

Suppose that your SCT has an objective with a focal length of 2,000 mm. If you have a 40-mm eyepiece, a focal reducer/corrector shortens the effective focal length to 1,260 mm. This reduces the magnification from 50✕ to a little more than 30✕. It also increases the absolute field of view by a factor of about 1.6. The corrector feature helps to ensure proper focus throughout the apparent field of view.

SOLAR FILTER

You can use a telescope to look at the Sun, but there are some precautions you must take to avoid damage to your telescope, your eyesight, or both. Before you point a telescope toward the Sun, get a *solar filter* that fits over the entire skyward opening of the telescope. The filter must be as large in diameter as the objective and is called a *full-aperture solar filter*. With such a filter, direct sunlight does not fall on any of the telescope optics. Only certain types of filters are acceptable; these block ultraviolet (UV) rays that otherwise could damage your eyes even if the image is not uncomfortably bright. The brand-name telescope manufacturers such as Celestron supply excellent solar filters. They're not cheap, but neither are your telescope or your eyesight.

If you have a finder that uses lenses, such as a Keplerian refractor with cross hairs, cover it before aiming the telescope at the Sun. Otherwise, you risk damage to the finder's eyepiece and cross hairs.

Never use a "sun filter" that screws into telescope eyepieces! Such a device is at the prime focus of the telescope objective, so it will heat up. Such filters have been known to melt or crack. If one of these "filters" fails while you're looking at the Sun, you will remember the experience for the rest of your life. You'll be lucky if your retina is not injured permanently.

Have you heard that you can aim a telescope at the Sun without a solar filter, with the eyepiece installed, and let the brilliant light shine onto a white piece of paper or a screen to see details of the Sun's surface? In theory, this scheme works, and you can in fact get a decent image without risking damage to your eyes. Several people can view the image at the same time. But this is a bad idea. It subjects the eyepiece to direct focused sunlight, which can permanently damage the eyepiece. Besides this, as the Sun moves in the sky or as you move the telescope around while locating the Sun, the focused spot will strike and heat up interior components of the telescope.

Think of the focused, unfiltered rays of the Sun as the business end of a blowtorch. Would you turn a hot flame on anything you value? Of course not. So follow the universal rule: Always filter sunlight before it gets into a telescope (Fig. 20-9). Treat your telescope as kindly as you treat your own eyes.

MOON (LUNAR) FILTER

If you have a telescope whose objective lens or mirror is larger than about 10 cm (4 in), the Moon will appear extremely bright at low magnification when it is near the full phase. In fact, at the lowest obtainable magnifications with SCTs using focal reducer/correctors, the full Moon can appear so brilliant that it hurts your eyes to look at it. A *Moon filter*, attached to the eyepiece, renders the Moon's image tolerable under these conditions.

A Moon filter, also called a *lunar filter*, is tinted grayish, gray-green, or brownish, like the lenses in a high-quality pair of sunglasses. It is mounted in a threaded ring that screws into the objective side of the eyepiece. This place-

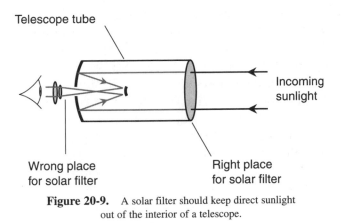

Figure 20-9. A solar filter should keep direct sunlight
out of the interior of a telescope.

ment is all right; the Moon's light is not intense enough to cause damage to
a telescope's interior components or to an eyepiece. You'll know when you
need a lunar filter and when you don't. At high magnification levels or when
the Moon is a thin crescent, you won't want one. At low magnification, after
sundown, and when the Moon is gibbous or full, you will.

OTHER FILTERS

It is not easy for most people to find places where the nighttime sky is not
polluted by airglow. Airglow doesn't interfere very much with viewing of
the Moon or the planets, although dust and particulate pollution, along with
convection currents rising from the day-heated land that roil the evening
air, can blur even these images and reduce their contrast. If you want to see
nebulae, globular clusters, and galaxies, you will have trouble with airglow
unless you take some measures to reduce it.

A *light-pollution-reduction (LPR) filter* reduces the effects of airglow at
night. Most big outdoor lamps are sodium-vapor devices that emit most of
their radiation at well-defined wavelengths in the yellow part of the visible
spectrum. Mercury-vapor lamps are less common, but they too emit most
of their light at certain discrete wavelengths. A *line-type LPR filter* is
designed to transmit light at all visible wavelengths except specific ones. In
this way, the airglow from sodium-vapor and mercury-vapor lamps can be
attenuated, whereas light at other wavelengths passes through the filter
unaffected. Other LPR filters include *narrowband* and *broadband* types.
The particular filter that will work best in a given situation must be found

by trial and error. The folks in your local astronomy club can give you advice based on their own experiences. All LPR filters, like Moon filters, are designed to be screwed into the objective side of an eyepiece.

Planetary filters are simple color filters that are screwed into eyepieces in the same manner as are Moon filters and LPR filters. They are available in almost any tint you can imagine. You can use an orange filter to look at Mars, a yellow-green filter to look at Jupiter, or a red filter to look at Venus. You can even use these filters (*in addition to a full-aperture solar filter— never all by itself!*) to look at the Sun. Experimentation is the key. Try all the filters you can find. Look at anything you want with them. See if you can borrow some from friends, so that you don't spend a lot of money unnecessarily. You're bound to see some interesting things.

Supports, Mounts, and Drives

Telescopes can be supported, mounted, and driven in various ways. Some systems are designed for simplicity and convenience; others are intended for ease of tracking once the telescope has been aimed. Any system can be driven by a motor that keeps it aimed at an object as the Earth rotates. The most sophisticated systems can locate objects using computer programs and can follow them so precisely that they remain in the field of view for an hour or so.

TRIPOD VERSUS PEDESTAL

With the exception of the *Dobsonian mount* (described below), the most common amateur telescope support is the *tripod*. A telescope tripod resembles the tripods used for photography or video recording, but the telescope design is sturdier and more resistant to vibration. A typical telescope tripod can be adjusted in height from approximately 1 m (40 in) to 1.5 m (60 in). As its name implies, the tripod has three legs, the lengths of which are independently adjustable. This allows you to level the instrument even if the surface is somewhat irregular.

The *pedestal support* is preferred by some people for use with large telescopes, especially those using the *German equatorial mount* (described below). A pedestal consists of a single, massive, thick vertical post. The post can be supported on a flat base, or it can be driven and cemented into the ground. Pedestal supports are less portable than tripods (and of course, not portable at all if permanently secured to the surface). This type of support is sturdier than most tripods. Care must be exercised to ensure that the pedestal is perfectly plumb (vertical).

AZ-EL MOUNT

The simplest, and generally the cheapest, set of bearings you can get for a telescope is the *azimuth-elevation (az-el) mount*. It goes by other names too, such as *altitude-azimuth*, *altazimuth*, or *alt-az*. Figure 20-10 is a simplified drawing of a refractor employing this system. The *azimuth bearing* turns 360 degrees in the horizontal plane. The *elevation bearing* rotates in the vertical plane for as far as the telescope will allow. Theoretically, only 90 degrees of elevation range is necessary, from the horizon to the zenith. Using the bearings in combination, the telescope can be pointed to any object in the sky. Because of the construction of the particular az-el system shown in Fig. 20-10, it is sometimes called a *fork mount*.

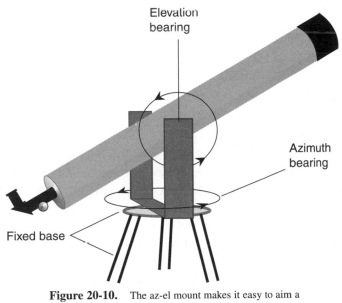

Figure 20-10.   The az-el mount makes it easy to aim a
telescope at any point in the sky.

The az-el mount, while convenient for casual telescope users, has limitations. As Earth rotates, objects move across the sky in paths parallel to the celestial equator. Near the celestial equator, this motion is from east to west; near the celestial poles it is in circles, counterclockwise in the northern hemisphere and clockwise in the southern. In order to follow an object across the sky over a period of time, you'll have to adjust both the azimuth and the elevation settings in an az-el mount (unless you happen to be at either the north or the south geographic pole). It would be much easier if you only had to move one of the bearings. This is possible with a simple modification of the az-el system.

FORK MOUNT/WEDGE

If an az-el mount is tilted so that the plane of the "horizon" corresponds with the celestial equator rather than with the actual horizon, a telescope can track objects in the sky by continuous adjustment of only one bearing. The 360 degree azimuth bearing from the az-el system becomes a *right-ascension (RA) bearing*. When it is rotated, the telescope moves east and west in celestial longitude. The range of the elevation bearing from the az-el system must be extended to cover 180 degrees, and it becomes a *declination bearing*. When it is adjusted, the telescope moves north and south in celestial latitude.

The proper tilt for the converted az-el system is accomplished by means of a wedge, constructed or set at an angle that corresponds to the terrestrial latitude where the telescope is located. The *fork mount*, which gets its name from its shape, lends itself readily to this scheme (Fig. 20-11). This system is popular among SCT users.

The *fork mount/wedge* requires alignment to work properly. The *RA axis* (the axis of the right-ascension bearing) must point precisely at the north celestial pole. A slight misalignment will result in improper tracking, especially over long periods of time. To ensure that the alignment is correct, the wedge is adjustable. You should determine your latitude down to the minute of arc. (There are several Web sites that can provide you with this information if you live in a town of at least medium size. Because the Web page locations change constantly, the best way to find them is to enter the phrase *latitude and longitude* into a well-known search engine such as *google.com*.)

DOBSONIAN MOUNT

Large Newtonian telescopes—those over 25 cm (10 in) in diameter—present a special challenge when it comes to mounting them and viewing through them. The *Dobsonian mount*, named after its inventor, is an az-el system that sits directly on the ground or pavement (Fig. 20-12).

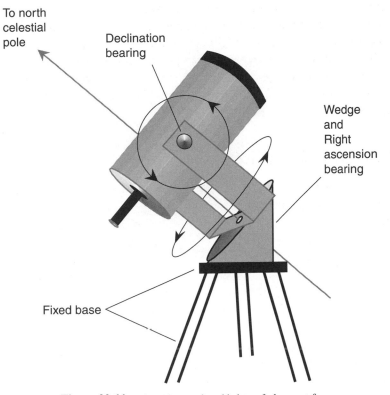

To north
celestial
pole

Declination
bearing

Wedge
and
Right
ascension
bearing

Fixed base

Figure 20-11.   A wedge can be added to a fork mount for
easy tracking of celestial objects.

Dobsonian mounts usually are constructed from plywood with Teflon
bearings. The plywood helps to dampen vibrations transmitted through the
ground, such as can be caused by heavy trucks on nearby streets. The
Teflon bearings provide ease and smoothness of movement. Because the
telescope sits lower to the surface than is the case with a tripod or pedestal
mount, the eyepiece can be reached with less difficulty. However, with
extremely large Newtonian reflectors (those over 40 cm across and/or with
high *f*-ratios), a ladder is necessary for viewing objects near the zenith.

The limitations of the Dobsonian mount are similar to those of the az-el
mount. Tracking can be inconvenient because both the azimuth and the elevation bearings must be moved. Special equatorial mounting tables are
available for Dobsonian telescopes. The table is sloped, and the slope can
be adjusted; it performs the same function as the wedge in the fork
mount/wedge system. If you happen to live in the tropics, the slope of the

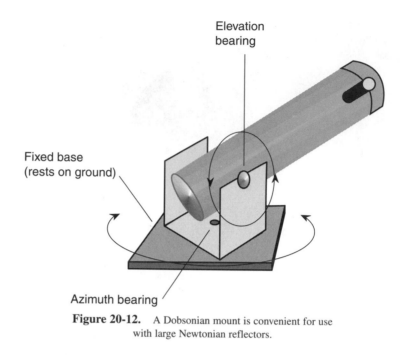

Figure 20-12.   A Dobsonian mount is convenient for use
with large Newtonian reflectors.

table can interfere with the full range of movement of the Dobsonian
mount. However, at higher latitudes, including most of Europe and North
America, the equatorial table is a convenient option.

GERMAN EQUATORIAL MOUNT

One of the best-known sets of telescope bearings is found in the *German
equatorial mount*. It can be recognized by its unique configuration and counterweight (Fig. 20-13). It, like the fork mount/wedge, moves along RA and
declination coordinates. The RA axis is adjustable and must be aimed at the
north celestial pole. The most sophisticated German equatorial mounts are
equipped with sighting scopes that make them fairly easy to align. Once the
mount is adjusted properly and the telescope is aimed at an object in the sky,
the object can be followed by moving only the RA bearing.

Some people find the German equatorial mount unnatural and awkward,
especially when observing things in the circumpolar region. However, a
well-made and well-adjusted system of this type is superior to any other for
serious use with massive and bulky telescopes. Not surprisingly, the best
German equatorial mounts are expensive. Some of the largest ones, including a heavy-duty pedestal base, cost more than $2,000.

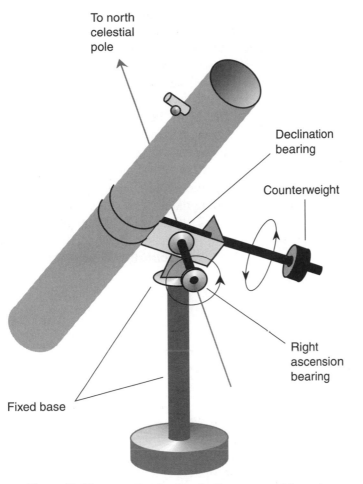

To north
celestial
pole

Declination
bearing

Counterweight

Right
ascension
bearing

Fixed base

Figure 20-13.   Simplified drawing of a German equatorial mount.

CLOCK DRIVE

If you want to gaze at celestial objects through a telescope for a long time, a *clock drive* is a great convenience. This is especially true if you have a group of people, each of whom must look through the telescope in turn and not all of whom are experts at aiming it. The higher the magnification, the more quickly a celestial object will "drift" out of the field of view unless some compensation is made. Objects near the celestial equator drift faster than objects near the celestial pole.

The heart of the clock drive is a slow motor that makes one complete 360-degree revolution per sidereal day, that is, every 23 hours and 56 minutes.

This is accomplished by gearing-down and precise regulation of the actual motor speed. Clock drives are typically designed to work with either a fork mount/wedge or a German equatorial mount because only the RA bearing needs to be connected to the motor. Clock drives can function with az-el drive systems, but they are more complicated and more expensive because both bearings must be adjusted by the device as time passes.

With any clock drive, the quality of the gears is particularly important. If there is significant play in the gears, the telescope will wobble with the slightest disturbance. This effect is particularly annoying at higher magnification and takes away much of the advantage of the clock drive. If you plan to do any time-exposure astrophotography with your telescope, you will need a clock drive, an equatorial mount, and a guiding device that makes minute corrections based on a *guide star* that you select within the field of view. Under these circumstances, telescope vibration and gear play cannot be tolerated at all.

Clock drives require a source of electrical power. This can be a battery, or it can be the household utility current. Batteries tend to wear out fast in most clock drives, and they never go dead when it's convenient. (No time is convenient for a power failure!) Use of the utility power requires an extension cord.

COMPUTERIZED DRIVES

In recent years, a number of amateur telescopes, particularly SCTs, have been made available equipped with microcomputers that automatically guide the instrument to any of several thousand objects in the sky. These telescopes are manufactured by Celestron, Inc., among others. The most sophisticated models incorporate Global Positioning System (GPS) receivers so that the alignment process is automatic. You don't even have to know where on Earth you are to set up such a telescope and use it.

A computerized drive, like the venerable videocassette recorder (VCR), requires the user to climb a learning curve. The system must be programmed and objects selected according to a certain sequence of entries. The information is entered on a keypad, and the system status is displayed on a screen. You also can locate objects manually using a set of up/down and left/right (or north/south and east/west) buttons. The *slew rate* (the speed at which the telescope turns as you hold down one of the buttons) can be selected over a range from slowest to fastest, for example, from 1 (very slow) to 9 (quite fast).

Once an object has been located in the sky, the telescope can be programmed to follow it. This scheme can work with az-el or equatorial systems, but if you have astrophotography in mind, the equatorial system is a must.

Celestial objects rotate in the field of view over time when an az-el clock drive is used because an az-el–mounted telescope doesn't maintain a constant *attitude* (orientation) with respect to celestial coordinates as the heavens sweep around the celestial pole. This will blur a time-exposure photograph, even if the bearings and the clock drive itself are aligned perfectly.

For a computerized drive to work correctly, the telescope must be aligned with great accuracy. If it doesn't have the GPS feature, this means that it has to be set by using two or three reference stars. These stars must be far apart from one another in the sky, and they must all be above the horizon at the time of alignment. You will have to know your terrestrial latitude and longitude down to the minute of arc or better. In addition, you'll have to know the exact time. This can be found at the following Web site (as of the time of this writing):

http://www.time.gov

It will take some practice to get good at aligning a computerized drive system unless it has GPS built in. You will have to learn to carry out the whole process within a few moments. With each tick of the clock, your reference stars move across the heavens by several seconds of arc. Once you have the telescope aligned accurately, you can select the object you want by navigating the computer menu, and the telescope will aim itself. It's best to use the lowest available magnification initially because this gives your telescope the greatest margin for error. You can then fine-tune the position of the telescope if you want more magnification.

 Quiz

Refer to the text if necessary. A good score is 8 correct. Answers are in the back of the book.

1. A telescope has an *f*-ratio of *f*/5. The magnification is 100×. What is the focal length?
    (a) 500 mm
    (b) 20 cm
    (c) 100 mm
    (d) It cannot be determined from this information.

2. A telescope has an objective focal length of 1,000 mm, and the eyepiece has a focal length of 25 mm. A 2× Barlow lens is used. What is the magnification?
(a) 20×
(b) 40×
(c) 80×
(d) It cannot be determined from this information.

3. Some telescopes have a tendency to produce blurring near the outer edge of the field of view, even when things are in perfect focus at the center. This is called
(a) coma.
(b) dispersion.
(c) chromatic aberration.
(d) depth of field.

4. The optimal position of a solar filter in an SCT is
(a) in front of the objective.
(b) between the objective and the secondary mirror.
(c) between secondary mirror and the eyepiece.
(d) between the eyepiece and the observer's eye.

5. You see a pair of binoculars in a surplus store. You see "7 × 70" stamped on them. The diameter of the objective lenses is
(a) 10 cm.
(b) 70 cm.
(c) 0.1 m.
(d) 70 mm.

6. Suppose that an SCT has a magnification of 200× with a particular eyepiece. A focal reducer/corrector that cuts the *f*-ratio by 37 percent is installed. What is the resulting magnification if the same eyepiece is used?
(a) 63×
(b) 74×
(c) 126×
(d) 200×

7. The bearings of a properly adjusted fork mount/wedge move the telescope along coordinates of
(a) compass direction and elevation.
(b) celestial latitude and celestial longitude.
(c) azimuth and altitude.
(d) right ascension and elevation.

8. The light-gathering area of a refracting telescope is
(a) proportional to the objective diameter.
(b) proportional to the square of the objective diameter.
(c) proportional to the *f*-ratio.
(d) proportional to the magnification.

9. A filter that allows all light to pass through, with the exception of light at a single wavelength, is called
   (a) broadband.
   (b) line-type.
   (c) planetary.
   (d) lunar.

10. A prism that turns the path of the light rays entirely by means of total internal reflection is called
   (a) a dispersive prism.
   (b) a reflective prism.
   (c) a flint prism.
   (d) a porro prism.

Test: Part Five

Do not refer to the text when taking this test. A good score is at least 30 correct. Answers are in the back of the book. It is best to have a friend check your score the first time so that you won't memorize the answers if you want to take the test again.

1. A telescope can be made using
   (a) a concave objective lens and a convex eyepiece lens.
   (b) a convex objective lens and a concave eyepiece lens.
   (c) a convex objective mirror and a concave eyepiece lens.
   (d) a concave objective lens and a concave eyepiece lens.
   (e) any of the above.

2. Which of the following will not present a problem for interstellar travelers?
   (a) Cosmic radiation
   (b) Boredom
   (c) Meteoroids
   (d) Maintaining a supply of food and water
   (e) All of the above will be cause for concern.

3. Which of the following statements is false?
   (a) A convex lens can bring parallel light rays to a focus.
   (b) A concave mirror can collimate light rays from a point source.
   (c) A convex mirror can be used as the main objective in a reflecting telescope.
   (d) A concave lens makes close-up objects appear smaller.
   (e) A convex lens can be used as the main objective in a refracting telescope.

4. What color of light is best for referring to star charts and other data when observing the heavens on a moonless night?
   (a) Red
   (b) Green
   (c) Blue
   (d) White
   (e) It doesn't matter.

5. A cloud chamber can be used to detect
   (a) cosmic particles.
   (b) IR radiation.
   (c) UV radiation.
   (d) microwaves.
   (e) all of the above.

6. In a good pair of binoculars, the lenses
   (a) reflect most of the light internally.
   (b) have aluminized surfaces.
   (c) are polarized.
   (d) are coated to minimize reflection of light.
   (e) are tinted to filter out glare.

7. One of the most annoying problems of a Galilean refractor, which uses a concave lens as the eyepiece, is the fact that
   (a) the image appears upside-down.
   (b) the image appears laterally reversed.
   (c) the apparent field of view is narrow.
   (d) focusing is critical.
   (e) the objective mirror is difficult to adjust.

8. What is a telechir?
   (a) A long-distance radio transmitter for interplanetary communication
   (b) A system for remotely controlling a space probe
   (c) A shuttle craft that can be used for transportation between a large space ship and the surface of a planet, moon, or asteroid
   (d) A remotely controlled robot
   (e) A virtual-reality computer program

9. A rich-field Newtonian reflector
   (a) has relatively high magnification.
   (b) has a relatively large objective mirror.
   (c) has a relatively small f-ratio.
   (d) requires a Barlow lens.
   (e) requires a focal reducer.

10. Visible-light wavelengths are commonly denoted in
    (a) light-years.
    (b) kilometers.
    (c) meters.
    (d) centimeters.
    (e) nanometers.

11. Suppose that a certain material transmits light at a speed of 150,000 km/s. What is its index of refraction, accurate to three significant figures?
    (a) 0.500

(b) 0.805
(c) 1.00
(d) 1.24
(e) 2.00

12. A Keplerian refracting telescope has an objective lens with a focal length of 1,000 mm. A 10-mm eyepiece is used in conjunction with a 2× Barlow lens. The magnification of this telescope is
(a) 20×.
(b) 50×.
(c) 100×.
(d) 200×.
(e) impossible to calculate without more information.

13. One of the most serious challenges that will face the designers and operators of matter-antimatter spacecraft propulsion systems is
(a) keeping the antimatter contained while it is stored.
(b) preventing the ship from exceeding the speed of light.
(c) the relative inefficiency of matter-antimatter reactions.
(d) keeping the temperature high enough for reactions to occur.
(e) building an entire space ship out of antimatter.

14. The interferometer is
(a) a special high-resolution radio telescope.
(b) a device for splitting light into colors.
(c) an instrument for detecting x-rays.
(d) a means of minimizing radio noise.
(e) an antenna for transmitting signals through the ionosphere.

15. On an interplanetary journey, food will come mainly from
(a) supplies carried or delivered from Earth.
(b) mineral matter gathered from meteoroids, asteroids, and comets.
(c) extraterrestrial sources.
(d) water and vitamin pills.
(e) intravenous feeding.

16. The wavelength of a light beam is inversely proportional to its
(a) energy.
(b) bandwidth.
(c) intensity.
(d) frequency.
(e) period.

17. A disadvantage of using robots in place of human astronauts for long-distance journeys into space is, arguably, the fact that
(a) the electronic circuits in robots are sensitive to cosmic radiation.
(b) sending robots to celestial destinations is not romantic.

(c) robots are too heavy and bulky to transport into space.

(d) robots require electrical power to function, but humans do not.

(e) robots could bring computer viruses back to Earth.

18. An SCT has an objective mirror with an effective focal length of 2,000 mm. A 40-mm eyepiece is used in conjunction with a focal reducer/corrector that reduces the *f*-ratio by 37 percent. The magnification of this configuration is

  (a) 18.5×.

  (b) 31.5×.

  (c) 50×.

  (d) 68.5×.

  (e) impossible to determine without more information.

19. Which of the following statements is false?

  (a) A glass prism bends green light more than it bends orange light.

  (b) The focal length of a simple glass convex lens is shorter for green light than for orange light.

  (c) Dispersion occurs when white light passes through a simple glass lens.

  (d) The index of refraction of glass depends on the color of the light shining through it.

  (e) All of the above statements are true.

20. Fill in the blank to make the following sentence true: "The _____ temperature of a celestial object is determined by examining its radiation intensity at various wavelengths."

  (a) infrared

  (b) convective

  (c) spectral

  (d) apparent

  (e) Fahrenheit

21. A fork mount and wedge consists of

  (a) an az-el mount that is tilted so that its azimuth axis points at the celestial pole.

  (b) an az-el mount that is tilted so that its elevation axis points at the celestial pole.

  (c) a German equatorial mount that is tilted so that its right-ascension axis points at the zenith.

  (d) a German equatorial mount that is tilted so that its declination axis points at the zenith.

  (e) none of the above.

22. The maximum diameter of a refracting telescope is limited, in practice, by

  (a) lens sag.

  (b) spherical aberration.

  (c) paraboloidal aberration.

  (d) focal length.

  (e) dispersion.

23. Suppose that you travel into space at a speed arbitrarily close to the speed of light for a vast distance and then return to Earth. Which of the following scenarios is impossible?
    (a) Your parents have not yet been born.
    (b) All the classmates you knew in high school are years older than you.
    (c) The climate of the Earth has changed.
    (d) Humans have become extinct.
    (e) The Sun has died.

24. In order to observe the Cosmos at x-ray wavelengths, it is necessary to
    (a) get above Earth's atmosphere.
    (b) use special color filters.
    (c) use a telescope with a large *f*-ratio.
    (d) find a location far away from city lights.
    (e) find a location where there is little rf interference.

25. To keep the rotation of Earth from causing celestial objects to drift out of a telescope's field of view, you can use
    (a) a Dobsonian mount.
    (b) a right-ascension mount.
    (c) a German equatorial mount.
    (d) an equatorial mounting table.
    (e) a clock drive.

26. Imagine a solid sphere of glass, perfectly transparent and perfectly uniform, with a spherical hollow space in the exact center. Imagine a lightbulb, call it Lamp A, with a point-source filament located at the center of the spherical hollow space and therefore also at the center of the whole sphere of glass. Imagine a second lightbulb in the open air, also with a point-source filament; call it Lamp B. How do the rays of light from the two lamps compare in their behavior?
    (a) The rays from both lamps radiate outward in straight lines and in exactly the same way.
    (b) The rays from Lamp A are reflected totally inside the cavity within the sphere of glass, but the rays from Lamp B radiate outward in straight lines.
    (c) The rays from Lamp A converge to a point somewhere outside the sphere of glass, but the rays from Lamp B radiate outward in straight lines.
    (d) The rays from Lamp A diverge more when they emerge from the sphere of glass compared with the rays from Lamp B that do not have to pass through the glass.
    (e) It is impossible to say without more information.

27. A device that bends light, making refractors and SCTs easier to look through when observing objects high in the sky, is called
    (a) a Kellner eyepiece.
    (b) a finder.
    (c) a Barlow lens.

(d) a focal reducer.

(e) a star diagonal.

28. Which of the following instruments can be used to analyze celestial objects in the UV part of the spectrum?

(a) A radar telescope

(b) A spectrophotometer

(c) An interferometer

(d) A Keplerian refractor

(e) A Galilean refractor

29. Most telescope eyepieces have focal lengths ranging from approximately

(a) 0.4 to 4 mm.

(b) 4 to 40 mm.

(c) 40 to 400 mm.

(d) 40 cm to 4 m.

(e) 4 to 40 m.

30. A Cassegrain reflector

(a) has a concave objective mirror and a flat secondary mirror.

(b) has a concave objective mirror and a convex secondary mirror.

(c) has a longer tube than an equivalent Newtonian reflector.

(d) suffers from lens sag if the diameter of the objective is too large.

(e) has the eyepiece mounted in the side of the tube.

31. Fill in the blank to make the following sentence correct: "For best viewing, the _____ of a pair of binoculars should be the same as the diameter of the pupils of the observer's eyes when adjusted to the darkness."

(a) objective-lens diameter

(b) eyepiece diameter

(c) magnification

(d) exit pupil

(e) focal length

32. One theory concerning the nature of x-ray stars suggests that they consist of

(a) diffuse gas and dust congealing into new stars.

(b) remnants of supernovae.

(c) binary systems in which a neutron star pulls matter from a normal star.

(d) cool stars with intense magnetic fields.

(e) pairs of neutron stars in mutual orbit.

33. In an interplanetary or intergalactic spacecraft, the oxygen supply can be supplemented by

(a) burning flammable materials.

(b) electrolysis of water.

(c) combining water with hydrogen.

(d) hydrogen fusion.

(e) none of the above.

34. A Cassegrain reflector has an objective measuring 25 cm in diameter. The magnification of the telescope is 250×. What is the *f*-ratio of this instrument?
    (a) *f*/25
    (b) *f*/250
    (c) *f*/10
    (d) *f*/0.10
    (e) It cannot be calculated from this information.

35. Which, if any, of the following (a, b, c, or d) has *not* been suggested as a good reason for humanity to venture into space?
    (a) Extraterrestrial objects contain natural resources we can use.
    (b) We should strive to encounter and learn from extraterrestrial beings.
    (c) Relativistic space travel will allow us to travel back in time and correct our past mistakes.
    (d) The Sun will eventually die and we will need to find a new home.
    (e) All of the above have been suggested as good reasons for humanity to venture into space.

36. Radar astronomy has proven valuable in mapping the surface of
    (a) Neptune.
    (b) Uranus.
    (c) Saturn.
    (d) Jupiter.
    (e) Venus.

37. The Bussard ramjet is a proposed form of
    (a) aircraft propulsion system.
    (b) hydrogen-fusion propulsion system.
    (c) chemical propulsion system.
    (d) matter-antimatter propulsion system.
    (e) laser-beam propulsion system.

38. When light from the Sun shines on a spherical, reflective object such as a steel ball bearing, the reflected rays
    (a) are all parallel.
    (b) diverge.
    (c) converge.
    (d) focus to a point hot enough to start fires.
    (e) behave unpredictably.

39. When an astronaut lives for a long time at zero-*g*,
    (a) the body becomes stronger because it doesn't have to constantly work against the pull of gravity.
    (b) the bones lose calcium, which is excreted in the urine; this weakens the skeleton and can cause kidney stones.
    (c) the lung capacity increases because the thin air requires that they grow larger to get enough oxygen.

(d) the body gradually loses its need for sleep so that by the end of a long journey, space travelers can stay awake and alert indefinitely.

(e) nothing in particular happens to the body.

40. Karl Jansky was the first experimenter to discover radio noise coming from outer space. He found especially high levels of rf energy when his antennas were aimed at

(a) the Moon.

(b) the north celestial pole.

(c) the center of the galaxy.

(d) the horizon.

(e) the zenith.

Final Exam

Do not refer to the text when taking this exam. A good score is at least 75 correct. Answers are in the back of the book. It is best to have a friend check your score the first time so that you won't memorize the answers if you want to take the test again.

1. As seen from temperate or tropical latitudes, the positions of the constellations shift gradually westward in the sky from night to night because
    (a) the Earth rotates on its axis.
    (b) the Earth revolves around the Sun.
    (c) the tilt of the Earth's axis changes.
    (d) their right ascensions constantly change.
    (e) their declinations constantly change.

2. A star cluster in which the individual stars do not show a striking central concentration nor an orderly pattern or structure is called
    (a) a thin cluster.
    (b) an open cluster.
    (c) a nebulous cluster.
    (d) a globular cluster.
    (e) a random cluster.

3. An average star such as our Sun will, it is believed, eventually end up as
    (a) a blue dwarf.
    (b) a black hole.
    (c) a proton star.
    (d) a pulsar.
    (e) a black dwarf.

4. an imaginary line connecting any planet with the Sun
    (a) revolves around the Sun at a constant angular rate.
    (b) is always the same length.
    (c) is longest when the planet moves fastest.
    (d) sweeps out equal areas in equal periods of time.
    (e) is constantly getting longer.

5. A radio galaxy
   (a) is more radioactive than a typical galaxy.
   (b) is a galaxy that can be observed only at radio wavelengths.
   (c) is a galaxy that emits an unusually large amount of energy at radio wavelengths.
   (d) is a galaxy that emits radio signals that sound like they come from an intelligent civilization.
   (e) is a galaxy to which we have sent radio signals in the hope of contacting extraterrestrial civilizations.

6. Total internal reflection can occur for a light beam
   (a) striking a pane of glass from the outside.
   (b) passing through a pane of glass at a right angle.
   (c) striking the surface of a prism at a grazing angle from the inside.
   (d) traveling from one place to another through a vacuum.
   (e) under no circumstances.

7. The distance from the Earth to the Moon is approximately
   (a) 3 times Earth's diameter.
   (b) 10 times Earth's diameter.
   (c) 30 times Earth's diameter.
   (d) 100 times Earth's diameter.
   (e) 300 times Earth's diameter.

8. The nanometer is commonly used to denote
   (a) image resolution in a telescope.
   (b) diameters of telescope objectives.
   (c) focal lengths of telescope eyepieces.
   (d) wavelengths of longwave radio signals.
   (e) wavelengths of visible light.

9. Why do meteors seem to come from a particular spot in the sky during a meteor shower?
   (a) Because they are traveling perpendicular to the Earth's orbit around the Sun.
   (b) Because they move at extreme speed relative to Earth, differences in their actual direction don't make a significant difference.
   (c) It is an illusion; we see them from a certain perspective, although they fall toward Earth in more or less parallel paths.
   (d) For good reason! They actually originate in the constellation that seems to produce them.
   (e) The premise is wrong. Meteors seem to come from all over the sky during a shower.

10. The celestial meridians all intersect each other at the
    (a) celestial equator.
    (b) celestial parallels.
    (c) celestial poles.

   (d) zenith.

   (e) horizon.

11. The solar orbit of Mars lies between the solar orbits of

   (a) Mercury and Venus.

   (b) Venus and Earth.

   (c) Earth and Jupiter.

   (d) Jupiter and the asteroids.

   (e) Saturn and the asteroids.

12. Ganymede, the largest moon of Jupiter, keeps one face toward the planet at all times because

   (a) Jupiter's powerful magnetic field interacts with Ganymede's magnetic field to keep Ganymede's rotation rate in sync with its orbital period.

   (b) Ganymede is slightly out of round, and Jupiter's gravitation takes advantage of this to keep Ganymede's rotation rate in sync with its orbital period.

   (c) Ganymede lies in the plane of Jupiter's equator, and this sort of orbit always produces synchronization of a satellite's rotation rate with its orbital period.

   (d) of a sheer coincidence.

   (e) nothing! Ganymede does not keep the same face toward Jupiter at all times.

13. Distances to some of the galaxies outside the Milky Way have been estimated by observing and measuring the relative magnitudes of

   (a) Cepheids.

   (b) globular clusters.

   (c) blue dwarfs.

   (d) neutron stars.

   (e) red giants.

14. In an electromagnetic field, the direction of wave travel (propagation) is

   (a) parallel to the magnetic lines of flux.

   (b) parallel to the electrical lines of flux.

   (c) parallel to both the magnetic and the electrical lines of flux.

   (d) parallel to neither the magnetic nor the electrical lines of flux.

   (e) dependent on the wavelength.

15. The elevation of an object in the sky is its

   (a) angle in degrees with respect to the zenith.

   (b) angle in degrees with respect to true (geographic) north.

   (c) angle in degrees with respect to the meridian.

   (d) angle in degrees above the horizon.

   (e) distance in kilometers from the Earth.

16. Long-distance space travel will require

   (a) an advanced or even an as-yet unknown propulsion system.

   (b) patience and cooperation among the spacecraft crew.

   (c) a means of replenishing or maintaining food, air, and water supplies.

    (d) a shelter against stellar and cosmic radiation.

    (e) all of the above.

17. Dust storms on Mars

    (a) are too small to be seen by Earth-based telescopes.

    (b) occur only near the poles.

    (c) never last more than a few hours.

    (d) sometimes cover the whole planet.

    (e) do not occur. Mars has no atmosphere and therefore no wind.

18. On Earth, there are certain places where the Sun stays above the horizon for 24 hours a day on June 21. These places all lie between

    (a) the equator and 23.5°N. latitude.

    (b) the equator and 66.5°N. latitude.

    (c) the equator and 23.5°S. latitude.

    (d) the equator and 66.5°S. latitude.

    (e) none of the above.

19. By definition, living things generally

    (a) act against entropy.

    (b) distribute energy throughout the Cosmos.

    (c) produce ionizing radiation.

    (d) convert oxygen into carbon dioxide.

    (e) give off energy in the form of heat.

20. Suppose that you are located at 45° north latitude. It is late January. At exactly 6:00 P.M. local time, you see a star at the zenith. Where will the star be at 6:00 A.M., 12 hours later?

    (a) On the southern horizon

    (b) On the northern horizon

    (c) On the western horizon

    (d) On the eastern horizon

    (e) Below the horizon, at the nadir

21. The rate at which time "flows" depends on

    (a) absolute position in space.

    (b) the intensity of the gravitational field.

    (c) the speed of light.

    (d) the intensity of Earth's magnetic field.

    (e) all of the above.

22. The Kuiper belt contains an abundance of

    (a) radiation from Earth's magnetic field.

    (b) radiation from Jupiter's magnetic field.

    (c) high-speed particles from the Sun.

    (d) comets.

    (e) meteors and asteroids.

23. The surface of Venus is
    (a) cold, dark, and icy.
    (b) an ocean of liquid methane and ammonia.
    (c) hot, dry, and under enormous pressure.
    (d) a mystery because of the clouds that shroud the planet.
    (e) nonexistent; it is a "gas ball" like Jupiter or Saturn.

24. We do not see impact craters on the surface of Jupiter because
    (a) Jupiter has no solid surface on which craters can form.
    (b) Jupiter's moons take all the hits from incoming meteors.
    (c) Jupiter's gravitational field protects it from meteors.
    (d) Jupiter's magnetic field deflects meteors.
    (e) Jupiter is too far away for our telescopes to resolve craters there.

25. What do Mercury and Venus have in common?
    (a) They are both larger than the Earth.
    (b) They both lack atmospheres.
    (c) They both have retrograde orbits around the Sun.
    (d) They both have solar orbits inside the Earth's solar orbit.
    (e) Neither of them ever appears in a crescent phase.

26. In every respect, acceleration force manifests itself in precisely the same way as
    (a) the force caused by gravitation.
    (b) the force caused by time dilation.
    (c) the force caused by high speeds.
    (d) the force caused by constant relative motion.
    (e) none of the above.

27. As seen through telescopes on Earth, Jupiter appears as a crescent
    (a) when it is at or near opposition.
    (b) when it is at or near quadrature.
    (c) when it is at or near conjunction.
    (d) when it is at or near perihelion.
    (e) at no time.

28. Distances between stars in our galaxy are generally expressed in
    (a) meters.
    (b) kilometers.
    (c) miles.
    (d) light-miles.
    (e) light-years.

29. A robotic remote-control system that employs feedback to give the operator a sense of "being the robot" is called
    (a) remote synchronization.
    (b) telemetry.
    (c) telepresence.

(d) virtual control.

(e) time dilation.

30. Fill in the blank to make the following sentence true: "The path that the Sun follows from west to east against the background of stars during the course of the year is a circle in the sky called the _____."

(a) celestial equator

(b) celestial meridian

(c) solar circle

(d) ecliptic

(e) prime parallel

31. According to the special theory of relativity, a space ship can exceed the speed of light

(a) if it turns into antimatter.

(b) if time is made to run backward.

(c) if antigravity propulsion systems are used.

(d) if space is curved in the vicinity of the vessel.

(e) under no circumstances.

32. Suppose that you live in Dallas, Texas. You see the Moon just after sunset on an evening approximately four days before the new Moon. Which of the following is true?

(a) The Moon appears gibbous and waxing.

(b) The Moon appears gibbous and waning.

(c) The Moon appears crescent and waxing.

(d) The Moon appears crescent and waning.

(e) You must be dreaming. In reality, the Moon is below the horizon.

33. Fill in the blank in the following sentence to make it true: "The special theory of relativity is concerned with relative motion, and the general theory of relativity is concerned with _____."

(a) black holes and white dwarfs

(b) acceleration and gravitation

(c) space travel

(d) absolute motion

(e) the gravitational collapse of the Universe

34. Which of the following is *not* a feature of the surface of Mars today?

(a) Mountains of volcanic origin

(b) Canyons that apparently once were rivers

(c) Polar ice caps

(d) Impact craters caused by meteorites

(e) Liquid oceans

35. The number of other advanced civilizations in our galaxy

(a) is probably diminishing with time.

(b) can only be guessed at.

    (c) must be small because none of them have paid us a visit.

    (d) must be large because there are so many stars.

    (e) must be small because we have heard no signals from them.

36. An extremely massive star can collapse until it is so dense that nothing, not even light rays, can escape from its immediate vicinity. This type of object is called

    (a) a black dwarf.

    (b) a black hole.

    (c) a neutron star.

    (d) a pulsar.

    (e) nothing! There can be no such object.

37. At the north celestial pole, the circumpolar constellations

    (a) fill the sky; every constellation you see is circumpolar.

    (b) do not exist; no constellation you see is circumpolar.

    (c) rise in the east and set in the west, just as they do anywhere else.

    (d) never rise above the horizon.

    (e) remain fixed in the sky at all times.

38. Virtually all Earth's weather takes place within the

    (a) torrid zone.

    (b) temperate zone.

    (c) troposphere.

    (d) stratosphere.

    (e) exosphere.

39. In the Search for Extraterrestrial Intelligence (SETI) program, the 21-cm hydrogen wavelength has received particular attention because

    (a) scientists think that alien civilizations would stay away from the corresponding frequency.

    (b) scientists think that alien civilizations would be likely to listen for and transmit signals near the corresponding frequency.

    (c) that wavelength contains large amounts of EM energy.

    (d) the corresponding range of frequencies is vast.

    (e) the premise is false. The 21-cm hydrogen wavelength has not received any special attention.

40. The southern pole star is

    (a) Canopus.

    (b) Crux.

    (c) Polaris.

    (d) Regulus.

    (e) none of the above; there is no southern pole star.

41. The Earth's axis "wobbles" over long periods, like the axis of a spinning top. This is called

    (a) retrograde motion.

    (b) libration.

    (c) albedo.
    (d) precession.
    (e) convolution.

42. When the first pulsar was detected with a radio telescope, some people got the idea that the emissions were coming from
    (a) an extraterrestrial civilization.
    (b) galaxies blowing up.
    (c) Jupiter's magnetic field.
    (d) the Moon.
    (e) the Sun.

43. A type E0 galaxy is shaped like
    (a) a sphere.
    (b) a flat disk.
    (c) an irregular blob of stars.
    (d) a football.
    (e) a cigar.

44. From where on Earth's surface is every point on the celestial sphere above the horizon at one time or another as Earth makes one complete rotation relative to the stars? (Assume that the terrain is flat so that the horizon is not obstructed by hills, buildings, or anything else.)
    (a) Nowhere
    (b) Either geographic pole
    (c) The geographic equator
    (d) Anywhere between the Tropic of Cancer and the Tropic of Capricorn
    (e) Anywhere

45. Hubble and Humason believed that the red shifts of distant galaxies are caused by expansion of the Universe. Based on this assumption, they found that the speed-versus-distance function can be graphed as
    (a) a bell-shaped curve.
    (b) a curve similar to the Main Sequence.
    (c) a straight line.
    (d) a logarithmic curve.
    (e) an exponential curve.

46. Which of the following planets cannot be seen from Earth with the unaided eye?
    (a) Neptune
    (b) Saturn
    (c) Mercury
    (d) Canopus
    (e) All the planets can be seen from Earth with the unaided eye.

47. A telescope has an objective lens with a diameter of 250 mm. An eyepiece with a focal length of 10 mm is used. What is the magnification?
    (a) $25\times$

(b) 2500×

(c) 250×

(d) 10×

(e) It cannot be calculated from this information.

48. The diameter of Earth is roughly the same as the diameter of
    (a) Mercury.
    (b) Venus.
    (c) Mars.
    (d) Jupiter.
    (e) Ganymede.

49. The "hairy" glowing region of a visible a comet, immediately surrounding the core, is known as
    (a) the dust tail.
    (b) the coma.
    (c) the gas tail.
    (d) the nucleus.
    (e) the atmosphere.

50. Lunar libration is responsible for
    (a) the fact that the tides are extreme at certain times and moderate at other times.
    (b) the fact that the Moon goes through phases rapidly at certain times of the year and slowly at other times of the year.
    (c) the fact that, over time, we can see slightly more than half the Moon's surface.
    (d) the fact that the Moon is sometimes a little closer to Earth than at other times.
    (e) the fact that the Moon affects certain people's moods, particularly when it is at perigee and it is full.

51. Which of the following stars or constellations can be seen in the southern sky on winter evenings by people who live in the northern hemisphere at temperate latitudes?
    (a) Ursa minor
    (b) Ursa major
    (c) Polaris
    (d) Cassiopeia
    (e) None of the above

52. In order to read star charts on a dark night without desensitizing your eyes too much, you can use
    (a) candles.
    (b) a dim source of red light.
    (c) a dim source of blue light.
    (d) a dim source of black light (ultraviolet).
    (e) a cathode-ray-tube (CRT) computer monitor.

53. A runaway greenhouse effect is believed responsible for the hellish environment of
    (a) the Moon.
    (b) Mars.
    (c) Jupiter.
    (d) Venus.
    (e) Pluto.

54. Neptune has a moon called
    (a) Phobos.
    (b) Callisto.
    (c) Europa.
    (d) Triton.
    (e) Charon.

55. Which of the following planets or moons has a specific gravity less than that of liquid water?
    (a) The Moon
    (b) Mars
    (c) Saturn
    (d) Titan
    (e) Ganymede

56. The nuclei of atoms heavier than hydrogen contain
    (a) protons, neutrons, and electrons.
    (b) protons and neutrons.
    (c) neutrons and electrons.
    (d) protons and electrons.
    (e) protons only.

57. A total eclipse of the Sun occurs when
    (a) the Moon passes between Earth and the Sun and the Moon is at apogee.
    (b) The Moon passes between Earth and the Sun and the Moon is at perigee.
    (c) Earth passes between the Moon and the Sun and the Moon is at apogee.
    (d) Earth passes between the Moon and the Sun and the Moon is at perigee.
    (e) Earth, the Moon, and the Sun line up at any time.

58. In an H-R diagram, most stars are found along or in the
    (a) bell-shaped curve.
    (b) median curve.
    (c) Main Sequence.
    (d) lower-left corner.
    (e) upper-right corner.

59. The density of a pulsar is approximately the same as the density of
    (a) Earth.
    (b) the Moon.
    (c) Jupiter.

    (d) the Sun.

    (e) none of the above.

60. The moons of Mars

    (a) are nonexistent; Mars has no moons.

    (b) are tiny compared with the planet.

    (c) are significant in size compared with the planet.

    (d) are volcanically active.

    (e) revolve around the planet in polar orbits.

61. What is the right ascension of the Sun at the summer solstice in the northern hemisphere?

    (a) 0 h

    (b) 6 h

    (c) 12 h

    (d) 18 h

    (e) It cannot be determined without more information.

62. The volcanoes on Io, one of Jupiter's moons, occur because

    (a) Io is under constant tidal stress because of Jupiter's gravitation.

    (b) Io is bombarded by subatomic particles because of Jupiter's magnetic field.

    (c) Io is subjected to powerful magnetic forces from Jupiter.

    (d) Io was just "born that way."

    (e) Io is less dense than most other celestial objects.

63. The orbit of the Pluto-Charon system crosses the orbit of the Neptune-Triton system. However, astronomers believe that Pluto-Charon and Neptune-Triton will never collide because

    (a) space is just too vast; the probability is in effect zero.

    (b) the orbital periods of the two systems are in resonance, so they always miss each other.

    (c) the orbits of the two systems are not in the same plane.

    (d) the solar orbit of the Pluto-Charon system is retrograde, whereas the solar orbit of the Neptune-Triton system is prograde.

    (e) nothing! Astronomers believe the two systems will collide.

64. According to Ptolemy's geocentric theory, the orbits of the planets consisted of two elements called

    (a) radius and circumference.

    (b) celestial latitude and celestial longitude.

    (c) right ascension and declination.

    (d) deferent and epicycle.

    (e) nothing! Ptolemy never formulated a geocentric theory.

65. The Sun derives its energy from

    (a) combustion.

    (b) nuclear fission.

    (c) nuclear fusion.

    (d) matter-antimatter reactions.

    (e) all of the above.

66. According to geocentric theories,
    (a) Earth revolves around the Sun.
    (b) Earth is the center of the Universe.
    (c) Earth's motion is relative.
    (d) the center of Earth is the source of all energy.
    (e) the stars are mere illusions.

67. The objective in a reflecting telescope is
    (a) a convex mirror.
    (b) a concave mirror.
    (c) a convex lens.
    (d) a concave lens.
    (e) a planoconcave lens.

68. The star Rigel is a feature of the evening sky in January for viewers at 45° north latitude; it is fairly high in the southern sky. For viewers at 45° south latitude,
    (a) Rigel is seen in the evening sky in July; it is fairly high in the north.
    (b) Rigel is seen in the evening sky in July; it is fairly high in the south.
    (c) Rigel is seen in the evening sky in January; it is fairly high in the north.
    (d) Rigel is seen in the evening sky in January; it is fairly high in the south.
    (e) Rigel is never seen at all.

69. Massive black holes are believed to be common
    (a) in the cores of stars.
    (b) in orbits around planets.
    (c) at the centers of galaxies.
    (d) at the centers of open star clusters.
    (e) in the arms of spiral galaxies.

70. According to the theory of natural selection, life forms, if given enough time, will evolve in such a way that they
    (a) get more and more biologically simple.
    (b) have longer and longer individual life spans.
    (c) become better suited to the environment in which they live.
    (d) all converge toward a single, universal species.
    (e) become more and more hostile to each other.

71. A simple objective lens refracts orange light to a slightly different extent than it refracts blue light. This is observed in a telescope as
    (a) partial internal refraction.
    (b) selective refraction.
    (c) chromatic aberration.
    (d) coma.
    (e) astigmatism.

72. On Mars, your weight would be
    (a) less than your weight on Earth but more than your weight on Jupiter.
    (b) less than your weight on Earth but more than your weight on an asteroid.
    (c) more than your weight on Earth but less than your weight on Neptune.
    (d) more than your weight on Earth but less than your weight on Saturn.
    (e) the same as your weight on Earth.

73. In order to obtain the speeds necessary for interstellar travel, nuclear bombs have been suggested as a method of producing thrust. A practical problem with this scheme is the fact that
    (a) nuclear bombs would be too heavy to carry.
    (b) the impulse would occur in uncomfortable bursts.
    (c) a source of oxygen would be needed.
    (d) great quantities of water would be required.
    (e) the explosions would irradiate the food supply.

74. Which of the following theories concerning the origin of the Solar System is most commonly accepted?
    (a) The planets formed from a black hole that sucked gas and dust into orbit around itself from a large region of interstellar space.
    (b) The planets formed when a passing star ripped matter out of the Sun.
    (c) The planets and Sun formed from a condensing disk of interstellar gas and dust.
    (d) The planets formed from comets that fell into the Sun, causing solar matter to be ejected into space.
    (e) The planets and Sun formed from a white hole, that is, a place where new matter was introduced from another universe.

75. The velocity (speed and direction) of a meteor as it enters the Earth's atmosphere can be measured accurately using
    (a) a high-powered telescope.
    (b) binoculars.
    (c) radar.
    (d) an interferometer.
    (e) a spectrograph.

76. If intelligent civilizations in the Cosmos always become warlike, we should not expect to get any visits from them because
    (a) they look at us and think we are too weak to bother with.
    (b) they are afraid of us because we are so barbaric.
    (c) they have already conquered us; we just don't realize it yet.
    (d) they annihilate themselves before they become capable of visiting us.
    (e) none of the above; we ought to expect them any time.

77. A converging lens
    (a) brings parallel light rays to a focus.
    (b) spreads parallel light rays out.

(c) makes converging light rays parallel.

(d) can do any of a, b, and c.

(e) can do none of a, b, or c.

78. Which of the following planets has no known moons?

(a) Mercury

(b) Mars

(c) Uranus

(d) Neptune

(e) Pluto

79. The carbon dioxide produced by human respiration in an interstellar space ship could be used by

(a) the nuclear propulsion system.

(b) the radiation shield.

(c) the water-electrolysis equipment.

(d) plants grown for food.

(e) nothing; it could serve no useful purpose.

80. The interferometer is a form of

(a) refracting telescope.

(b) spectrograph.

(c) high-resolution radio telescope.

(d) x-ray camera.

(e) device for measuring red shift.

81. Which of the following planets lacks a ring system?

(a) Jupiter

(b) Saturn

(c) Uranus

(d) Neptune

(e) All of the above planets have ring systems.

82. Approximately how far is the Moon from Earth?

(a) 1.3 light-seconds

(b) 250,000 light-seconds

(c) 400,000 light-seconds

(d) 8.2 light-minutes

(e) 4.3 light-years

83. An angle that measures 120 seconds of arc is the same as an angle that measures

(a) 2 minutes of arc.

(b) 2 degrees of arc.

(c) 7,200 minutes of arc.

(d) one-third of a circle.

(e) two complete circles.

84. Meteorites are always found
    (a) in solar orbit between Mars and Jupiter.
    (b) in solar orbit.
    (c) in Earth orbit.
    (d) in lunar orbit.
    (e) on Earth's surface.

85. The *f*-ratio of a telescope is the
    (a) diameter of the objective divided by the diameter of the eyepiece in the same units.
    (b) eyepiece focal length divided by the objective focal length in the same units.
    (c) objective focal length divided by the eyepiece focal length in the same units.
    (d) objective focal length divided by the objective diameter in the same units.
    (e) none of the above.

86. If you observe the sky from the southern tip of South America, which of the following constellations will you never see?
    (a) Crux
    (b) Carina
    (c) Grus
    (d) Ursa Minor
    (e) You will never see any of the above.

87. A Barlow lens can be useful for
    (a) increasing the number of magnification levels obtainable with a given set of telescope eyepieces.
    (b) obtaining extreme magnification for looking at the details of distant stars and galaxies.
    (c) reducing the effective focal length of a telescope's objective.
    (d) eliminating chromatic aberration in reflecting telescopes.
    (e) increasing the apparent field of view for a particular telescope eyepiece.

88. As the wavelengths of x-rays become shorter, it gets increasingly difficult to direct and focus them because
    (a) they become fainter.
    (b) their penetrating power increases.
    (c) their frequency decreases.
    (d) they are increasingly affected by magnetic fields.
    (e) nothing! The premise is false. It gets easier to focus x-rays as their wavelengths become shorter.

89. A full-aperture solar filter
    (a) screws into the barrel of a telescope eyepiece.
    (b) is used with telescope finders only.
    (c) attaches to the secondary mirror in a reflecting telescope.
    (d) keeps direct, unfiltered sunlight entirely out of a telescope.
    (e) should never be used with any telescope.

90. Under what circumstances is the shortest distance between two points a curve rather than a straight line?
    (a) Under no circumstances
    (b) When space is "flat"
    (c) When two observers are moving relative to each other at constant speed
    (d) In the presence of a powerful gravitational field
    (e) Whenever clocks are not synchronized

91. A Moon filter can improve the image of
    (a) a crescent Moon viewed at high magnification.
    (b) a full Moon viewed at low magnification.
    (c) a crescent Moon viewed with a Barlow lens.
    (d) a new Moon viewed with a focal reducer/corrector.
    (e) the Moon at any phase viewed with a short-focal-length eyepiece.

92. The full phase of Mercury, as seen through telescopes on Earth, takes place when the planet is near
    (a) opposition.
    (b) inferior conjunction.
    (c) quadrature.
    (d) superior conjunction.
    (e) perihelion.

93. If the Pluto-Charon system had been discovered today rather than many years ago, some astronomers would classify it as
    (a) a pair of large asteroids.
    (b) a pair of huge comets.
    (c) a pair of tiny, unborn stars.
    (d) a pair of errant moons of Neptune.
    (e) a pair of objects outside the Solar System.

94. Fill in the blank in the following sentence to make it true: "Quasars exhibit _____ in their spectral lines."
    (a) stability
    (b) blurring
    (c) blue shifts
    (d) red shifts
    (e) alternating red and blue shifts

95. Titan, the largest moon of Saturn, has an atmosphere that is
    (a) much colder than that of Earth.
    (b) much warmer than that of Earth.
    (c) plagued by constant dust storms.
    (d) rich in hydrogen and filled with blue-white clouds of water vapor.
    (e) similar to the atmosphere of Earth.

96. Earth is closest to the Sun during the month of
    (a) January.
    (b) May.
    (c) July.
    (d) September.
    (e) November.

97. Suppose that the spectral lines in the light from a star shift at a regular rate toward the red, then toward the blue, then back toward the red, and so on. What can we surmise about this star?
    (a) It is receding from us.
    (b) It is approaching us.
    (c) It is in orbit around or is in mutual orbit with something.
    (d) Its temperature varies on a regular basis.
    (e) Its brightness varies on a regular basis.

98. Fill in the blank to make the following sentence true: "According to Kepler, each planet follows an elliptical orbit around the Sun, with the Sun _____."
    (a) at the center of the ellipse
    (b) outside the ellipse
    (c) on the minor axis of the ellipse
    (d) at one focus of the ellipse
    (e) between the two foci of the ellipse

99. The brightest star in the entire sky is
    (a) Canis Minor.
    (b) Andromeda.
    (c) Polaris.
    (d) Rigel.
    (e) Sirius.

100. A person living in Perth, Australia, gets the most daylight during the month of
    (a) December.
    (b) March.
    (c) June.
    (d) September.
    (c) Any of the above; they're all the same.

Answers to Quiz, Test, and Exam Questions

Chapter 1

1. D	2. C	3. D	4. A	5. D
6. A	7. D	8. C	9. B	10. B

Chapter 2

1. B	2. A	3. D	4. C	5. A
6. D	7. B	8. D	9. C	10. B

Chapter 3

1. D	2. D	3. C	4. C	5. C
6. B	7. D	8. B	9. A	10. B

Chapter 4

1. B	2. C	3. B	4. D	5. D
6. B	7. A	8. C	9. C	10. C

Test: Part One

1. A	2. A	3. E	4. C	5. A
6. A	7. A	8. A	9. A	10. C
11. E	12. D	13. B	14. C	15. A
16. E	17. B	18. C	19. D	20. B
21. D	22. E	23. E	24. C	25. A
26. E	27. D	28. A	29. C	30. C
31. A	32. D	33. C	34. E	35. E
36. B	37. B	38. E	39. D	40. D

Chapter 5

1. A	2. D	3. D	4. B	5. C
6. B	7. D	8. A	9. C	10. D

Chapter 6

1. C	2. C	3. D	4. A	5. B
6. B	7. A	8. B	9. A	10. D

Chapter 7

1. C	2. A	3. D	4. C	5. A
6. C	7. D	8. A	9. B	10. A

Chapter 8

1. A	2. D	3. A	4. D	5. C
6. A	7. C	8. A	9. C	10. D

Test: Part Two

1. C	2. B	3. C	4. B	5. A
6. A	7. A	8. B	9. C	10. D
11. A	12. C	13. E	14. C	15. B
16. C	17. A	18. E	19. D	20. E
21. C	22. D	23. E	24. A	25. C
26. D	27. C	28. C	29. C	30. A
31. E	32. D	33. A	34. E	35. E
36. D	37. A	38. D	39. A	40. E

Chapter 9

1. B	2. A	3. A	4. C	5. D
6. B	7. D	8. B	9. D	10. C

Chapter 10

1. A	2. A	3. B	4. B	5. C
6. D	7. D	8. C	9. A	10. C

Chapter 11

1. C	2. C	3. B	4. C	5. A
6. C	7. B	8. D	9. B	10. A

Chapter 12

1. C	2. A	3. A	4. D	5. B
6. A	7. D	8. D	9. C	10. C

Test: Part Three

1. E	2. C	3. D	4. D	5. A
6. B	7. A	8. B	9. A	10. A
11. D	12. D	13. E	14. B	15. E
16. A	17. A	18. D	19. E	20. B
21. D	22. C	23. C	24. B	25. E
26. C	27. C	28. C	29. D	30. A
31. A	32. C	33. C	34. A	35. A
36. B	37. E	38. E	39. E	40. E

Chapter 13

1. D	2. A	3. A	4. C	5. D
6. B	7. C	8. D	9. B	10. B

Chapter 14

1. D	2. C	3. B	4. B	5. C
6. D	7. D	8. A	9. C	10. A

Chapter 15

1. A	2. C	3. B	4. A	5. B
6. D	7. D	8. D	9. D	10. C

Chapter 16

1. D	2. C	3. A	4. B	5. C
6. C	7. D	8. C	9. B	10. A

Test: Part Four

1. C	2. B	3. E	4. D	5. E
6. B	7. C	8. A	9. A	10. A
11. B	12. A	13. A	14. C	15. D
16. B	17. D	18. D	19. B	20. E
21. C	22. C	23. D	24. B	25. D
26. E	27. A	28. C	29. A	30. A
31. C	32. C	33. A	34. E	35. E
36. C	37. C	38. B	39. A	40. C

Chapter 17

1. C	2. B	3. A	4. C	5. D
6. D	7. A	8. B	9. B	10. D

Chapter 18

1. A	2. B	3. D	4. D	5. C
6. C	7. B	8. C	9. D	10. C

Chapter 19

1. B	2. C	3. B	4. B	5. C
6. C	7. A	8. D	9. D	10. A

Chapter 20

1. D	2. C	3. A	4. A	5. D
6. C	7. B	8. B	9. B	10. D

Test: Part Five

1. B	2. E	3. C	4. A	5. A
6. D	7. C	8. D	9. C	10. E
11. E	12. D	13. A	14. A	15. A
16. D	17. B	18. B	19. E	20. C
21. A	22. A	23. A	24. A	25. E
26. A	27. E	28. B	29. B	30. B
31. D	32. C	33. B	34. E	35. C
36. E	37. B	38. B	39. B	40. C

Final Exam

1. B	2. B	3. E	4. D	5. C
6. C	7. C	8. E	9. C	10. C
11. C	12. B	13. A	14. D	15. D
16. E	17. D	18. E	19. A	20. B
21. B	22. D	23. C	24. A	25. D
26. A	27. E	28. E	29. C	30. D
31. E	32. E	33. B	34. E	35. B
36. B	37. A	38. C	39. B	40. E
41. D	42. A	43. A	44. C	45. C
46. A	47. E	48. B	49. B	50. C
51. E	52. B	53. D	54. D	55. C
56. B	57. B	58. C	59. E	60. B
61. B	62. A	63. B	64. D	65. C
66. B	67. B	68. C	69. C	70. C
71. C	72. B	73. B	74. C	75. C
76. D	77. A	78. A	79. D	80. C
81. E	82. A	83. A	84. E	85. D
86. D	87. A	88. B	89. D	90. D
91. B	92. D	93. B	94. D	95. A
96. A	97. C	98. D	99. E	100. A

Suggested Additional Reference

Books

Charles, Jeffrey R., *Practical Astrophotography*, Springer-Verlag, New York, 2000.

Harrington, Philip S., *StarWare*, Wiley, New York, 1998.

Jastrow, Robert, *Journey to the Stars*, Bantam Books, New York, 1989.

Moche, Dinah L., *Astronomy: A Self-Teaching Guide*, Wiley, New York, 2000.

Moore, Patrick, *Astronomy*, NTC/Contemporary Publishing, New York, 1995.

Web Sites

Encyclopedia Britannica Online, www.britannica.com
Eric's Treasure Troves of Science, www.treasure-troves.com
Sky and Telescope, www.skypub.com
Weather Underground, www.wunderground.com

INDEX

NUMBERS

ABOUT THE AUTHOR

Stan Gibilisco is one of McGraw-Hill's most diverse and best-selling authors. Known for his clear, user-friendly, and entertaining writing style, Mr. Gibilisco's depth of knowledge and ease of presentation make him an excellent choice for a book such as *Astronomy Demystified.* His previous titles for McGraw-Hill include: *The TAB Encyclopedia of Electronics for Technicians and Hobbyists, Teach Yourself Electricity and Electronics,* and the *Illustrated Dictionary of Electronics. Booklist* named his book, *The McGraw-Hill Encyclopedia of Personal Computing* one of the Best References of 1996.